METHODS IN MOLECULAR BIOLOGY

Series Editor
John M. Walker
School of Life and Medical Sciences
University of Hertfordshire
Hatfield, Hertfordshire, UK

For further volumes:
http://www.springer.com/series/7651

For over 35 years, biological scientists have come to rely on the research protocols and methodologies in the critically acclaimed *Methods in Molecular Biology* series. The series was the first to introduce the step-by-step protocols approach that has become the standard in all biomedical protocol publishing. Each protocol is provided in readily-reproducible step-by-step fashion, opening with an introductory overview, a list of the materials and reagents needed to complete the experiment, and followed by a detailed procedure that is supported with a helpful notes section offering tips and tricks of the trade as well as troubleshooting advice. These hallmark features were introduced by series editor Dr. John Walker and constitute the key ingredient in each and every volume of the *Methods in Molecular Biology* series. Tested and trusted, comprehensive and reliable, all protocols from the series are indexed in PubMed.

Embryo Models In Vitro

Methods and Protocols

Edited by

Magdalena Zernicka-Goetz

Division of Biology and Biological Engineering, California Institute of Technology, Pasadena, CA, USA

Kursad Turksen

Ottawa, ON, Canada

Humana Press

Editors
Magdalena Zernicka-Goetz
Division of Biology and Biological Engineering
California Institute of Technology
Pasadena, CA, USA

Kursad Turksen
Ottawa, ON, Canada

ISSN 1064-3745 ISSN 1940-6029 (electronic)
Methods in Molecular Biology
ISBN 978-1-0716-3685-5 ISBN 978-1-0716-3686-2 (eBook)
https://doi.org/10.1007/978-1-0716-3686-2

Cover Illustration Caption: Artwork created by Dr. Kursad Turksen.

This Humana imprint is published by the registered company Springer Science+Business Media, LLC, part of Springer Nature.
The registered company address is: 1 New York Plaza, New York, NY 10004, U.S.A.

Preface

Successes in the use of pluripotent cells to generate very early embryo-like structures in vitro are among the most exciting and innovative advances in developmental biology in recent times. That the resultant 3D embryo-like structures in vitro phenocopy many of the developmental landmarks and 3D architectures seen normally in vivo has opened multiple new and important avenues of investigation.

In this volume, we have collected a series of protocols that cover diverse aspects of the topic and exemplify some of the developments and improvements in the field.

Once again, the protocols gathered here are faithful to the mission statement of the Methods in Molecular Biology series: They are well-established and described in an easy-to-follow, step-by-step fashion so as to be valuable for not only experts but also novices in the stem cell field. That goal is achieved because of the generosity of the contributors who have carefully described their protocols in this volume, and we are very grateful for their efforts.

Our thanks as well go to Dr. John Walker, the Series Editor of the Methods in Molecular Biology series, for giving us the opportunity to create this volume and for supporting us along the way.

We are also grateful to Patrick Marton, the Executive Editor of Methods in Molecular Biology and the Springer Protocols collection, for his continuous support from idea to completion of this volume.

A special thank you goes to Anna Rakovsky, Assistant Editor for Methods in Molecular Biology, for her support from the beginning to the end of this project.

We would also like to thank David C. Casey, Senior Editor of Methods in Molecular Biology, for his outstanding editorial work during the production of this volume.

Finally, we would like to thank the production crew for their work in putting together an outstanding volume.

Pasadena, CA, USA — ***Magdalena Zernicka-Goetz***
Ottawa, ON, Canada — ***Kursad Turksen***

Contents

Contributors

GIANLUCA AMADEI • *Department of Biology, University of Padua, Padua, Italy*

ASHA SHAJI ANTONY • *Department of Genomic Science, Central University of Kerala, Tejaswini Hills, Kerala, India*

SHARON ARCURI • *Laboratory of Biomedical Embryology and Tissue Engineering, Department of Veterinary Medicine and Animal Sciences, Centre for Stem Cell Research, Università degli Studi di Milano, Lodi, Italy*

BRADLEY BALATON • *Department of Development and Regeneration, Leuven Stem Cell Institute, Leuven Institute for Single-cell Omics (LISCO), KU Leuven-University of Leuven, Leuven, Belgium*

SAVANA BIONDIC • *Département de Médecine, Université de Montréal, Montréal, QC, Canada; Centre de Recherche du Centre Hospitalier de l'Université de Montréal, Axe Immunopathologie, Montréal, QC, Canada*

TIZIANA A. L. BREVINI • *Laboratory of Biomedical Embryology and Tissue Engineering, Department of Veterinary Medicine and Animal Sciences, Centre for Stem Cell Research, Università degli Studi di Milano, Lodi, Italy*

JAN J. BROSENS • *Division of Biomedical Sciences, Warwick Medical School, University of Warwick, Clinical Sciences Building, University Hospital Coventry, Coventry, UK; Tommy's National Centre for Miscarriage Research, University Hospitals Coventry and Warwickshire NHS Trust, Coventry, UK*

JESICA CANIZO • *Département de Médecine, Université de Montréal, Montréal, QC, Canada; Centre de Recherche du Centre Hospitalier de l'Université de Montréal, Axe Immunopathologie, Montréal, QC, Canada*

A. S. DEVIKA • *Department of Genomic Science, Central University of Kerala, Tejaswini Hills, Kerala, India*

MARGARETE DIAZ-CUADROS • *Department of Genetics, Harvard Medical School, Boston, MA, USA; Department of Molecular Biology, Massachusetts General Hospital, Boston, MA, USA*

FULVIO GANDOLFI • *Department of Agricultural and Environmental Sciences – Production, Landscape, Agroenergy, Centre for Stem Cell Research, Università degli Studi di Milano, Milan, Italy*

MICHAEL HAGEMANN-JENSEN • *Department of Cell and Molecular Biology, Karolinska Institutet, Stockholm, Sweden*

CHARLOTTE E. HANDFORD • *Division of Biology and Biological Engineering, California Institute of Technology, Pasadena, CA, USA; Department of Physiology, Development and Neuroscience, University of Cambridge, Cambridge, UK*

TIAM HEYDARI • *School of Biomedical Engineering, University of British Columbia, Vancouver, BC, Canada; Michael Smith Laboratories, University of British Columbia, Vancouver, BC, Canada*

ROWAN M. KARVAS • *Department of Developmental Biology, Washington University School of Medicine, St. Louis, MO, USA; Center of Regenerative Medicine, Washington University School of Medicine, St. Louis, MO, USA*

LAURA KEROSUO • *Neural Crest Development and Disease Unit, National Institute of Dental and Craniofacial Research, Intramural Research Program, Neural Crest Development and Disease Unit, National Institutes of Health, Bethesda, ML, USA*

Sherif Khodeer • *Department of Development and Regeneration, Leuven Stem Cell Institute, Leuven Institute for Single-Cell Omics (LISCO), Leuven Cancer Institute, KU Leuven – University of Leuven, Leuven, Belgium*
Sergio Ledda • *Department of Veterinary Medicine, University of Sassari, Sassari, Italy*
Katherine Vandal Lenghan • *Molecular Biology Program, Université de Montréal, Montreal, QC, Canada; Centre de Recherche du Centre Hospitalier de l'Université de Montréal, Montreal, QC, Canada*
Tianqing Li • *State Key Laboratory of Primate Biomedical Research, Institute of Primate Translational Medicine, Kunming University of Science and Technology, Kunming, Yunnan, China*
Clifford L. Librach • *CReATe Fertility Centre, Toronto, ON, Canada; University of Toronto, Department of Obstetrics and Gynecology, Toronto, ON, Canada; University of Toronto, Department of Physiology, Toronto, ON, Canada; Sunnybrook Research Institute, Toronto, ON, Canada*
Kun Liu • *Colorado Center for Reproductive Medicine, Lone Tree, CO, USA; Paul M. Rady Department of Mechanical Engineering, University of Colorado, Boulder, CO, USA*
Deirdre Logsdon • *Colorado Center for Reproductive Medicine, Lone Tree, CO, USA*
Emma S. Lucas • *Division of Biomedical Sciences, Warwick Medical School, University of Warwick, Clinical Sciences Building, University Hospital Coventry, Coventry, UK*
Yu-Xin Luo • *Beijing Key Laboratory of Reproductive Endocrinology and Assisted Reproductive Technology and Key Laboratory of Assisted Reproduction, Ministry of Education, Center of Reproductive Medicine, Department of Obstetrics and Gynecology, Peking University Third Hospital, Beijing, China*
Komal Makwana • *Division of Biomedical Sciences, Warwick Medical School, University of Warwick, Clinical Sciences Building, University Hospital Coventry, Coventry, UK*
Yuchuan Miao • *Department of Genetics, Harvard Medical School, Boston, MA, USA; Department of Pathology, Brigham and Women's Hospital, Boston, MA, USA*
Ikbal B. Mohammad Choudhury • *Division of Biology and Biological Engineering, California Institute of Technology, Pasadena, CA, USA*
Zachary T. Olmsted • *State University of New York Polytechnic Institute, College of Nanoscale Science and Engineering, Nanobioscience, Albany, NY, USA; University of California Los Angeles, Ronald Reagan UCLA Medical Center, Department of Neurosurgery, Los Angeles, CA, USA*
Ceren Pajanoja • *Neural Crest Development and Disease Unit, National Institute of Dental and Craniofacial Research, Intramural Research Program, Neural Crest Development and Disease Unit, National Institutes of Health, Bethesda, ML, USA; Faculty of Medicine, University of Helsinki, Helsinki, Finland*
Janet L. Paluh • *State University of New York Polytechnic Institute, College of Nanoscale Science and Engineering, Nanobioscience, Albany, NY, USA*
Amitesh Panda • *Department of Development and Regeneration, Leuven Stem Cell Institute, Leuven Institute for Single-Cell Omics (LISCO), Leuven Cancer Institute, KU Leuven – University of Leuven, Leuven, Belgium*
Pallavi Panda • *Division of Biology and Biological Engineering, California Institute of Technology, Pasadena, CA, USA*
Maria Belen Paredes-Espinosa • *State University of New York Polytechnic Institute, College of Nanoscale Science and Engineering, Nanobioscience, Albany, NY, USA*
Vincent Pasque • *Department of Development and Regeneration, Leuven Stem Cell Institute, Leuven Institute for Single-Cell Omics (LISCO), Leuven Cancer Institute,*

KU Leuven – University of Leuven, Leuven, Belgium; Department of Development and Regeneration, Leuven Stem Cell Institute, Leuven Institute for Single-cell Omics (LISCO), KU Leuven-University of Leuven, Leuven, Belgium

GEORGIA PENNAROSSA • *Laboratory of Biomedical Embryology and Tissue Engineering, Department of Veterinary Medicine and Animal Sciences, Centre for Stem Cell Research, Università degli Studi di Milano, Lodi, Italy*

SOPHIE PETROPOULOS • *Département de Médecine, Université de Montréal, Montréal, QC, Canada; Centre de Recherche du Centre Hospitalier de l'Université de Montréal, Axe Immunopathologie, Montréal, QC, Canada; Department of Clinical Science, Intervention and Technology, Karolinska Institutet, and Division of Obstetrics and Gynecology, Karolinska. Universitetssjukhuset, Stockholm, Sweden*

THI XUAN AI PHAM • *Department of Development and Regeneration, Leuven Stem Cell Institute, Leuven Institute for Single-Cell Omics (LISCO), Leuven Cancer Institute, KU Leuven – University of Leuven, Leuven, Belgium*

CARLOS A. PINZON-ARTEAGA • *Department of Molecular Biology, University of Texas Southwestern Medical Center, Dallas, TX, USA*

OLIVIER POURQUIÉ • *Department of Genetics, Harvard Medical School, Boston, MA, USA; Department of Pathology, Brigham and Women's Hospital, Boston, MA, USA; Harvard Stem Cell Institute, Harvard University, Cambridge, MA, USA*

THOMAS M. RAWLINGS • *Division of Biomedical Sciences, Warwick Medical School, University of Warwick, Clinical Sciences Building, University Hospital Coventry, Coventry, UK*

STEWART J. RUSSELL • *CReATe Fertility Centre, Toronto, ON, Canada*

GAOHUI SHI • *State Key Laboratory of Primate Biomedical Research, Institute of Primate Translational Medicine, Kunming University of Science and Technology, Kunming, Yunnan, China*

SMITA SUDHEER • *Department of Genomic Science, Central University of Kerala, Tejaswini Hills, Kerala, India*

DEBORAH M. TAYLOR • *Centre for Reproductive Medicine, University Hospitals Coventry and Warwickshire NHS Trust, Coventry, UK*

THOROLD W. THEUNISSEN • *Department of Developmental Biology, Washington University School of Medicine, St. Louis, MO, USA; Center of Regenerative Medicine, Washington University School of Medicine, St. Louis, MO, USA*

SAN KIT TO • *Department of Development and Regeneration, Leuven Stem Cell Institute, Leuven Institute for Single-cell Omics (LISCO), KU Leuven-University of Leuven, Leuven, Belgium*

MARIA TRYFONOS • *Division of Biomedical Sciences, Warwick Medical School, University of Warwick, Clinical Sciences Building, University Hospital Coventry, Coventry, UK*

KATHERINE VANDAL • *Département de Médecine, Université de Montréal, Montréal, QC, Canada; Centre de Recherche du Centre Hospitalier de l'Université de Montréal, Axe Immunopathologie, Montréal, QC, Canada*

HONGMEI WANG • *State Key Laboratory of Stem Cell and Reproductive Biology, Institute of Zoology, Chinese Academy of Sciences, Beijing, China; Institute for Stem Cell and Regeneration, Chinese Academy of Sciences, Beijing, China; University of Chinese Academy of Sciences, Beijing, China; Beijing Institute for Stem Cell and Regenerative Medicine, Beijing, China*

YUE WANG • *State Key Laboratory of Stem Cell and Reproductive Biology, Institute of Zoology, Chinese Academy of Sciences, Beijing, China; Institute for Stem Cell and Regeneration, Chinese Academy of Sciences, Beijing, China; University of Chinese Academy of Sciences, Beijing, China; Beijing Institute for Stem Cell and Regenerative Medicine, Beijing, China*

HAO WU • *State Key Laboratory of Stem Cell and Reproductive Biology, Institute of Zoology, Chinese Academy of Sciences, Beijing, China; Institute for Stem Cell and Regeneration, Chinese Academy of Sciences, Beijing, China; University of Chinese Academy of Sciences, Beijing, China; Beijing Institute for Stem Cell and Regenerative Medicine, Beijing, China*

JUN WU • *Department of Molecular Biology, University of Texas Southwestern Medical Center, Dallas, TX, USA; Cecil H. and Ida Green Center for Reproductive Biology Sciences, University of Texas Southwestern Medical Center, Dallas, TX, USA; Hamon Center for Regenerative Science and Medicine, University of Texas Southwestern Medical Center, Dallas, TX, USA*

LIFENG XIANG • *Department of Reproductive Medicine, The First People's Hospital of Yunnan Province, Kunming, China; State Key Laboratory of Primate Biomedical Research, Institute of Primate Translational Medicine, Kunming University of Science and Technology, Kunming, Yunnan, China*

YU YIN • *State Key Laboratory of Primate Biomedical Research, Institute of Primate Translational Medicine, Kunming University of Science and Technology, Kunming, Yunnan, China*

LEQIAN YU • *Department of Molecular Biology, University of Texas Southwestern Medical Center, Dallas, TX, USA; State Key Laboratory of Stem Cell and Reproductive Biology, Institute of Zoology, Chinese Academy of Sciences, Beijing, China; Institute for Stem Cell and Regeneration, Chinese Academy of Sciences, Beijing, China*

YANG YU • *Beijing Key Laboratory of Reproductive Endocrinology and Assisted Reproductive Technology and Key Laboratory of Assisted Reproduction, Ministry of Education, Center of Reproductive Medicine, Department of Obstetrics and Gynecology, Peking University Third Hospital, Beijing, China; Clinical Stem Cell Research Center, Peking University Third Hospital, Beijing, China*

YE YUAN • *Colorado Center for Reproductive Medicine, Lone Tree, CO, USA*

PETER W. ZANDSTRA • *School of Biomedical Engineering, University of British Columbia, Vancouver, BC, Canada; Michael Smith Laboratories, University of British Columbia, Vancouver, BC, Canada*

MAGDALENA ZERNICKA-GOETZ • *Division of Biology and Biological Engineering, California Institute of Technology, Pasadena, CA, USA; Department of Physiology, Development and Neuroscience, University of Cambridge, Cambridge, UK*

CHENG ZHAO • *Department of Clinical Science, Intervention and Technology, Karolinska Institutet, and Division of Obstetrics and Gynecology, Karolinska. Universitetssjukhuset, Stockholm, Sweden*

YI ZHENG • *Department of Biomedical and Chemical Engineering, Syracuse University, Syracuse, NY, USA; BioInspired Syracuse: Institute for Material and Living Systems, Syracuse University, Syracuse, NY, USA*

Methods in Molecular Biology (2024) 2767: 1–18
DOI 10.1007/7651_2023_485

Published online: 24 June 2023

Generation of Human Blastoids from Naive Pluripotent Stem Cells

Yulei Wei, Kun Liu, Carlos A. Pinzon-Arteaga, Deirdre Logsdon, Leqian Yu, Ye Yuan, and Jun Wu

Abstract

Under certain culture conditions, naive human pluripotent stem cells can generate human blastocyst-like structures (called human blastoids). Human blastoids serve as an accessible model for human blastocysts and are amenable for large-scale production. Here, we describe a detailed step-by-step protocol for the robust and high-efficient generation of human blastoids from naive human pluripotent stem cells.

Key words Human blastoids, Human extraembryonic cell types, Trophoblast cells, Hypoblast cells, Human implantation model, Integrated models of human embryos, Naive human pluripotent stem cells, The generation of blastocyst-like structures

1 Introduction

Few days after fertilization, a single-celled zygote progresses through several cell divisions and develops into a blastocyst—a spherical structure composed of an outer layer of trophectoderm cells encasing an inner cell mass and a fluid-filled cavity. Blastocysts contain all founding tissues of a developing conceptus. Following implantation, blastocysts will generate extraembryonic tissues including placenta and extraembryonic membranes that support the development of the entire fetus. Human blastocysts are the gold standard for the study of human peri-implantation development. However, human blastocysts are ethically challenging to study, difficult to acquire, and more difficult to perturb. As a consequence, the cellular and molecular mechanisms underlying human embryogenesis and implantation remain unconfirmed or unknown. The recent development of stem-cell-based embryo models opens new avenues for studying early human development in vitro [2]. Here, we elaborate on a robust and efficient protocol for the generation of human blastocyst-like structures (termed blastoids) from naive human pluripotent stem cells (hPSCs). Human blastoids are easily accessible, scalable, and amenable to genetic and chemical manipulations, and thus represent a surrogate 3D model to human blastocysts to study early human development, understand early pregnancy loss, and gain insights into early developmental defects [3, 4].

2 Materials

2.1 Reagents

1. Geltrex™ LDEV-Free Reduced Growth Factor Basement Membrane Matrix (Gibco, cat. No. A1413201).
2. Matrigel hESC-Qualified Matrix, LDEV-free (Corning, cat. no. 354277).
3. Dulbecco's phosphate-buffered saline (1×), no calcium, no magnesium (DPBS) (Gibco, cat. no. 14190-144).
4. DMEM/F12 (Gibco, cat. no. 11320-033).
5. DMEM, high glucose, no glutamine (Gibco, cat. no. 11960044).
6. Neurobasal™ medium (Gibco, cat. no. 21103-049).
7. N2 supplement (100×) (Gibco, cat. no. 17502-048). *Note: Store at −20 °C and protect from light.*
8. B27 supplement (50×) (Gibco, cat. no. 17504-044). *Note: Store at −20 °C and protect from light.*
9. GlutaMAX (100×) (Gibco, cat. no. 35050-061).
10. MEM Non-Essential Amino Acids (100×) (Gibco, cat. no. 11140-050).
11. 2-Mercaptoethanol (1000×) (Gibco, cat. no. 21985-023).
12. Insulin-Transferrin-Selenium-Ethanolamine (ITS-X) (100×) (Gibco, cat. no. 51500-056).
13. Fetalgro EX Bovine Growth Serum (rmbio, cat. no. FGX-BBT).
14. Penicillin-Streptomycin (Gibco, cat. no. 15140-122).
15. TrypLE™ Express (Gibco, cat. no. 12604-013).
16. Knockout Serum Replacement (KSR) (Gibco, cat. no. 10828-028).
17. Bovine Serum Albumin (BSA) (Sigma-Aldrich, cat. no. A3059-50G).
18. Anti-Adherence Rinsing Solution (STEMCELL Technologies, cat. no. 07010).
19. Gelatin solution (0.1% in water) (STEMCELL Technologies, cat. no. 07903).
20. DNase I, RNase free (Thermo Scientific, cat. no EN0521).

21. ROCK inhibitor Y-27632 (ReproCell, cat. no. 04-0012). *Note: Store aliquots at −20 °C for up to 1–6 months.*
22. Human FGF2 (PeproTech, cat. no. 100-18B). *Note: Store aliquots at −20 °C for up to 1–6 months.*
23. Activin A (Peprotech, cat. no. 120-14E). *Note: Store aliquots at −20 °C for up to 1–6 months.*
24. CHIR99021 (APExBIO, cat. no. A8396). *Note: Store aliquots at −20 °C for up to 1–6 months.*
25. PD0325901 (ReproCell, cat. no. 04-0006). *Note: Store aliquots at −20 °C for up to 1–6 months.*
26. IM-12 (MedChemExpress, cat. no. HY-12292). *Note: Store aliquots at −20 °C for up to 1–6 months.*
27. SB590885 (Tocris, cat. no. 2650). *Note: Store aliquots at −20 °C for up to 1–6 months.*
28. WH-4-023 (MedChemExpress, cat. no. HY-12299). *Note: Store aliquots at −20 °C for up to 1–6 months.*
29. A83-01 (MedChemExpress, cat. no. HY-10432). *Note: Store aliquots at −20 °C for up to 1 month. Once thawed, do not refreeze again.*
30. Human Leukemia Inhibitory Factor (hLIF) (MilliporeSigma, cat. no. LIF1050). *Note: Maintain at 2–8 °C for 6–12 months.*
31. CEPT cocktail [50 nM Chroman 1 (MedChemExpress, cat. no. HY-15392), 5 μM Emricasan (MedChemExpress, cat. no. HY-10396), 1× Polyamine Supplement (Sigma-Aldrich, cat. no. P8483-5ML), and 0.7 μM Trans-ISRIB (Tocris, cat. no. 5284)]. *Note: Store aliquots at −20 °C for up to 1–6 months.*
32. 1-Oleoyl lysophosphatidic acid (LPA) (STEMCELL Technologies, cat. no. 72694). *Note: Store aliquots at −20 °C for up to 1–6 months.*
33. Trypan Blue Stain (0.4%) (Gibco, cat. no. 15250-061).
34. HyClone Embryonic Stem Cell Screened Fetal Calf Serum (FCS) (Cytiva, cat. no. SH30070.02E).
35. β-Estradiol (E2) (Sigma-Aldrich, cat. no. E2257).
36. Pure Ethyl Alcohol (EtOH) (Millipore Sigma, cat. no. E7023).
37. Progesterone (P4) (Sigma-Aldrich, cat. no. P8783).
38. Sodium Pyruvate (Sigma-Aldrich, cat. no. P4562).
39. Gentamicin (CellGro, cat. no. 30-005-CR).
40. Donkey serum (Sigma-Aldrich, cat. no. D9663).
41. Paraformaldehyde (PFA) (Electron Microscopy Sciences, cat. no. 15710). ! CAUTION As a formaldehyde releasing agent, paraformaldehyde is a potential carcinogen and a suspected reproductive hazard. Handle PFA with safety precautions.

42. Triton X-100 (Sigma-Aldrich, cat. no. X100-100ML).
43. Polyvinylpyrrolidone (PVP) (Sigma-Aldrich, cat. no. P0930).
44. Polysorbate 20 (Tween-20) (Fisher Scientific, cat. no. BP337-100).
45. Glycine (Sigma-Aldrich, cat. no. G5417-100G).
46. Mouse anti-GATA-6 monoclonal antibody (R&D Systems, cat. no. MAB1700).
47. Goat anti-GATA-3 polycolonal antibody (R&D Systems, cat. no. AF2605).
48. Rabbit anti-OCT-4 polycolonal antibody (Cell Signaling, cat. no. 2750S). *Note: Store at −20 °C. Do not aliquot as recommended by the manufacturer.*
49. Donkey anti-goat IgG Alexa Fluor™ Plus 647 (Invitrogen, cat. no. PIA32849).
50. Donkey anti-rabbit IgG Alexa Fluor™ 488 (Invitrogen, cat. no. A21206).
51. Goat anti-mouse IgG Alexa Fluor™ 555 (Invitrogen, cat. no. A21424).
52. NucBlue™ Live Cell Stain ReadyProbes Reagent (Hoechst 33342) (Invitrogen, cat. no. R37605).

2.2 Equipment

1. Hypoxic (5% O_2, 5% CO_2) incubator with humidity and gas control.
2. Normoxic (20% O_2, 5% CO_2) incubator with humidity and gas control.
3. Biological safety cabinet.
4. AggreWell 400 plates (STEMCELL Technologies, cat. no. 34415).
5. Centrifuge.
6. Water bath.
7. Inverted microscope.
8. Confocal microscope.
9. Stereomicroscope.
10. Conical tubes, 15 mL.
11. 6-Well Flat-Bottom Plate, Tissue Culture-Treated
12. Hemocytometer.
13. Syringe filter, 0.2 μm (Sigma-Aldrich, cat. no. SLGVR33RB).
14. Cell strainers, 37 μm and 70 μm (STEMCELL Technologies, cat. no. 27215, 27216), 20 μm (Pluristrainer, 43-50020-03).
15. μ-Slide 8 Well ibiTreat dishes (Ibidi, cat. no. 80826),

16. Pipettes (Eppendorf, P1000, P200, and P20)
17. −20 °C freezer.
18. −80 °C freezer.
19. Glass mouth pipette.
20. Common consumables (serological pipettes, centrifuge tubes, pipette tips, and blades).
21. OptiPrep™ (STEMCELL Technologies, cat. no. 07820).

2.3 Reagent Setup

2.3.1 β-Estradiol (E2)

Dissolve 1 mg in 1 mL of 100% EtOH to prepare a 1 mg/mL initial stock solution. Dilute initial stock solution with 9 mL of DMEM/F12 to make a 0.367 mM final stock solution that can be aliquoted and stored for up to 6 months at −80 °C.

2.3.2 Progesterone (P4)

Dissolve 0.2 mg in 1 mL of MilliQ H_2O to prepare a 200 μg/mL stock solution.

2.3.3 Sodium Pyruvate

Dissolve 22 mg in 10 mL of DMEM/F12 to prepare a 40 mM stock solution.

2.3.4 Gentamicin

Dilute 400 μL of 50 mg/mL Gentamicin solution with 3.6 mL of MilliQ H_2O to prepare a 5 mg/mL stock solution that can be stored until expiration date at 4 °C.

2.3.5 MEF Medium

To prepare 200 mL of MEF medium, mix 188 mL of Advanced DMEM with 10 mL of Fetalgro EX bovine growth serum and 2 mL of GlutaMAX (100×). Store it for up to 4 weeks at 4 °C. Alternatively, Advanced DMEM can be replaced with DMEM high glucose and Fetalgro EX can be replaced for other fetal bovine serum (FBS) alternatives at a final concentration of 10% (vol/vol).

2.3.6 Immortalized Stromal Cell Culture Medium

To prepare 100 mL of the medium, mix 93 mL of Advanced DMEM with 5 mL of Fetalgro EX bovine growth serum, 1 mL of ITS-X, and 1 mL of GlutaMAX (100×). Store it for up to 4 weeks at 4 °C. Alternatively, the Fetalgro EX can be replaced with 20% (vol/vol) FBS commercial alternatives.

2.3.7 N2B27 Basal Medium

Prepare the medium by adding 1× N2 supplement, 1× B27 supplement, 1× GlutaMAX, 1× NEAA, and 2-mercaptonethanol (final concentration 0.1 mM) to 1:1 (vol/vol) mixture of DMEM/F12 and neurobasal medium. Store it for up to 4 weeks at 4 °C.

2.3.8 Washing Medium

Prepare the medium by supplementing DMEM/F12 with 0.1% (wt/vol) BSA. Store it for up to 4 weeks at 4 °C.

2.3.9 4.5i/L/A Medium

Prepare the medium by adding 0.5% penicillin–streptomycin, BSA (final concentration 50 μg/mL), 0.5% knockout serum replacement, and the following small molecules and cytokines: PD0325901 (final concentration 1 μM), IM-12 (final concentration 0.5 μM), SB590885 (final concentration 0.5 μM), WH-4-023 (final concentration 1 μM), human LIF (final concentration 20 ng/mL), activin A (final concentration 20 ng/mL), and Y-27632 (final concentration 5 μM) to N2B27 basal medium. Store at 4 °C for up to 1 week.

2.3.10 eHDM Medium

Prepare the medium by adding 0.5% penicillin–streptomycin, FGF2 (final concentration 20 ng/mL), activin A (final concentration 20 ng/mL), CHIR99021 (final concentration 3 μM), and CEPT cocktail [Chroman 1 (final concentration 50 nM), Emricasan (final concentration 5 μM), 1× Polyamine Supplement, and Trans-ISRIB (final concentration 0.7 μM)] to N2B27 basal medium. Prepare the medium in advance to allow at least 4 h equilibration at 37 °C under hypoxic conditions (5% O_2, 5% CO_2). Store at 4 °C for up to 1 week.

2.3.11 eTDM Medium

To prepare eTDM medium, add 0.25× N2 supplement, 0.25× B27 supplement, 0.5× GlutaMAX, 0.5× NEAA, 2-mercaptonethanol (final concentration 0.1 mM), 0.5% ITS-X, 0.5% penicillin–streptomycin, and the following small molecules and cytokines: PD0325901 (final concentration 1 μM), A83-01 (final concentration 1 μM), SB590885 (final concentration 0.5 μM), WH-4-023 (final concentration 1 μM), human LIF (final concentration 10 ng/mL), LPA (final concentration 0.5 μM), and CEPT cocktail [Chroman 1 (final concentration 50 nM), Emricasan (final concentration 5 μM), 1× Polyamine Supplement, and Trans-ISRIB (final concentration 0.7 μM)] to 3:1 (vol/vol) mixture of DMEM/F12 and neurobasal medium. Prepare the medium in advance to allow at least 4 h equilibration at 37 °C under hypoxic conditions (5% O_2, 5% CO_2). Store at 4 °C for up to 1 week.

2.3.12 EBC1 Medium

Prepare the medium by adding 15% FCS, E2 (final concentration 8 nM), P4 (final concentration 200 ng/mL), sodium pyruvate (final concentration 1 mM), Gentamicin (final concentration 10 μg/mL), and CEPT cocktail [Chroman 1 (final concentration 50 nM), Emricasan (final concentration 5 μM), 1× Polyamine Supplement, and Trans-ISRIB (final concentration 0.7 μM)] to N2B27 basal medium. Prepare the medium in advance to allow at least 4 h equilibration at 37°C under normoxic conditions (20% O_2, 5% CO_2). Store at 4 °C for up to 1 week.

2.3.13 Fixation Buffer

Prepare fixation buffer by adding 0.05 g PVP (0.1% (wt/vol)) in 50 mL DPBS. Dilute 16% PFA with fixation buffer to a final concentration of 4% PFA solution. Filter through 0.2 μm syringe filter. Store at 4 °C for up to 6 months.

2.3.14 Holding Buffer

Prepare holding buffer by adding 0.25 g BSA (0.5% (wt/vol)) in 50 mL DPBS add 0.1% Triton X-100. Filter through 0.2 μm syringe filter. Store at 4 °C for up to 6 months.

2.3.15 Washing Buffer

Prepare washing buffer by adding 500 μL of Tween-20 (0.1% (vol/vol)) and 0.5 g PVP (0.1% (wt/vol)) in 500 mL DPBS. Filter through 0.2 μm syringe filter. Store at 4 °C for up to 6 months.

2.3.16 Permeabilization Buffer

Prepare permeabilization buffer by adding 150 μL of Triton X-100 (1–0.3% (vol/vol)) in 50 mL DPBS. Filter through 0.2 μm syringe filter. Store at 4 °C for up to 6 months.

2.3.17 Blocking Buffer

Prepare blocking buffer by adding 50 μL of Tween-20 (0.1% (vol/vol)), 0.5 g BSA (1.0% (wt/vol)), and 0.3754 g glycine (final concentration 100 mM) in 50 mL DPBS. Filter through 0.2 μm syringe filter. Store at 4 °C for up to 6 months.

2.3.18 Antibody Buffer

Prepare antibody buffer by adding 50 μL of Tween-20 (0.1% (vol/vol)) and 0.5 g BSA (1.0% (wt/vol)) in 50 mL DPBS. Filter through 0.2 μm syringe filter. Store at 4 °C for up to 6 months.

3 Methods

3.1 Culturing of Naive hPSCs

3.1.1 Preparation of iMEF Feeder Layers—Timing 1 h

CRITICAL (*See* **Notes 1–3**)

1. Prewarm MEF medium at 37 °C.
2. Prepare a gelatin-coated six-well culture plate by adding 1 mL of 0.1% (wt/vol) gelatin solution per well.
3. Incubate the plate at 37 °C for 1–2 h.
4. Thaw the iMEFs in a 37 °C water bath by shaking the cryovial occasionally until no ice clump presents (*see* **Note 24**).
5. Wipe the cryovial with 70% (vol/vol) ethanol to sterilize before bringing it into the biological safety cabinet.
6. Place the cell mixture in the bottom of a 15 mL conical tube, and slowly add 10 mL of prewarmed washing medium to avoid osmotic shock in the cells.
7. Centrifuge the cells at 200 × g for 5 min.
8. Aspirate the supernatant from the tube and resuspend the cells in 1 mL of prewarmed MEF medium.

9. Before seeding the iMEFs to wells, prepare the plates of gelatin by removing gelatin solution and adding 2 mL of the warm MEF medium to each well.
10. Count cells using a hemacytometer or other cell counting device according to manufacturer protocols and transfer the iMEFs at a density of 1×10^6 cells per well.
11. Distribute the cells evenly by moving the plate in back-and-forth and side-to-side motions.
12. Incubate the iMEFs at 37 °C under normoxic conditions (20% O_2, 5% CO_2) in a humidified environment.

3.1.2 Thawing 4.5i/L/A Naive hPSCs—Timing 30 min

CRITICAL (*See* **Note 4**)

1. Thaw an aliquot of Geltrex on ice. Dilute Geltrex Matrix with prechilled DMEM/F12 medium (final concentration 1% (vol/vol)). Keep on ice until use (*see* **Note 5**).
2. Prewarm 4.5i/L/A medium and washing medium at 37 °C.
3. Prepare the plates of iMEFs by aspirating MEF medium and washing the iMEFs twice with DPBS. Add 2 mL of 4.5i/L/A medium supplemented with CEPT cocktail for each well and equilibrate the plate at 37 °C under hypoxic conditions (5% O_2, 5% CO_2) (*see* **Notes 25** and **26**).
4. Thaw a cryovial of naive hPSCs in a 37 °C water bath by shaking the cryovial occasionally until no ice clump presents.
5. Wipe the cryovial with 70% (vol/vol) ethanol in distilled H_2O to sterilize before bringing it into the biological safety cabinet.
6. Place the cell mixture in the bottom of a 15 mL conical tube, and slowly add 10 mL of prewarmed washing medium to avoid osmotic shock in the cells.
7. Centrifuge the cells at 200 × g for 5 min.
8. Aspirate the supernatant from the tube and resuspend the cells gently in 1 mL of prewarmed 4.5i/L/A medium supplemented with CEPT cocktail.
9. Count cells and transfer the naive hPSCs onto the iMEFs at a density of 3×10^4 cells/cm^2. Add diluted Geltrex solution with a final concentration of 0.5 μL/cm^2 for each well to promote cell attachment.
10. Distribute the cells evenly by moving the plate in back-and-forth and side-to-side motions.
11. Incubate the naive hPSCs at 37 °C under hypoxic conditions (5% O_2, 5% CO_2) in a humidified environment.
12. Perform a daily medium refreshment with 2 mL equilibrated 4.5i/L/A medium per well.
13. Once naive hPSCs are stabilized and expanded, passage the cells every 3 to 4 days onto fresh iMEF-coated plates (*see* **Note 6**).

3.1.3 Confirmation of the Extraembroynic Lineage Plasticity of 4.5i/L/A Naive hPSCs—Timing 2 h

1. Prepare a gelatin-coated six-well culture plate by adding 1 mL of 0.1% (wt/vol) gelatin solution per well and incubating the plate at 37 °C for at least 2 h.
2. Prewarm 4.5i/L/A medium and washing medium at 37 °C.
3. Aspirate the medium from naive hPSCs and wash with DPBS twice.
4. Dislodge the naive hPSCs from the well with 500 μL TrypLE per well at 37 °C for 3 min.
5. Use a P1000 pipette to dissociate the colonies into single cells and add 1 mL of prewarmed washing medium to quench dissociation.
6. Pass the cells through a 37 μm cell strainer to remove any large pieces of iMEF sheet.
7. Centrifuge the cells at 200 × g for 3 min.
8. Remove the supernatant from the tube and resuspend the pellet gently in 2 mL of prewarmed 4.5i/L/A medium supplemented with CEPT cocktail.
9. Aspirate the gelatin solution from the gelatin-coated plate and transfer the cell suspension into wells. Incubate the plate at 37 ° C under hypoxic conditions (5% O_2, 5% CO_2) for 40–60 min to deplete iMEF cells. To ensure complete removal repeat this twice.
10. Collect the supernatant containing the naive hPSCs and replate the cells onto a Matrigel-coated 12-well plate at a density of 1×10^5 cells per well in eHDM media with 1× CEPT (*see* **Note 7**).
11. Incubate the naive hPSCs at 37 °C under hypoxic conditions (5% O_2, 5% CO_2) in a humidified environment for 24 h.
12. Prepare and equilibrate eHDM medium on the next day, warm, and carefully perform a full medium change with equilibrated eHDM.
13. Day 3–11: Prepare and equilibrate eTDM medium on the day of usage. Culture the cells in eTDM and refresh medium every 2 days.
14. Fix and perform immunofluorescent staining (protocol below) to confirm the lineage marker expression (*see* **Note 8**).

3.2 Generation of Human Blastoid—Timing 7–9 Days

3.2.1 Preparation of the AggreWell Plates—Timing 10–20 min

1. Wipe the sleeve containing an AggreWell plate with 70% (vol/-vol) ethanol to sterilize before bringing it into the biological safety cabinet.
2. Open the plate in the biological safety cabinet.

3. Add 500 μL Anti-Adherence Rinsing Solution per well. Note: Dispense Anti-Adherence Rinsing Solution slowly without introducing air bubbles to wells.
4. Centrifuge the plate at 1500 × g for 5 min.
5. Keep the plate treated with Anti-Adherence Rinsing Solution at room temperature until use (at least 5 min).

3.2.2 Aggregation of Naive hPSCs in the AggreWell Plates—Timing 2–3 h

1. Prepare a gelatin-coated six-well culture plate by adding 1 mL of 0.1% (wt/vol) gelatin solution per well and incubating the plate at 37 °C for at least 2 h.
2. Prepare and equilibrate sufficient 4.5i/L/A medium supplemented with CEPT cocktail and eHDM medium at 37 °C under hypoxic conditions (5% O_2, 5% CO_2). Prewarm washing medium.
3. Aspirate the medium from naive hPSCs and wash with DPBS twice.
4. Dislodge the naive hPSCs from the well with 500 μL TrypLE per well at 37 °C for 3 min (*see* **Note 9**).
5. Use a P1000 pipette to dissociate the colonies into single cells and add 1 mL prewarmed washing medium to quench dissociation. CRITICAL (*see* **Note 10**).
6. Centrifuge the cells at 200 × g for 3 min.
7. Aspirate the supernatant and resuspend the cell pellet in 4.5i/L/A medium supplemented with CEPT cocktail.
8. Add DNase to the cell suspension at a final concentration at 10 μL/mL to dissolve DNA meshes formed from dead cells, maintain during **step 9** incubation.
9. Aspirate the gelatin solution from the gelatin-coated plate and transfer the cell suspension into wells. Incubate the plate at 37 °C under hypoxic conditions (5% O_2, 5% CO_2) for 40–60 min to remove iMEF cells.
10. Carefully collect the supernatant containing naive hPSCs without disturbing the attached iMEFs.
11. Pass the supernatant through a 70 μm cell strainer to exclude any large cell clumps.
12. Perform another round of iMEF depletion by passing the cell suspension through a 40 μm cell strainer.
13. Gently pipette up and down 2–3 times to mix the cell content thoroughly. Count the cells manually using a hemocytometer and adjust cell concentration accordingly (*see* **Notes 11–13**).
14. For the naive WIBR3 human ES cells, 18 ± 4 cells should be seeded per microwell (~100 wells) (*see* **Note 14**).
15. Centrifuge the cells at 200 × g for 3 min.

16. Aspirate the supernatant and resuspend the cell pellet in equilibrated eHDM.
17. Aspirate Anti-Adherence Rinsing Solution from the wells of the AggreWell plate and rinse the wells twice with DPBS.
18. Add 1 mL of the suspension of cells with the corresponding concentration in eHDM to each well of a pretreated AggreWell plate.
19. Mix the cell suspension evenly by gently pipetting in zig-zag motions 2–3 times using a P200 pipette.
20. Centrifuge the AggreWell plate at 1200 × g for 2 min. CRITICAL (*see* **Note 15**).
21. Incubate the AggreWell plate at 37 °C under hypoxic conditions (5% O_2, 5% CO_2) in a humidified environment. The day of cell plating is designated as day 0.

3.2.3 Blastoid Development—Timing 7–8 Days

1. Day 1: Prepare eHDM medium (without CEPT cocktail). Aliquot sufficient medium (500 μL per well) and store at 4 °C for up to 1 week.
2. Supplement aliquoted eHDM medium with CEPT cocktail and equilibrate the medium at 37 °C under hypoxic conditions (5% O_2, 5% CO_2) 4 h before use.
3. Carefully remove half (500 μL) of wasted medium from the wells and add 500 μL of fresh medium by leaning P200 tips against the wall of the wells and pipetting slowly.
4. Day 2: Repeat **steps 2–3** to perform a half-medium change with eHDM.
5. Day 3: Prepare eTDM medium (without CEPT cocktail and A83-01). Aliquot sufficient eTDM medium (750 μL per well) store at 4 °C for up to 1 week.
6. Supplement aliquoted eTDM medium with fresh CEPT cocktail and A83-01 and equilibrate the medium at 37 °C under hypoxic conditions (5% O_2, 5% CO_2) before use.
7. Carefully remove as much eHDM as possible and add 1.5 mL of eTDM (*see* **Note 16**).
8. Day 4–8: Equilibrate eTDM medium on the day of usage. Perform a half-medium refreshment with eTDM every 2 days.
9. Incubate the AggreWell plate at 37 °C under hypoxic conditions (5% O_2, 5% CO_2) in a humidified environment.
10. The human blastoids usually mature after approximately 5 to 6 days of incubation in eTDM and can be harvested for immunostaining or implantation assay.

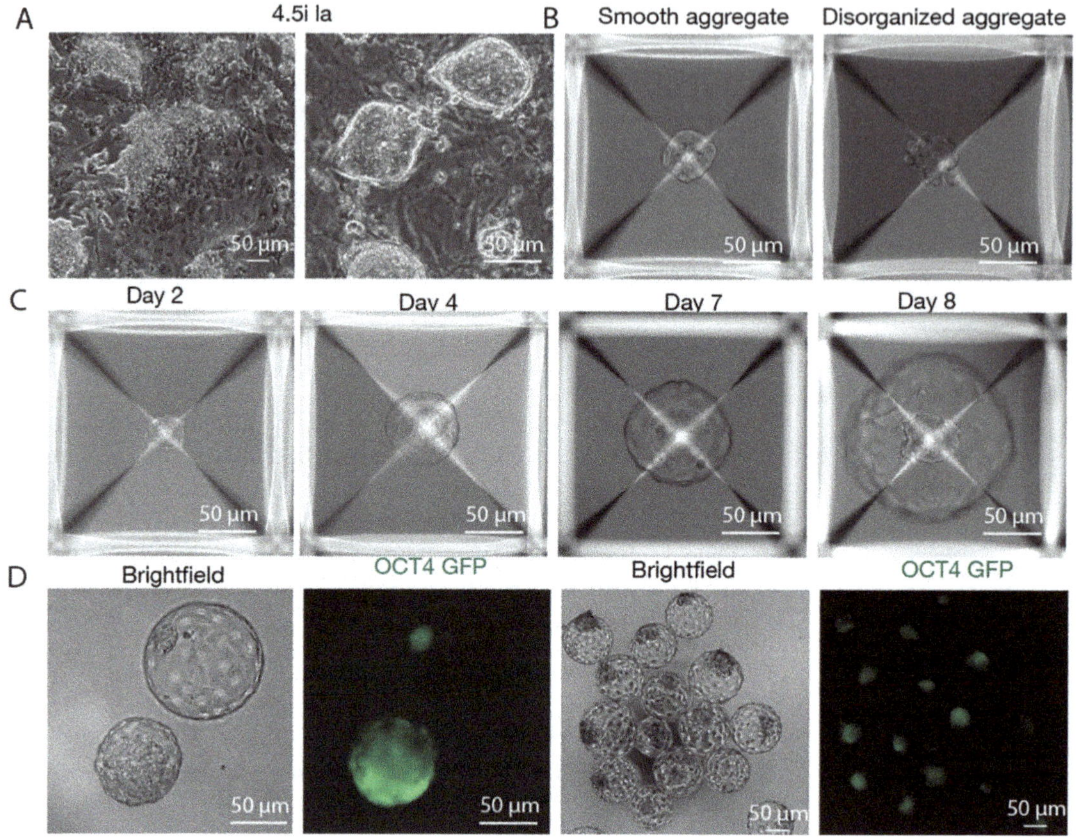

Fig. 1 Generation of human blastoids from 4.5i/L/A ΔPE–OCT4–GFP WIBR3 hESCs [1] that contain an OCT4-GFP reporter. (**a**) Representative bright field images of hESCs grown in 4.5i/L/A. (**b**) Representative bright field images of an optimal aggregate (left) with smooth, polarized edges, and a suboptimal disorganized aggregate (right). Notice the presence of cell debris. (**c**) Representative bright field images showing blastoid generation at different time points. (**d**) Example blastoids batch. Left panels show expanded blastoid and early blastoid. Notice early trophectoderm has not silenced OCT4 expression

3.3 Human Blastoid Handling—Timing 30 min

1. Widen the caliber of a 1 mL pipette tip by cutting 2–3 mm from the thin tip end to minimize the shearing force (*see* **Note 17**).
2. After blastoid is generated, gently pipette up and down 2–3 times to suspend the structures and transfer the suspension to a well of six-well culture plate using the cut tips.
3. Prepare and preload the swan-neck of a glass mouth pipette with eTDM medium. Pick up intact blastoids displaying standard blastocyst morphology (Fig. 1b) using the glass mouth pipette under a stereomicroscope with a heating stage and proceed to the following experiments.

3.4 Differential Immunofluorescent Staining of Human Blastoids—Timing 3 Days

3.4.1 Human Blastoid Fixation—Timing 30 min

1. Prepare a fresh 4% (vol/vol) PFA solution by mixing fixation buffer and 16% PFA in a 3:1 ratio.
2. Transfer blastoids into a well containing 4% PFA solution and incubate for 20 min at room temperature to fix the samples (*see* **Note 18**).
3. Transfer fixed samples to holding buffer and maintain at 5 °C until proceeding to immunostaining. Cover holding buffer containing the samples with a thin layer of mineral oil to minimize evaporation or store in 1.5 mL tubes.

3.4.2 Incubation with Primary Antibodies—Timing 2–3 h

1. Wash samples three times with washing buffer for 5 min per wash.
2. Transfer samples into a well containing permeabilization buffer and incubate for 30 min at room temperature.
3. Wash samples once with washing buffer for 5 min.
4. Prepare a complete blocking solution by adding 10% donkey serum to blocking buffer. Incubate samples in the complete blocking solution at room temperature for 2 h on a shaker set at low speed.
5. Wash samples once with washing buffer for 5 min.
6. Prepare a primary antibody solution using antibody buffer with optimal dilutions suggested by the manufacturer. Incubate samples in the primary antibody solution at 5 °C overnight on a shaker set at low speed.

3.4.3 Incubation with Secondary Antibodies—Timing 2 h

1. Wash samples three times with washing buffer for 5 min per wash.
2. Prepare a secondary antibody solution using antibody buffer with optimal dilutions suggested by the manufacturer. Add NucBlue Live Cell Stain (2 drops/mL) directly to the secondary antibody solution (*see* **Note 19**).
3. Incubate samples in the secondary antibody solution at room temperature for 1 h on a shaker set at low speed.
4. Wash samples once with washing buffer for 5 min.
5. Wash samples once with 1:1 (vol/vol) mixture of washing buffer and antibody buffer for 5 min (*see* **Note 20**).
6. Wash samples once with washing buffer for 5 min.
7. Place samples in a 250 μL washing buffer on μ-Slide 8 well imaging slide.
8. Perform fluorescence imaging by confocal microscope (Fig. 2).

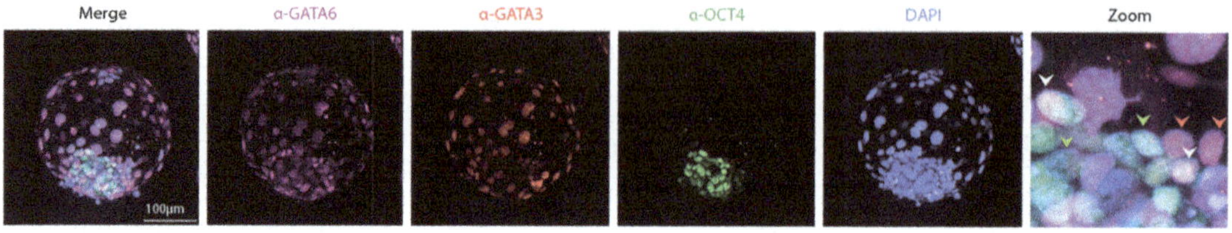

Fig. 2 Immunostaining of human blastoids. Immunofluorescence staining and confocal imaging of human basltoid with GATA6 (magenta), GATA3 (red), and OCT4 (green). White arrows indicate GATA6 and OCT4 double-positive cells, hypoblast cells. Green arrows indicate OCT4-only positive cells, epiblast cells. Red arrows indicate GATA6- and GATA3-positive cells, trophectoderm cells

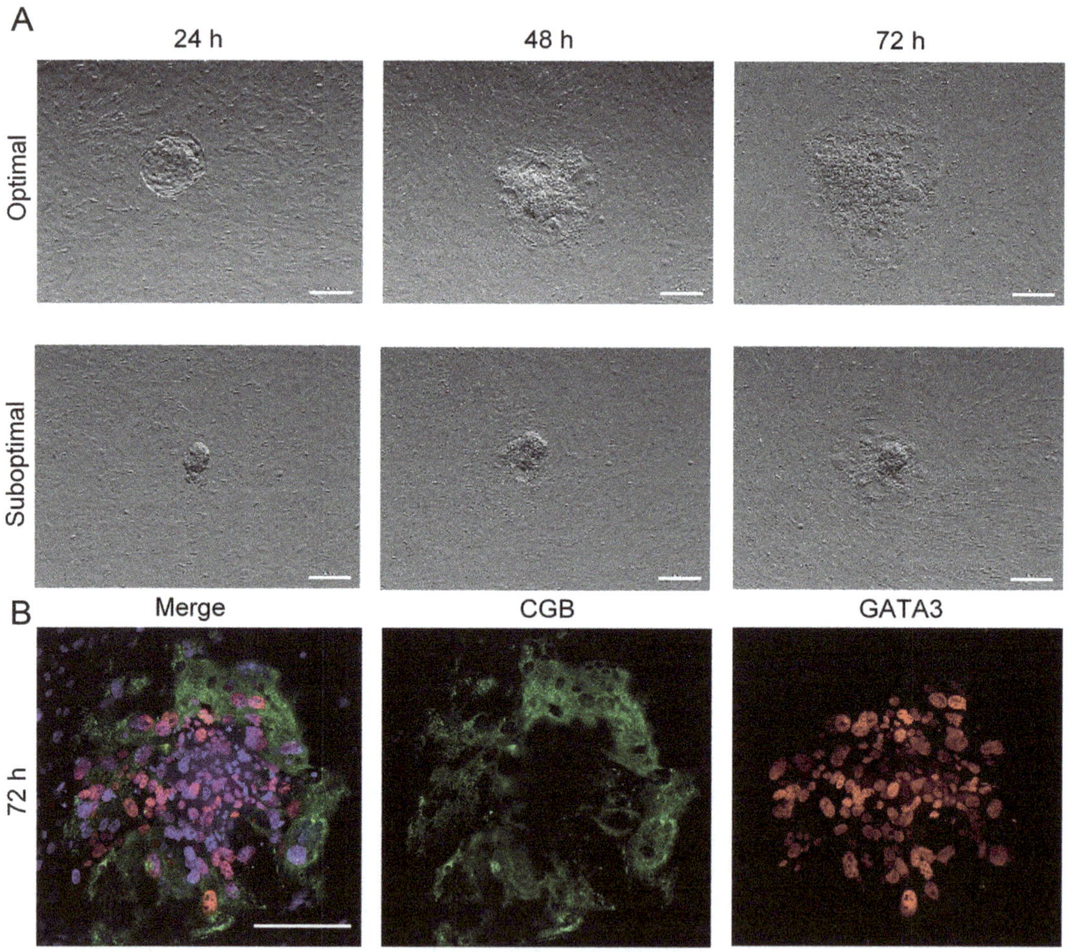

Fig. 3 Blastoid development during the peri-implantation period. (**a**) Optimal (top panel) and suboptimal (bottom panel) growth over 24 h, 48 h, and 72 h in extended co-culture on immortalized stromal cells. (**b**) Representative immunoflourescent staining of human chorionic gonadotropin subunit beta (CGB), and GATA3 after 72 h of co-culture with immortalized stromal cells. Scale bar = 200 μm

3.5 Peri-implantation Modeling of Human Blastoids (Fig. 3)—Timing 3–4 Days

3.5.1 Preparation of the Extended Culture Plates—Timing 30 min

CRITICAL (*See* **Note 21**)

1. Prewarm immortalized stromal cell medium and washing medium at 37 °C.
2. Wipe the sleeve containing an eight-well ibidi dish with 70% (vol/vol) ethanol to sterilize before bringing it into the biological safety cabinet.
3. Open the dish in the biological safety cabinet while taking care not to touch the wells.
4. Retrieve the culture plate culturing stromal cells and detach cells using TrypLE at 37 °C for 3 min.
5. Use a P1000 pipette to dislodge the cells from the plate and add prewarmed washing medium to quench dissociation.
6. Centrifuge the cells at 200 × g for 4 min.
7. Aspirate the supernatant from the tube and resuspend the cells in 1 mL of prewarmed immortalized stromal cell medium.
8. Transfer the stromal cells at a density of 6.9×10^4 cells/cm^2 to the ibidi dish.
9. Distribute the cells evenly by gently pipetting up and down 2–3 times.
10. Incubate the ibidi dish at 37 °C under hypoxic conditions (5% O_2, 5% CO_2) in a humidified environment.

3.5.2 Blastoid Extended culture [6]—Timing 3 Days

1. Prepare EBC medium and equilibrate the medium at 37 °C under normoxic conditions (20% O_2, 5% CO_2) for at least 4 h.
2. Retrieve the ibidi dish coated with stromal cells to the biological safety cabinet. Discard all wasted stromal cell medium and add 300 μL of equilibrated EBC medium to each well (*see* **Note 22**).
3. Return the ibidi dish to the incubator until blastoid transfer.
4. Prepare a center-well tissue culture dish with 500 μL of equilibrated EBC medium. Transfer the selected blastoids to the center-well culture dish for a brief wash (*see* **Note 23**).
5. Carefully move blastoids to each well of the ibidi dish and maintain blastoid identification if necessary.
6. Incubate the ibidi dish at 37°C under normoxic conditions (20% O_2, 5% CO_2) in a humidified environment for 2 days.
7. Day 2: Prepare EBC medium and equilibrate the medium at 37 °C under normoxic conditions (20% O_2, 5% CO_2) for at least 4 h.
8. Assess the attachment of blastoids each day by gently tapping the ibidi dish and examining whether the blastoid has securely attached to the stromal cells under microscope.
9. Perform a half-medium change with equilibrated EBC medium without disturbing the attached blastoids.

10. Incubate the ibidi dish at 37 °C under normoxic conditions (20% O_2, 5% CO_2) in a humidified environment for one additional day.
11. After 3 days in stromal co-culture, samples can be used for immunostaining or single cell digestion.

4 Notes

1. CRITICAL Successful maintenance of naive hPSCs in 4.5i/L/A medium requires high-quality inactivated mouse embryonic fibroblasts (iMEFs) as feeder cells. MEFs derived from either the CF-1 or DR4 mouse strains at embryonic day (E) 12.5 are recommended.
2. CRITICAL The density of feeder layers is another crucial parameter for maintenance of naive hPSCs in 4.5i/L/A medium. In our experience, 4.5i/L/A medium is toxic to iMEFs. The density of iMEFs typically used for culturing naive hPSCs often decreases in exposure of 4.5i/L/A medium, which will lead to increased spontaneous differentiation and loss of naive pluripotency characteristics (e.g., appearance of a flat morphology rather than dome-shaped morphology, ambiguous colony border integrity without refractive edges). Thus, we recommend plating iMEFs at high density (at a confluency of approximately 80–90%) to achieve optimal feeder layer density for 4.5i/L/A cultured naive hPSCs.
3. CRITICAL Ideally, the iMEFs should be plated for 24–48 h before use.
4. CRITICAL We recommend culturing naive hPSCs on iMEF feeders at 37 °C under hypoxic conditions (5% O_2, 5% CO_2) as previously reported [5].
5. Geltrex Matrix diluted below 9 mg/mL does not form a gel. Store aliquots at −80 °C to −20 °C and avoid multiple freeze/thaw cycles. When working from a 5-mL diluted solution, keep on ice if it is not used within 5 min.
6. Do not let cells become overly confluent. Keep the colonies sparse to avoid any spontaneous differentiation.
7. The plating cell number will need to be increased if the cell viability/proliferation is low.
8. We suggest using the following antibodies to verify lineage differentiation: epiblast (SOX2, OCT4, and KLF17), trophoblast (GATA3, GATA2, or TFAP2C), and hypoblast (GATA6, GATA4, and SOX17). After 9 days of differentiation usually >20% of cells are either GATA3$^+$ or GATA6$^+$.
9. The passage number of naive hPSCs is crucial for successful blastoid generation. It is better to use the early passage naive hPSCs (below **Note 20**).

10. CRITICAL Excessive force during the single cell dissociation and subsequent steps will compromise naive hPSCs viability and reduce blastoid generation efficiency.
11. A small amount of iMEFs may still remain even after a sequential iMEFs exclusion, 20 μm cell strainer can be used if many iMEFs remain. Only count the cells with small 16–20 μm. iMEFs normally present larger than hPSCs.
12. CRITICAL Too much cell debris can negatively affect the blastoid formation.
13. OptiPrep™ is a sterile endotoxin tested solution of 60% iodixanol in water with a density of 1.32 g/ml that can be used to remove nonviable cells. Cells that are nonviable have a compromised membrane integrity and the iodixanol will enter the cytosolic compartment of the cells generating a slightly higher density. Adjust the density of the cell suspension to 1.15–1.16 g/ml using OptiPrep™, up to a volume of 2 mL and overlay slowly at a 45° angle the solution with 2 mL of a solution of 1.12 g/mL, and 1 mL of PBS on top. After centrifugation at 700 × g for 30 min at 90° angle, the viable cells will form a band at the top of the upper layers (just below the PBS) while nonviable cells and contaminants remain in the load zone. Human ESCs viability of up to 98% has been achieved with no visible cell debris. As an alternative to OptiPrep™, cells can be centrifuged at low speed 300 rpm for 10 minutes to recover only viable cells in the pellet.
14. To reach the maximum efficiency of blastoid formation, optimal starting cell number needs to be titrated, as cell proliferation rate can vary across different hPSC lines.
15. CRITICAL Ensure equal distribution of the cells within the AggreWell. Repeat **steps 19–20** or adjust the centrifuge as necessary to ensure no angle differences are present.
16. The aggregates may fuse together if the well is disturbed, which will dramatically affect the final blastoid formation efficiency.
17. Wide-bore filtered pipette tips are also commercially available.
18. Do not use a mouth pipette to move blastoids in and out of PFA.
19. In order to avoid photo bleaching, protect the secondary antibody solution and samples from light in all subsequent steps.
20. BSA in the antibody buffer helps to prevent samples from sticking to the glass pipette.
21. CRITICAL We use endometrial stromal cells obtained from patient biopsy and immortalized by telomerase reverse transcriptase (TERT) overexpression for extended culture. In brief,

isolated primary stromal cells are infected with Lentivirus particles for hTERT overexpression driven by EEF1A1 promoter (GeneCopoeia, LP730-025) as described by manufacturer. We recommend using immortalized stromal cells within 9 passages for extended culture, and thawing and plating cells 3 days before experiments for the recovery and expansion. Prepare extended culture plate 1 day prior to blastoid harvesting.

22. Use caution to avoid letting fluid touch the lid as this will increase the risk of contamination.
23. Use care to minimize oil carryover into the ibidi dish. Incompatible oil will crack the dishes and will disrupt downstream fixation and staining.
24. CRITICAL Successful maintenance of naive hPSCs in 4.5i/L/A medium requires high-quality inactivated mouse embryonic fibroblasts (iMEFs) as feeder cells. MEFs derived from either the CF-1 or DR4 mouse strains at embryonic day (E) 12.5 are recommended.
25. CRITICAL The density of feeder layers is another crucial parameter for maintenance of naive hPSCs in 4.5i/L/A medium. In our experience, 4.5i/L/A medium is toxic to iMEFs. The density of iMEFs typically used for culturing naive hPSCs often decreases on exposure to 4.5i/L/A medium, which will lead to increased spontaneous differentiation and loss of naive pluripotency characteristics (e.g., appearance of a flat morphology rather than dome-shaped morphology, ambiguous colony border integrity without refractive edges). Thus, we recommend plating iMEFs at high density (at a confluency of approximately 80–90%) to achieve optimal feeder layer density for 4.5i/L/A cultured naive hPSCs.
26. CRITICAL Ideally, the iMEFs should be plated for 24–48 h before use.

References

1. Gafni O et al (2013) Derivation of novel human ground state naive pluripotent stem cells. Nature 504:282–286. https://doi.org/10.1038/nature12745
2. Fu J, Warmflash A, Lutolf MP (2021) Stem-cell-based embryo models for fundamental research and translation. Nat Mater 20:132–144. https://doi.org/10.1038/s41563-020-00829-9
3. Yu L et al (2021) Blastocyst-like structures generated from human pluripotent stem cells. Nature 591:620–626. https://doi.org/10.1038/s41586-021-03356-y
4. Yu L, et al (2022) Large scale production of human blastoids amenable to modeling blastocyst development and maternal-fetal crosstalk. bioRxiv, 2022.2009.2014.507946. https://doi.org/10.1101/2022.09.14.507946
5. Theunissen TW et al (2014) Systematic identification of culture conditions for induction and maintenance of naive human pluripotency. Cell Stem Cell 15:471–487
6. Logsdon DM, Kile RA, Schoolcraft WB, Krisher RL, Yuan Y (2020) Single cell collection of trophoblast cells in peri-implantation stage human embryos. J Vis Exp. https://doi.org/10.3791/61476

Methods in Molecular Biology (2024) 2767: 19–25
DOI 10.1007/7651_2022_466

Published online: 15 December 2022

Human Pre-gastrulation Embryo Culture in 3D Condition

Lifeng Xiang, Yu Yin, Gaohui Shi, and Tianqing Li

Abstract

The development process of human embryo until blastocyst is well understood during the past 30 years, however, embryogenesis from blastocyst to pre-gastrulation was still remained a "black box". Limited by research materials and culture technologies, the "black box" is still unopened. We recently established an extended three-dimensional (3D) culture system of human blastocysts (Xiang et al., Nature 577(7791): 537–542, 2020). The 3D embryo culture system could enable human blastocyst growing up to early primitive streak anlage stage in vitro. Here, we introduce the detail protocol and notes of culturing human 3D embryos.

Key words Human embryo, Pre-gastrulation, Primitive streak, Three-dimensional (3D) culture

1 Introduction

Benefiting from the development of assisted reproductive technology, the culture system of pre-implantation human embryo has become very skilled [2, 3]. Based on this, studies of lineage separation, epigenetic regulation, and genome activation in human pre-implantation embryo development have been widely carried out [4–7]. However, the developmental process from pre- to post-implantation is still a mystery. Recently, two research teams established an in vitro post-implantation culture system using a two-dimensional (2D) platform and revealed some developmental landmarks of in vivo human embryos development [8, 9]. Nevertheless, during the culture process, these 2D-cultured embryos become flattened beyond 11–12 days post fertilization, indicating that the system is a suboptimal model for imitating embryonic development in vivo [10–12]. To solve the issue, we recently developed a new 3D culture system which allows human blastocysts growing up to early primitive streak anlage stage [1]. Under the 3D culture condition, we could be able to reproduce the lineage separation patterns of pre-gastrulation human embryo in vitro and observe the formation of specific structures such as embryonic disc, amnion, primary and secondary yolk sac, anterior–posterior axes, and primitive streak anlage.

2 Materials

2.1 Embryo Culture

1. mIVC1: Advanced DMEM/F12 supplemented with 20% (vol/vol) DFBS, 2 mM L-glutamine, 1 × ITS-X, 8 nM β-oestradiol (dissolved in DMSO), 200 ng/mL progesterone (dissolved in DMSO) and 25 μM N-acetyl-L-cysteine (dissolved in ddH_2O), 0.22% (vol/vol) sodium lactate, 1 mM sodium pyruvate (dissolved in ddH_2O), 10 μM Y27632 (dissolved in DMSO). Store at 4 °C for 2 weeks.
2. mIVC2: Advanced DMEM/F12 supplemented with 30% (vol/vol) KOSR, 2 mM L-glutamine, 1 × ITS-X, 8 nM β-oestradiol (dissolved in DMSO), 200 ng/mL progesterone (dissolved in DMSO) and 25 μM N-acetyl-L-cysteine (dissolved in ddH_2O), 0.22% (vol/vol) sodium lactate, 1 mM sodium pyruvate (dissolved in ddH_2O), 10 μM Y27632 (dissolved in DMSO). Store at 4 °C for 2 weeks.

2.2 Embryo Thawing and Zona Pellucida Removing

1. Kitazato Thawing Media Kit; store at 4 °C.
2. G-2; store at 4 °C.
3. Mineral oil; store at 4 °C.
4. Tyrode's solution; store at −20 °C.

2.3 Fixation and Immunostaining Medium

1. 4% Paraformaldehyde; store at −20 °C.
2. 0.5%/0.2%Triton X-100; store at −20 °C.
3. 0.05% Tween-20 in PBS; prepare when using.
4. 3% BSA; store at −20 °C.
5. PBS; store at 4 °C.

3 Methods

Carry out all procedures at room temperature unless otherwise specified.

3.1 Embryo Thawing and Zona Pellucida Removal

1. Prepare a 4-well plate, add 0.5 mL G-2 per well and then add 0.25 mL mineral oil on the surface of G-2, transfer the 4-well plate to the CO_2 incubator, and equilibrate the medium for at least 6 h (*see* **Note 1**).
2. Thaw human blastocysts (day 5 or 6 post-fertilization, d.p.f 5–6) using a Kitazato Thawing Media Kit by strictly following the manufacturer's instructions.
3. Embryos are cultured in equilibrated G-2 medium for 4 h, and then zona pellucida of embryos are removed by the acidic Tyrode's solution (*see* **Notes 2–6**).

3.2 Evaluation of Embryo Quality

Evaluation of blastocyst quality is according to the Gardner scoring system [13]. The grade of inner cell mass should be higher than B, and the blastocysts with above 3 of expansion status are included in this study. Otherwise, embryos are excluded in the research.

3.3 In Vitro Three-Dimensional Culture of Human Embryos

1. The CO_2 incubator setting is as follows: 6% CO_2, 37.2 °C and saturated humidity.
2. Prepare a low-attachment 96-well plate, add 150 μL/well mIVC1, and pre-equilibrate the medium for at least 6 h (*see* **Note 1**).
3. Wash blastocysts in pre-equilibrated mIVC1 3 times, and transfer blastocysts to a new well for each embryo (*see* **Note 6**).
4. Before d.p.f 8, the culture medium is mIVC1. On d.p.f 8, 50% mIVC2 is replaced per well. On d.p.f 9, each embryo is transferred to a new well with mIVC2 including 10% Matrigel (*see* **Note 7**).
5. Subsequently, 50% of mIVC2 is replaced with new pre-equilibrated medium including 10% Matrigel every day (*see* **Note 8** and Figs. 1 and 2).

3.4 Frozen Section of Embryo

1. Use 4% PFA to fix the embryos at room temperature.
2. Wash embryos 3 times in PBS at room temperature.
3. Dehydrate embryos with 15% sucrose for 1 min at room temperature (*see* **Note 9**).
4. Embed embryos in a plastic mold with O.C.T (*see* **Note 10**).
5. Use Leica frozen slicer to section the embedded embryos at − 20 °C (*see* **Note 11**).
6. Bake the slices on a heating plate for 40 min at 45 °C.
7. Store the slices in −20 °C or −80 °C refrigerator.

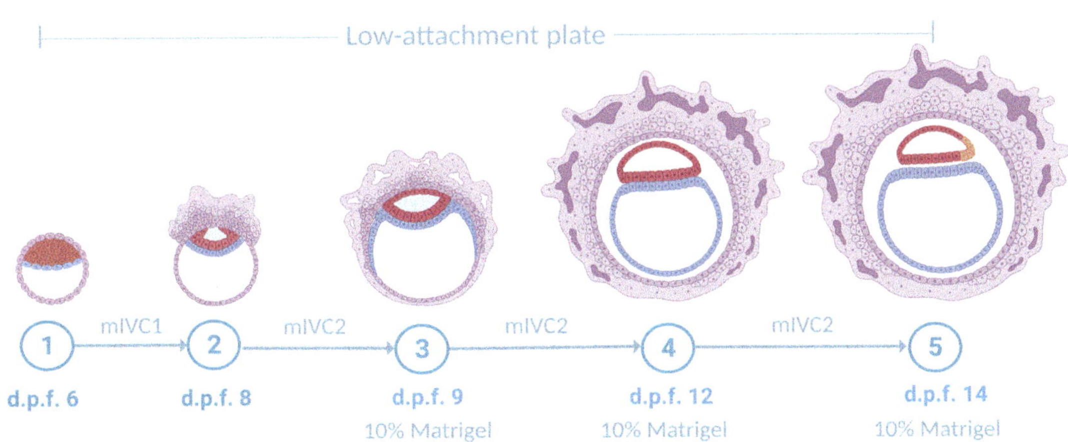

Fig. 1 Schematic diagram of embryo culture. (Figure partially created using BioRender)

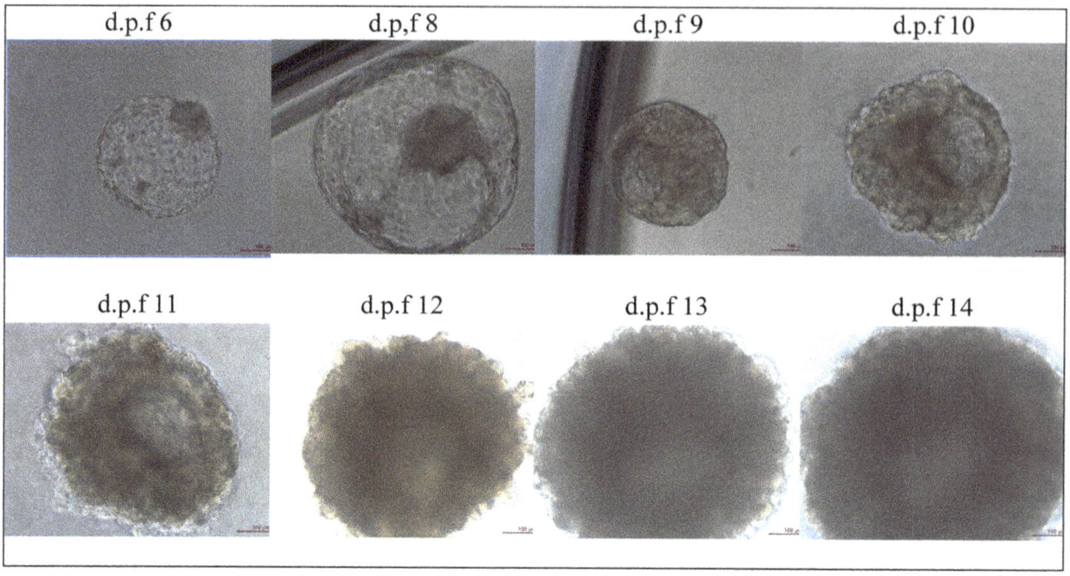

Fig. 2 Phase contrast images of human blastocysts developing through d.p.f 14. Scale bars,100 μm

3.5 Embryo Section Staining and Image Taking

1. Place embryo section slices in a dark box and wash with PBS 3 times.
2. Permeabilize the embryo sections with 0.2% Triton X-100 in PBS for 30 min at room temperature.
3. Wash the slices with PBS for one time.
4. Block samples by 3% BSA for 4 h at room temperature.
5. Wash the slices with PBS for one time.
6. Add primary antibodies with 1% BSA in PBS on the samples.
7. Incubate primary antibodies overnight in a 4 °C refrigerator (*see* **Note 12**).
8. Wash samples with 0.05% Tween-20 in PBS for 3 times, 3–5 min each time.
9. Add secondary antibodies dissolved in PBS on the samples for 2 h at room temperature (*see* **Note 12**).
10. Wash samples with 0.05% Tween-20 in PBS for 3 times, 3–5 min each time. Cover coverslips with 50% glycerine in PBS.
11. Use Confocal microscopy to take images.

3.6 Whole Embryo Staining and Photo Taking

1. Fix embryos by 4% PFA in a normal 96-well plate for 40 min in room temperature, 200 μL/well.
2. Transfer embryos to a new well that contained 200 μL PBS to wash out the PFA.

3. Transfer embryos to a new well that contained 200 μL 0.5% Triton X-100 for 2 h to permeabilize the membrane at room temperature.
4. Transfer embryos to a new well that contained 200 μL 3% BSA for 4 h to block the non-specific sites at room temperature.
5. Transfer embryos to a new well that contained 100 μL primary antibody diluted by 1% BSA.
6. Incubate primary antibodies overnight at the 4 °C refrigerator.
7. Wash embryos with 0.05% Tween-20 in PBS 3 times, 15 min each time.
8. Transfer embryos to a new well that contained 100 μL secondary antibody diluted by PBS.
9. Wash embryos with 0.05% Tween-20 in PBS for 3 times, 15 min each time.
10. Wash embryos with 50% glycerine and transferred to 8-well IbiTreat μ plates to take photos.
11. Embryos are mounted in 80% glycerol, and take photos with a multi-photon microscope.

4 Notes

1. Before use, the medium including G-2, mIVC1, and mIVC2 should be pre-equilibrated in CO_2 incubator for at least 6 h.
2. Suitable operation pipettes: to avoid damaging human embryos, the diameter of Pasteur's pipette should be larger than embryos.
3. Acidic Tyrode's solution should be warmed at 37 °C before use.
4. Not more than two embryos were operated in the acidic Tyrode's solution at the same time.
5. Once the zona pellucida is vanished, the embryos must be transferred to the G-2 medium as soon as possible.
6. When embryos are transferred to mIVC1, please make sure that the residue of Tyrode's solution is vanished completely.
7. All the operations related to Matrigel should be performed on ice.
8. Time of embryos exposed out of the incubator should be minimized when replacing medium or taking photos.
9. Before embedded in the mold, the embryo dehydration time could be extended appropriately if the sample does not sink.

10. Embryos should be washed in O.C.T 3 times before being transferred to the mold and then transferred to the −20 °C refrigerator.
11. Before frozen-slicing, make sure that the temperature of Leica frozen slicer is below −20 °C.
12. During the process of staining, samples must be kept moistly all the time and avoid drying out.

Acknowledgments

This work was supported by the National Natural Science Foundation of China (82160294), the Joint Special Funds for the Department of Science and Technology of Yunnan Province-Kunming Medical University (202101AY070001-260), and Open Project of Yunnan Provincial Reproductive and Obstetrics and Gynecology Clinical Medicine Center (zx2019-01-01and 2020LCZXKF-SZ04).

References

1. Xiang L, Yin Y, Zheng Y, Ma Y, Li Y, Zhao Z, Guo J, Ai Z, Niu Y, Duan K, He J, Ren S, Wu D, Bai Y, Shang Z, Dai X, Ji W, Li T (2020) A developmental landscape of 3D-cultured human pre-gastrulation embryos. Nature 577(7791):537–542. https://doi.org/10.1038/s41586-019-1875-y
2. Steptoe PC, Edwards RG, Purdy JM (1971) Human blastocysts grown in culture. Nature 229(5280):132–133. https://doi.org/10.1038/229132a0
3. Edwards RG, Steptoe PC, Purdy JM (1980) Establishing full-term human pregnancies using cleaving embryos grown in vitro. Br J Obstet Gynaecol 87(9):737–756. https://doi.org/10.1111/j.1471-0528.1980.tb04610.x
4. Yan L, Yang M, Guo H, Yang L, Wu J, Li R, Liu P, Lian Y, Zheng X, Yan J, Huang J, Li M, Wu X, Wen L, Lao K, Li R, Qiao J, Tang F (2013) Single-cell RNA-Seq profiling of human preimplantation embryos and embryonic stem cells. Nat Struct Mol Biol 20(9): 1131–1139. https://doi.org/10.1038/nsmb.2660
5. Guo H, Zhu P, Yan L, Li R, Hu B, Lian Y, Yan J, Ren X, Lin S, Li J, Jin X, Shi X, Liu P, Wang X, Wang W, Wei Y, Li X, Guo F, Wu X, Fan X, Yong J, Wen L, Xie SX, Tang F, Qiao J (2014) The DNA methylation landscape of human early embryos. Nature 511(7511): 606–610. https://doi.org/10.1038/nature13544
6. Smith ZD, Chan MM, Humm KC, Karnik R, Mekhoubad S, Regev A, Eggan K, Meissner A (2014) DNA methylation dynamics of the human preimplantation embryo. Nature 511(7511):611–615. https://doi.org/10.1038/nature13581
7. Petropoulos S, Edsgard D, Reinius B, Deng Q, Panula SP, Codeluppi S, Plaza Reyes A, Linnarsson S, Sandberg R, Lanner F (2016) Single-cell RNA-Seq reveals lineage and X chromosome dynamics in human preimplantation embryos. Cell 165(4):1012–1026. https://doi.org/10.1016/j.cell.2016.03.023
8. Shahbazi MN, Jedrusik A, Vuoristo S, Recher G, Hupalowska A, Bolton V, Fogarty NME, Campbell A, Devito LG, Ilic D, Khalaf Y, Niakan KK, Fishel S, Zernicka-Goetz M (2016) Self-organization of the human embryo in the absence of maternal tissues. Nat Cell Biol 18:700–708. https://doi.org/10.1038/ncb3347
9. Deglincerti A, Croft GF, Pietila LN, Zernicka-Goetz M, Siggia ED, Brivanlou AH (2016) Self-organization of the in vitro attached human embryo. Nature 533:251–254. https://doi.org/10.1038/nature17948
10. Rossant J (2016) Human embryology: implantation barrier overcome. Nature 533:182–183
11. Rossant J, Tam PPL (2017) New insights into early human development: lessons for stem cell derivation and differentiation. Cell Stem Cell

20(1):18–28. https://doi.org/10.1016/j.stem.2016.12.004

12. Wamaitha SE, Niakan KK (2018) Chapter Twelve - Human pre-gastrulation development. In: Plusa B, Hadjantonakis A-K (eds) Current topics in developmental biology, vol 128. Academic Press, pp 295–338. https://doi.org/10.1016/bs.ctdb.2017.11.004

13. Gardner DK, Lane M, Stevens J, Schlenker T, Schoolcraft WB (2000) Blastocyst score affects implantation and pregnancy outcome: towards a single blastocyst transfer. Fertil Steril 73(6): 1155–1158. https://doi.org/10.1016/s0015-0282(00)00518-5

Methods in Molecular Biology (2024) 2767: 27–41
DOI 10.1007/7651_2022_471

Published online: 08 February 2023

Protocol for the Generation of Human EPS-Blastoids Using a Three-Dimensional Two-Step Induction System

Yu-Xin Luo and Yang Yu

Abstract

Stem cell-derived embryos in vitro allow the exploration of the very early stages of human embryogenesis in vitro and are thus promising for widespread applications in developmental biology, related developmental disease modeling, and drug discovery. Several cell resources have been utilized, with different efficiencies and methods for generating human blastoids, a structure similar to natural blastocysts. Human EPS cells were reported to contribute to the embryonic and extraembryonic lineages and therefore can be a practical and efficient cell resource for constructing human blastoids. Here, we developed a three-dimensional, two-step induction system for generating human blastoids using human EPS cells. According to morphological and transcriptomic analysis, EPS-blastoids recapitulate the key developmental processes and cell lineages of human blastocysts. Moreover, in vitro extended culture for 8 and 10 days of EPS-blastoids can result in postimplantation embryonic structures. In this chapter, we describe a protocol that covers the generation, maintenance, and developmental phenocopying of human EPS blastoids.

Key words Blastoids, Embryogenesis, EPS cells, iPSCs, Stem cell

1 Introduction

The use of embryonic models derived from stem cells to study early embryonic development in mice and humans is an advance in the field of developmental biology. Blastoids are classical stem cell-derived embryonic models [1]. Blastoids can recapitulate key developmental events from preimplantation to peri-implantation, including cell compaction, polarization, lineage determination, cavitation, and implantation. Compared with the study of early development using human embryos, stem cell-derived blastoids have fewer ethical limitations and can be manufactured on a large scale. They are robust in vitro models for systematically interrogating early human embryogenesis and have widespread applications in related developmental diseases and drug discovery. It has been reported that human blastoids have been generated using different types of human stem cells and induction protocols [2–7], and these human blastoids can simulate the morphology and transcriptomic landscapes of natural blastocysts. Human EPS cells can effectively contribute to embryonic and extraembryonic lineages

[8]. Compared with hESCs [9–11], human EPS cells have the potential to develop into EPI (epiblast), TE (trophectoderm), and PE (primitive endoderm) and can better recapitulate the interaction between lineages of human embryo development. Therefore, human EPS cells are a good source of cells to generate blastoids. Mouse EPS cells have been tested to form blastoids that resemble mouse blastocysts in morphology and transcriptome and represent key events in early embryonic development [12]. Similarly, human blastoids using human EPS cells have been successfully constructed [7]. In contrast to the one-step induction system of mouse EPS-blastoids, the induction system of human EPs-blastoids requires two steps. First, human EPS cells are pretreated with BMP4 to induce TE-like cells. Then, human EPS cells and TE-like cells are mixed in a certain ratio and seeded into a 3D culture system to generate blastocyst-like structures, namely, human EPS blastoids. EPS-blastoids are similar to natural blastocysts in terms of morphology, cell lineages, and transcriptomic signatures at single-cell resolution, and extended culture in vitro can result in postimplantation morphology [7].

In this chapter, we will describe in detail the induction of human somatic cells into human EPS-blastoids, including the induction of somatic cells into hiPSCs, hiPSCs into human EPS cells, human EPS cells into EPS-blastoids in 3D culture systems and in vitro extended culture of EPS-blastoids, and further analysis.

2 Materials

2.1 Mouse Embryonic Fibroblast (MEF) Cell Medium

Reagents	Final concentration	Amount	Vendor, catalog number
DMEM/F12 medium	n/a	384 mL	Thermo Fisher Scientific, 11330-032
Fetal bovine serum	10%	50 mL	Thermo Fisher Scientific, 16000-044
Penicillin/streptomycin (100×)	1×	5 mL	Thermo Fisher Scientific, 15140-122
Total	n/a	500 mL	n/a

2.2 Human ESC Medium

Reagents	Final concentration	Amount	Vendor, catalog number
DMEM/F12 medium	n/a	384 mL	Thermo Fisher Scientific, 11330-032
KnockOut Serum Replacement	20%	100 mL	KSR, Thermo Fisher Scientific, A3181502
L-GlutaMAX (200 mM)	2 mM	5 mL	Thermo Fisher Scientific, 35050-061
NEAA (100×)	1×	5 mL	Thermo Fisher Scientific, 11140-050
2-Mercaptoethanol (55 mM)	0.1 mM	0.91 mL	Thermo Fisher Scientific, 21985-023
Penicillin/streptomycin (100×)	1×	5 mL	Thermo Fisher Scientific, 15140-122
Recombinant bFGF (10 μg/mL)	10 ng	500 μL	Thermo Fisher Scientific, PHG0261
Total	n/a	500 mL	n/a

2.3 Human EPS Cell Derivation Medium

Reagents	Final concentration	Amount	Vendor, catalog number
DMEM/F12 medium	n/a	225 mL	Thermo Fisher Scientific, 11330-032
Neurobasal medium	n/a	225 mL	Thermo Fisher Scientific, 21103-049
N2 supplement (100×)	0.5×	2.5 mL	Thermo Fisher Scientific, 17502-048
B27 supplement (50×)	0.5×	5 mL	Thermo Fisher Scientific, 12587-010
L-GlutaMAX (200 mM)	2 mM	5 mL	Thermo Fisher Scientific, 35050-061
NEAA (100×)	1×	5 mL	Thermo Fisher Scientific, 11140-050
2-Mercaptoethanol (55 mM)	0.1 mM	0.91 mL	Thermo Fisher Scientific, 21985-023
Penicillin/streptomycin (100×)	1×	5 mL	Thermo Fisher Scientific, 15140-122
KnockOut Serum Replacement or 20% BSA (optional)	5% or 0.1%	25 mL or 2.5 mL	KSR, Thermo Fisher Scientific, A3181502; BSA, Sigma, B2064
hLIF (100 μg/mL)	10 ng/mL	50 μL	R&D Systems, 7734
CHIR99021 (30 mM)	3 μM	50 μL	Tocris, 4423
(S)-(+)-Dimethindene meleate (20 mM)	2 μM	50 μL	Tocris, 1425
Minocycline hydrochloride (20 mM)	2 μM	50 μL	Selleckchem, S4226

(continued)

Reagents	Final concentration	Amount	Vendor, catalog number
IWR endo-1 (10 mM)	1 μM	50 μL	Selleckchem, S7086
Y-27632 (10 mM)	2 μM	100 μL	Selleckchem, S1049
Total	n/a	500 mL	n/a

2.4 TE-Like Cell Derivation Medium Using Human EPS Clls

Reagents	Final concentration	Amount	Vendor, catalog number
DMEM/F12 medium	n/a	225 mL	Thermo Fisher Scientific, 11330-032
Neurobasal medium	n/a	225 mL	Thermo Fisher Scientific, 21103-049
N2 supplement (100×)	0.5×	2.5 mL	Thermo Fisher Scientific, 17502-048
B27 supplement (50×)	0.5×	5 mL	Thermo Fisher Scientific, 12587-010
L-GlutaMAX (200 mM)	2 mM	5 mL	Thermo Fisher Scientific, 35050-061
NEAA (100×)	1×	5 mL	Thermo Fisher Scientific, 11140-050
2-Mercaptoethanol (55 mM)	0.1 mM	0.91 mL	Thermo Fisher Scientific, 21985-023
Penicillin/streptomycin (100×)	1×	5 mL	Thermo Fisher Scientific, 15140-122
KnockOut Serum Replacement or 20% BSA (optional)	5% or 0.1%	25 mL or 2.5 mL	KSR, Thermo Fisher Scientific, A3181502; BSA, Sigma, B2064
BMP4 (50 μg/mL)	25 ng/mL	250 μL	R&D Systems, 314-BP-010
Y-27632 (10 mM)	2 μM	100 μL	Selleckchem, S1049
Total	n/a	500 mL	n/a

2.5 IVC1 Culture Medium

Reagents	Final concentration	Amount	Vendor, catalog number
Advanced DMEM/F12 medium	n/a	382 mL	Thermo Fisher Scientific, 12634-010
Heat-inactivated fetal bovine serum	20%	100 mL	Thermo Fisher Scientific, 30044-333
L-GlutaMAX (200 mM)	2 mM	5 mL	Thermo Fisher Scientific, 35050-061
Penicillin/streptomycin (100×)	0.5×	2.5 mL	Thermo Fisher Scientific, 15140-122

(continued)

Reagents	Final concentration	Amount	Vendor, catalog number
ITS-X (100×)	1×	5 mL	Thermo Fisher Scientific, 51500-056
Sodium pyruvate (100×)	1×	5 mL	Thermo Fisher Scientific, 11360-070
β-Estradiol (8 μM)	8 nM	500 μL	Sigma-Aldrich, E8875
Progesterone (1 mg/mL)	200 ng/mL	100 μL	Sigma-Aldrich, P0130
N-acetyl-L-cysteine (50 nM)	25 μM	250 μL	Sigma-Aldrich, A7250
Total	n/a	500 mL	n/a

2.6 Human EPS-Blastoid Culture Medium [8]

Reagents	Final concentration	Amount	Vendor, catalog number
EPS medium	n/a	300 mL	n/a
IVC1 culture medium	n/a	200 mL	n/a
Total	n/a	500 mL	n/a

2.7 IVC2 Culture Medium [13, 14]

Reagents	Final concentration	Amount	Vendor, catalog number
Advanced DMEM/F12 medium	n/a	333 mL	Thermo Fisher Scientific, 12634-010
Knockout replacement serum	30%	150 mL	Thermo Fisher Scientific, 30044-333
L-GlutaMAX (200 mM)	2 mM	5 mL	Thermo Fisher Scientific, 35050-061
Penicillin/streptomycin (100×)	0.5×	2.5 mL	Thermo Fisher Scientific, 15140-122
ITS-X (100×)	1×	5 mL	Thermo Fisher Scientific, 51500-056
Sodium pyruvate (100×)	1×	5 mL	Thermo Fisher Scientific, 11360-070
β-Estradiol (8 μM)	8 nM	500 μL	Sigma-Aldrich, E8875
Progesterone (1 mg/mL)	200 ng/mL	100 μL	Sigma-Aldrich, P0130
N-acetyl-L-cysteine (50 nM)	25 μM	250 μL	Sigma-Aldrich, A7250
Y-27632 (10 mM)	2 μM	100 μL	Selleckchem, S1049
Total	n/a	500 mL	n/a

2.8 Cell Lines

Resource	Source	Vendor, catalog number
Human skin fibroblasts	Ddonation	n/a
Human iPSCs	Homemade	n/a
Human EPS cells	Homemade	n/a
ICR MEF cells	Stem cell bank of Chinese Academy of Sciences	n/a

2.9 Antibodies

Reagent	Dilution	Vendor	Catalog number
Anti-OCT4	1:200	Santa Cruz	sc5279
Anti-GATA6	1:1000	Cell Signaling Technology	5851S
Anti-SOX2	1:200	Abcam	ab171380
Anti-GATA2/3	1:200	Abcam	ab182747
Anti-CK8	1:200	Abcam	ab53280
Anti-E-cadherin	1:200	Abcam	ab1416
Anti-TFAP2C	1:200	Cell Signaling Technology	2320S
Anti-CK7	1:200	Abcam	ab192077
Anti-CK7	1:200	Invitrogen	OVTL12/30
Anti-TP63	1:200	Biotium	BUNB0375
Anti-CK18	1:100	Santa Cruz	sc-31700
Anti-GATA3	1:100	R&D Systems	634913
Anti-OCT4	1:100	Santa Cruz	sc-8628
Alexa Fluor 488 anti-Rabbit	1:500	Thermo Fisher Scientific	A-11008
Alexa Fluor 555 anti-Mouse	1:500	Cell Signaling Technology	4409S
Alexa Fluor 647 anti-Rabbit	1:500	Abcam	ab150083
Alexa Fluor 488 anti-Rabbit	1:500	Thermo Fisher Scientific	A-21206

2.10 Other Reagents

Reagent	Vendor	Catalog number
u-Slide 8-well	ibidi	80826
AggreWell#400 6-well plate, 5 pack	STEMCELL	34460
pCXLE-hOCT3/4-shp53	Addgene	27077
pCXLE-hSK	Addgene	27078
pCXLE-hUL	Addgene	27080

3 Methods

3.1 Generation of Human iPSCs from Human Skin Fibroblasts with Episomal Vectors

1. Human skin fibroblasts were isolated from the chest of a female aborted fetus that was obtained with informed written consent and approved by the Third Affiliated Hospital of Guangzhou Medical University.
2. Human skin fibroblasts were cultured in DMEM (Thermo Fisher Scientific, C11995500BT) with 10% FBS. Three micrograms of episomal vectors, including pCXLE-hOCT3/4-shp53-F, pCXLE-hSK, and pCXLE-hUL [15], was electroporated (4D-Nucleofector System, Lonza) into 6 × 105 fibroblast cells with a Microporator (Invitrogen) with a 100-μL kit according to the manufacturer's instructions (*see* **Note 1**).
3. The cells were trypsinized 7 d after transduction, and 1×10^5 cells were plated onto 100-mm dishes covered with a mitomycin C-treated MEF feeder layer. The culture medium was replaced the next day with human ESC medium.
4. The colonies were counted 26–32 d after plating, and colonies similar to human ESCs were selected for further cultivation and evaluation.

3.2 Preparation of MEF Feeders

3.2.1 Thawing of ICR MEF Cells

1. ICR mouse embryonic fibroblasts (Passage 0, P0) were obtained from the Stem Cell Bank of the Chinese Academy of Sciences. Thaw a vial of frozen cells in a 37 °C water bath with gentle agitation.
2. Transfer the cells into a 15-mL conical tube, gently add pre-warmed MEF cell medium, and centrifuge at 1000 rpm for 5 min at room temperature.
3. Aspirate the medium. Resuspend the cells with 10 mL of MEF cell medium. Transfer the cells to a 10-cm dish.

3.2.2 Maintenance of MEF Cells

1. Maintain the cells at 37 °C in a 5% CO_2 incubator. The MEF cell medium is changed every 2 days (*see* **Note 2**).
2. When the correct cell density is reached, the cells are passaged. Aspirate the medium and wash twice with 5 mL of PBS.
3. Add 1 mL of TrypLE Express and incubate at 37 °C for approximately 3 min. Add 5 mL of PBS and pipette the well to disperse into single cells.
4. Centrifuge at 1000 rpm for 5 min and remove the supernatant. Resuspend cells with MEF cell medium and split the cells at 1: 3–4 in a 10-cm dish with 10 mL of MEF cell medium.
5. Maintain the cells at 37 °C in a 5% CO_2 incubator. The MEF cell medium is changed every 2 days. Culture MEFs in MEF cell medium until P3 (*see* **Note 3**).

3.2.3 Preparation of Feeder Cells

1. To stop cell proliferation (*see* **Note 4**), treat MEF with 10 mg/mL mitomycin C (Sigma) for a final concentration of 10 μg/mL. Incubate at 37 °C in a 5% CO_2 incubator for 120–240 min.
2. Aspirate the medium and wash twice with 5 mL of PBS.
3. Detach MEFs from the plate with TrypLE Express, and centrifuge and resuspend the single cells with DMSO and FBS at 1:9 (*see* **Note 5**).
4. Measure the density of MEF cells. Freeze MEFs at 1.0×10^6 cells per vial.
5. Store MEFs at −80 °C overnight and then place them in liquid nitrogen for long-term storage.

3.2.4 Preparation of MEF-Coated Plates

1. Pregelatinize cell culture plates for 15 min with 1 mL of 0.1% (wt/vol) gelatine (*see* **Note 6**). Aspirate the gelatine before seeding the cells.
2. One day before planting the human iPSCs or human EPS cells, thaw a frozen vial of mitotically inactivated MEFs in a 37 °C water bath and transfer the contents into a 15-mL conical tube containing 5 mL of MEF cell medium.
3. Centrifuge the medium at 1000 rpm for 5 min at room temperature and aspirate the supernatant.
4. Resuspend the cell pellet in MEF medium and plate it at a density of $4–5 \times 10^5$ MEFs per well for a six-well plate and then incubate the plates overnight at 37 °C.

3.3 Generation of Human EPS Cells from Human iPSCs [8]

1. Human iPSCs are cultured on mitomycin C-treated MEF feeder cells in six-well plates with human ESC medium. Digest human iPSCs with TrypLE Express Enzyme (Thermo Fisher Scientific, 12604021) or Accutase (Stem Cell Technologies, #07920) (*see* **Note 7**).

2. Add PBS to dilute the digested solution and pipette the well to generate single cells.
3. Centrifuge at 1000 rpm for 5 min at room temperature and aspirate the supernatant.
4. Add 1 mL of human ESC medium to resuspend the cell pellet.
5. Aspirate the medium on the feeder cells and wash twice with PBS. Add 1.5 mL of human ESC medium to the plate.
6. Seed the single cells to the plate with fresh MEF feeder cells. The split ratio is 1:2–1:4. Add human ESC medium to 2 mL per well of 6-well plates.
7. Maintain the cells at 37 °C, 5% CO_2, and saturated humidity.
8. After 12 h of seeding, aspirate the medium and wash twice with PBS. Replace to prewarmed human EPS cell medium.
9. The EPS cell medium is changed every day with fresh EPS cell medium. Dome-shaped colonies emerge after 3–6 days.
10. Then, the cells are dissociated and passaged by TrypLE Express (Thermo Fisher Scientific, 12604-021) into single cells every 2–4 days.
11. Centrifuge the cells at 1000 rpm for 5 min at room temperature and aspirate the supernatant.
12. Add fresh human EPS cell medium. Resuspend the cells and pass twice to fresh gelatine-coated plates for 15–20 min each time.
13. The EPS cells are collected, centrifuged, and seeded into new six-well plates with fresh feeder cells. The split ratio is 1:3–1:10 (*see* **Note 8**).

3.4 The 3D, Two-Step Generation of Human EPS-Blastoids In Vitro [7]

3.4.1 Preparation of TE-Like Cells

1. Precoat the six-well plate with 1 mL of Matrigel for at least 30 min. Remove Matrigel solution from the plate before seeding the cells (*see* **Note 9**).
2. Digest human EPS cells into single cells by TrypLE Express and remove feeder cells. Collect the cells, count the cell number, and seed into a plate precoated with Matrigel in TE-like derivation medium.
3. After 3 days of differentiation, TE-like cells can be observed.

3.4.2 Prepare the AggreWell400 Culture Plate (See **Note 10**)

1. If using the plate for the first time, add a sufficient amount of anti-adherence solution (Stem Cell Technologies, #07010) to cover the well, spin down at 2000 × g for 1–2 min, and examine under a microscope to ensure bubbles are not present in the microwell. After 15 min, remove the solution and rinse with PBS twice.

2. The plate can be reused several times. Before reusing them, add TrypLE Express, incubate at 37 °C for 5 min, and wash with PBS. Then, add anti-adherence solution and incubate at room temperature for 15 min, remove solution, and rinse with PBS twice.

3.4.3 Generation of Human EPS-Blastoids (Fig. 1)

1. Then, the cells are dissociated into single cells with TrypLE Express. Human EPS cells are also dissociated into single cells with TrypLE Express.
2. Accurately determine the number of viable TE-like cells and human EPS cells after adding 0.1% trypan blue to stain dead cells.
3. Mix the TE-like cells (1.0×10^5 cells) and EPS cells (2.0×10^4) together following a 5:1 ratio to a total of 1.2×10^5 cells per well with 0.5 mL of human EPS-blastoid derivation medium (*see* **Note 11**).
4. Gently add the cell mixture to one well of 24-well AggreWell 400 culture plate and ensure the cells are evenly distributed to every microwell. The day of cell seeding is defined as Day 0.
5. The blastoid culturing conditions are 37 °C, 5% O_2, 5% CO_2 and saturated humidity (*see* **Note 12**).
6. On the next day, examine the cell culture under a microscope. The cells in each microwell should form a compact aggregate this time (*see* **Note 13**).
7. The EPS-blastoid derivation medium is slightly changed every day without disturbing aggregates in the V-shaped bottom. Human EPS-blastoids usually emerge from Day 3 to Day 6 (Fig. 2a, b).

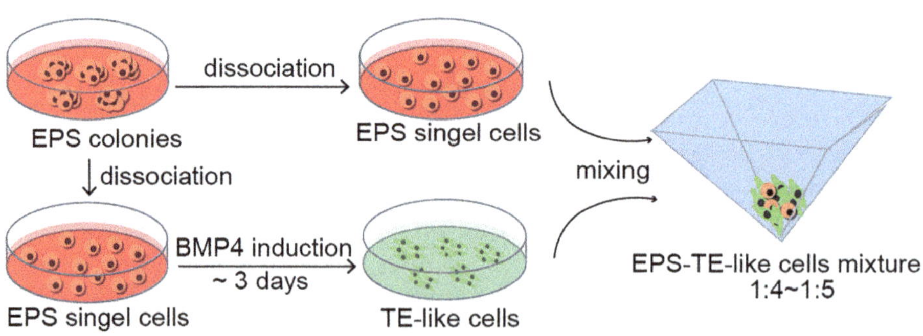

Fig. 1 Procedure of human blastoids generation using extended pluripotent stem cells. A part of extended pluripotent stem (EPS) cell colonies were dissociated to single cells and induced with BMP4 for 3 days, and these cells then displayed the characteristics of trophectoderm-like cells (TE-like cells). EPS cells and TE-like cells were mixed together in a ratio of 1:4 or 1:5 for blastoid induction. (Modified from ref. [7])

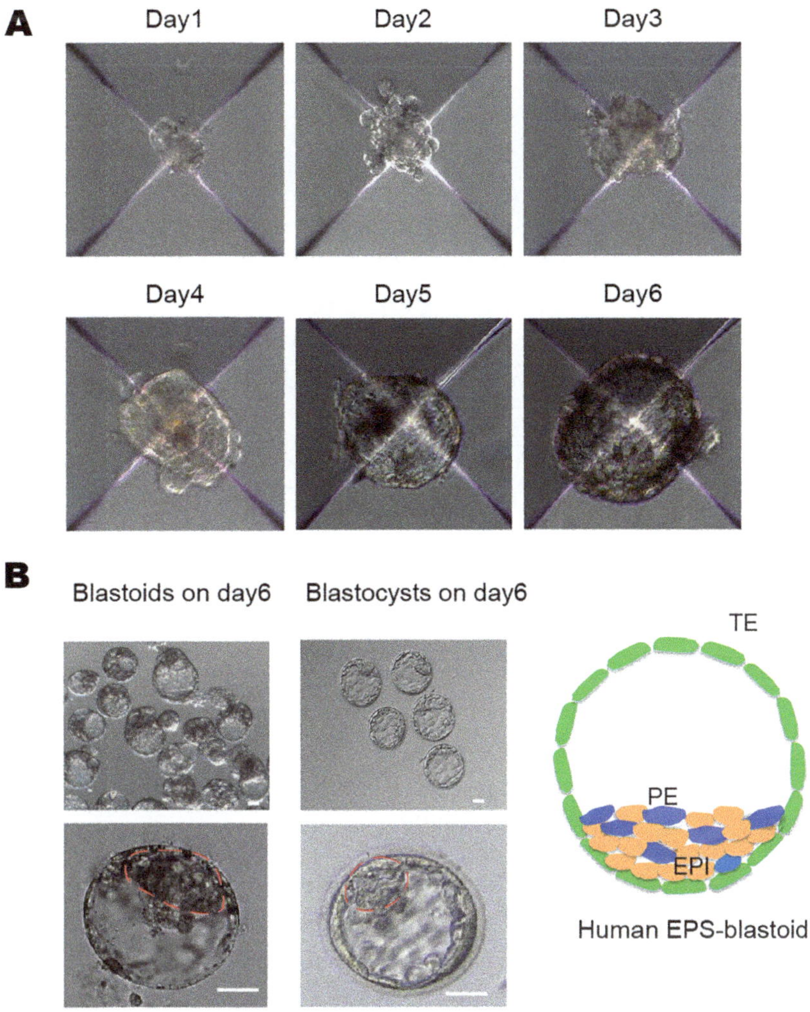

Fig. 2 Morphology characteristics of human EPS blastoids. (**a**) Phase-contrast images of human aggregates in the indicated days showing the formation of human blastoids from day 0 to day 6. (**b**) Phase-contrast image of human blastoids and human fertilized blastocysts on day 6. Red line indicates inner cell mass (ICM) of the structure and the outer layer cells represent trophoblast cells (TE). The diagram shows the distribution of three lineage cells in blastocysts on the right. (Modified from ref. [7])

3.4.4 Human EPS-Blastoid Collection

1. Use wide bore pipette tips to blow off the blastoid from the microwells by gently pipetting up and down.
2. Transfer the naive-blastoid culture into a 6-well plate for picking.
3. Place the stereomicroscope inside a biosafety cabinet and disinfect. Place the plate containing the naive-blastoid culture under the microscope (*see* **Note 14**).

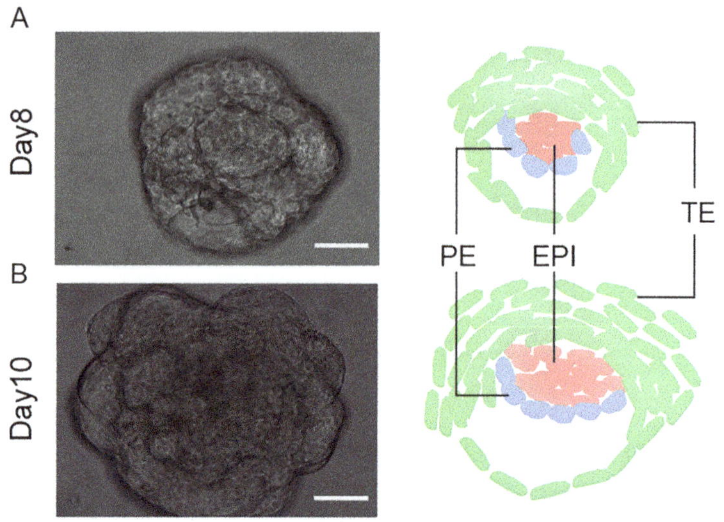

Fig. 3 The distribution of three lineage cells in early postimplantation embryo structures. (**a**) Phase-contrast and diagram images of human blastoids on (**a**) day 8 and (**b**) day 10. (Modified from ref. [7])

4. Use a mouth-controlled pipette with a hand-pulled glass microcapillary pipettes to pick up human EPS-blastoids.
5. Place the pooled EPS-blastoids into two drops of PBS successively for washing.

3.5 In Vitro Extended Culture of EPS-Blastoids for 8 and 10 Days

1. Precoat the eight-well plate with Matrigel for at least 30 min.
2. Transfer the human EPS-blastoids with a mouth-controlled pipette to the precoated eight-well plate. The blastoids are cultured with IVC1 for 2 days.
3. Then, on Day 8, some blastoids can be observed to be adherent to the bottle of the dish (Fig. 3). If the blastoids are adherent, replace the IVC1 with IVC2 for extended culture. If the blastoids are not adherent, do not replace the medium until Day 10.
4. Observe the morphology of the EPS-blastoids on Day 10 (Fig. 4) and collect the 96-well plates for further analysis.

3.6 Immunofluorescence of Human EPS-Blastoids (See Note 15)

1. The EPS-blastoids are fixed with 4% paraformaldehyde in PBS for 15 min at room temperature and then washed three times with PBS.
2. Permeabilize EPS-blastoids with 0.2% Triton X-100 in PBS for 30 min and then wash three times with PBS.
3. Block with 5% BSA in PBS for 2 h at room temperature.
4. Incubate EPS-blastoids with primary antibody diluted in blocking buffer overnight at 4 °C or 2 h at room temperature.

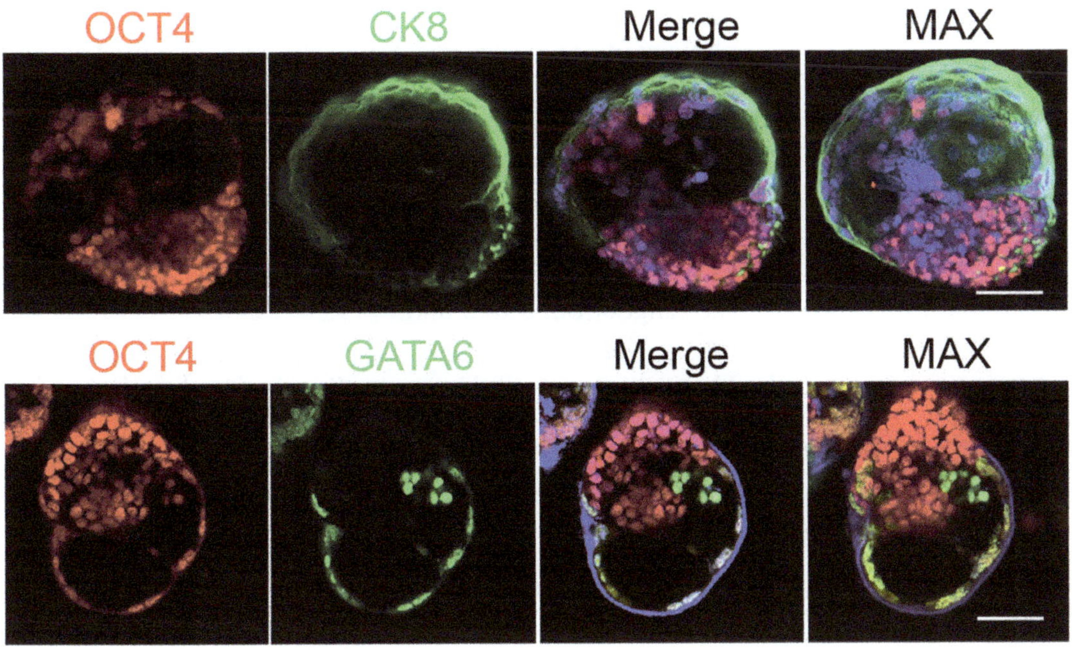

Fig. 4 Expression of lineage specific markers in human EPS blastoids. The expression of EPI lineage marker (OCT4), TE lineage markers (CK8), and PE lineage marker (GATA6) was detected in human EPS blastiods. (Modified from ref. [7])

5. After primary antibody incubation, wash three times with PBS containing 0.1% Tween 20 for 5–10 min each time.
6. Incubate blastoids with fluorescence-conjugated secondary antibodies diluted in blocking buffer at temperature for 2 h and then wash three times with PBS containing 0.1% Tween 20 for 5–10 min each time.
7. Nuclei are stained with Hoechst 33342 (Sigma, 94403) at 1 μg/mL.
8. Zeiss LSM 710 confocal microscopes and Fiji (ImageJ, V2.0.0) are used for imaging.

4 Notes

1. The plasmid mixtures used to reprogram human iPSCs are not limited to the mentioned combinations. More mixtures are shown in a previous report [15].
2. When the culture medium turns yellow or the cells grow fast, MEF cell medium can be changed daily.
3. To stop the proliferation of MEFs, we can treat MEFs with mitomycin C or irradiate the cells with 50 Gy (total).

4. The time of stopping proliferation is determined by the morphology of MEF cells, and MEF cells cannot be used at more than passage 3.
5. The freezing liquid contains 90% FBS and 10% DMSO, and the FBS can be replaced by MEF cell medium.
6. The concentration of gelatine ranges from 0.1% to 0.2% according to specific conditions.
7. The incubation time of digestion varies based on specific cells and ranges from 3 to 5 min.
8. When precoating Matrigel, ensure the whole plate is covered by Matrigel solution all the time until aspirating the solution. The time of Matrigel coating should not be less than 30 min. Maintain Matrigel in 4 °C no more than 1 week.
9. Some criteria are helpful to maintain human EPS cells in an undifferentiated state: avoid plating human EPS cells too sparsely, use the proper quantity of fresh feeder cells and avoid human EPS cell overgrowth. Additionally, remove the MEF cells when passaging as much as possible, for example, elongating the time of MEF attachment in gelatine-coated plates.
10. The treatment time of anti-adherence was at least 15 min, ensuring that the solution covered the bottom of the AggreWell 400 culture plate and bubbles are not present.
11. Create a uniform suspension of aggregates ~50–100 μm in size by gentle trituration with a pipette and add 10 μM ROCK inhibitor Y-27632 on the first day. The number of EPS cells is approximately 2.0×10^4 cells per well of a 6-well plate. Avoid cells that are too dense or too few.
12. The efficiency and morphology of human EPS-blastoids are affected by the concentration of oxygen. A high concentration of oxygen may harm the generation of human EPS blastoids.
13. At Day 1, the human EPS cells can form a compact aggregate. With time, the cell aggregates will grow and increase in size. On approximately Day 3, some cell aggregates undergo cavitation and continue to enlarge.
14. When selecting EPS-blastoids, keep the environment sterile and do not take too long.
15. The staining ratio of primary and secondary antibodies should be accurate, the incubation time of the second antibody should not be too long, attention should be given to prevent embryos from sticking to the oral pipette, and the oral pipette should be lubricated with 3% BSA.

Acknowledgments

This work was supported by the National Key Research and Development Program of China (2021YFC2700303) and the National Natural Science Funds (82192873 and 82225019).

References

1. Rivron NC, Frias-Aldeguer J, Vrij EJ, Boisset JC, Korving J, Vivié J et al (2018) Blastocyst-like structures generated solely from stem cells. Nature 557(7703):106–111. https://doi.org/10.1038/s41586-018-0051-0
2. Kagawa H, Javali A, Khoei HH, Sommer TM, Sestini G, Novatchkova M et al (2022) Human blastoids model blastocyst development and implantation. Nature 601(7894):600–605. https://doi.org/10.1038/s41586-021-04267-8
3. Yu L, Wei Y, Duan J, Schmitz DA, Sakurai M, Wang L et al (2021) Blastocyst-like structures generated from human pluripotent stem cells. Nature 591(7851):620–626. https://doi.org/10.1038/s41586-021-03356-y
4. Yanagida A, Spindlow D, Nichols J, Dattani A, Smith A, Guo G (2021) Naive stem cell blastocyst model captures human embryo lineage segregation. Cell Stem Cell 28(6):1016–22 e4. https://doi.org/10.1016/j.stem.2021.04.031
5. Sozen B, Jorgensen V, Weatherbee BAT, Chen S, Zhu M, Zernicka-Goetz M (2021) Reconstructing aspects of human embryogenesis with pluripotent stem cells. Nat Commun 12(1):5550. https://doi.org/10.1038/s41467-021-25853-4
6. Liu XD, Tan JP, Schroder J, Aberkane A, Ouyang JF, Mohenska M et al (2021) Modelling human blastocysts by reprogramming fibroblasts into iBlastoids. Nature 591(7851):627. https://doi.org/10.1038/s41586-021-03372-y
7. Fan Y, Min Z, Alsolami S, Ma Z, Zhang E, Chen W et al (2021) Generation of human blastocyst-like structures from pluripotent stem cells. Cell Discov 7(1):81. https://doi.org/10.1038/s41421-021-00316-8
8. Yang Y, Liu B, Xu J, Wang JL, Wu J, Shi C et al (2017) Derivation of pluripotent stem cells with in vivo embryonic and extraembryonic potency. Cell 169(2):243–257. https://doi.org/10.1016/j.cell.2017.02.005
9. Shao Y, Taniguchi K, Gurdziel K, Townshend RF, Xue X, Yong KMA et al (2017) Self-organized amniogenesis by human pluripotent stem cells in a biomimetic implantation-like niche. Nat Mater 16(4):419–425. https://doi.org/10.1038/nmat4829
10. Zheng Y, Xue X, Shao Y, Wang S, Esfahani SN, Li Z et al (2019) Controlled modelling of human epiblast and amnion development using stem cells. Nature 573(7774):421–425. https://doi.org/10.1038/s41586-019-1535-2
11. Shao Y, Taniguchi K, Townshend RF, Miki T, Gumucio DL, Fu J (2017) A pluripotent stem cell-based model for post-implantation human amniotic sac development. Nat Commun 8(1):208. https://doi.org/10.1038/s41467-017-00236-w
12. Li R, Zhong C, Yu Y, Liu H, Sakurai M, Yu L et al (2019) Generation of blastocyst-like structures from mouse embryonic and adult cell cultures. Cell 179(3):687–702.e18. https://doi.org/10.1016/j.cell.2019.09.029
13. Bedzhov I, Leung CY, Bialecka M, Zernicka-Goetz M (2014) In vitro culture of mouse blastocysts beyond the implantation stages. Nat Protoc 9(12):2732–2739. https://doi.org/10.1038/nprot.2014.186
14. Sozen B, Cox AL, De Jonghe J, Bao M, Hollfelder F, Glover DM et al (2019) Self-organization of mouse stem cells into an extended potential blastoid. Dev Cell 51(6):698. https://doi.org/10.1016/j.devcel.2019.11.014
15. Okita K, Matsumura Y, Sato Y, Okada A, Morizane A, Okamoto S et al (2011) A more efficient method to generate integration-free human iPS cells. Nat Methods 8(5):409–412. https://doi.org/10.1038/nmeth.1591

Methods in Molecular Biology (2024) 2767: 43–52
DOI 10.1007/7651_2022_463

Published online: 15 December 2022

Generation of Human Trophoblast Stem Cell-Dependent Placental In Vitro Models

Hao Wu, Yue Wang, and Hongmei Wang

Abstract

Currently, human trophoblast stem cell (hTSC) is considered to be the most promising laboratory model stimulating trophoblast criteria. Our group has established hTSCs allowing differentiation to syncytiotrophoblast (STB) and extravillous trophoblast (EVT). Further, hTSC-based three-dimensional (3D) trophoblast organoid (hTSC-organoid) provides a transformative model for studying human placental development and the interaction between trophoblast and maternal environment. Here, we present a protocol to obtain different types of placental trophoblast cells and trophoblast organoids using hTSCs. The generation of hTSC-organoids takes 6 days. hTSC-organoids permit passaging and can differentiate into EVT lineage.

Key words Human trophoblast stem cell, Human trophoblast organoid

1 Introduction

The placenta is a transient endocrine organ during pregnancy [1]. Besides producing hormones, placenta also serves as a pivot for nutrients/waste exchange and an immunological barrier to ensure normal fetal development [2]. Aberrant placental development threatens pregnancy outcomes and has profound effects on the lifelong health of the fetus and mother [3–5]. However, due to the complex anatomy structure of placenta and the lack of appropriate in vitro research models, the regulatory mechanisms of placental development are not well understood.

Mammalian development starts with cleavage divisions of zygote, and cells gradually differentiate and establish embryonic and extra-embryonic tissues which are important for embryo implantation. There are three lineages established in human peri-implantation blastocyst: the outer layer trophectoderm (TE), which will develop into the placenta; the epiblast (EPI), which will develop into embryo proper; and the hypoblast, which will develop into yolk sac [6].

Human trophoblast stem cells (hTSCs) of TE lineage can be isolated and maintained in vitro, which retain the potential for proliferation and differentiation [7–9]. Placental villi serve as the main units of placental structure and function, including floating villi and anchoring villi. In floating villi, the cytotrophoblast (CTB) fuses to form a syncytiotrophoblast (STB) covering the entire surface of villi, surrounding a core of connective tissue including fetal vessels, fibroblasts, and macrophages [5, 10–12]. In anchoring villi, the extravillous trophoblast cells (EVT) differentiated from CTB invade the uterine arterioles, destroy the media, and replace the endothelial cells, thus creating low-resistance, large-diameter blood vessels [13, 14].

For a long time, in vitro human placental models mainly include primary villous explants and trophoblast cells, various choriocarcinoma cell lines, and human embryonic stem cell (hESC)-derived trophoblast-like cells [15–22]. These models have limitations in accepting genetic manipulation or meeting primary trophoblast criteria. Recently, the hTSCs derived from blastocyst/first trimester villi are considered to be the best cell models modeling the development and function of human trophoblast cell [8, 9]. Further, 3D human trophoblast organoids make it possible to reconstruct the structure and microenvironment of the human placenta in a dish [23, 24]. In this chapter, we describe a step-by-step protocol to generate hTSC-organoids that can be long-term cultured and differentiate into both STB and EVT.

2 Materials

2.1 Reagents and Equipment

Reagents	Source	Identifier
DMEM/F12	Gibco	#11330032
BSA	MP biomedicals	#180728
ITS-X	Gibco	#51500056
L-Ascorbic acid	Sigma-Aldrich	#33034-100G
Valproic acid (VPA)	Wako	#227-01071
55mM 2-mercaptoethanol	Gibco	#M3148-250ML
FBS	Gibco	#10091148
Penicillin–streptomycin	Gibco	#15070063
DPBS	Hyclone	#SH30028.02
TrypLE	Gibco	#12605010
Collagen, type I	Corning	#354236
Recombinant human EGF	MedChemExpress	#HY-P7109
A83-01	MedChemExpress	#HY-10432

(continued)

CHIR-99021	MedChemExpress	#HY-10182
Y27632	MedChemExpress	#HY-10071
SB431542	MedChemExpress	#HY-10431
NRG1	Cell Signaling Technology	#5218SC
Knockout serum replacement	Gibco	#10828028
Matrigel	Corning	#354234
Forskolin	Sigma-Aldrich	#F3917-10MG
DMSO	Sigma-Aldrich	#V900090
Synth-a-Freeze CTS	Gibco	#A13713-01
Matrigel	Corning	#356231
Cell Recovery Solution	Corning	#354253
Equipment	**Source**	**Identifier**
6-well plate	Corning	#3516
Cryogenic storage container	Corning	#432001
8-well chamber	ibidi	#80826

2.2 Inhibitor Cocktail

Components	Final concentration
SB431542	1 mM
A83-01	0.5 mM
CHIR-99021	2 mM

Store up to 3 months at −20 °C.

2.3 hTSC Basal Medium

Combine DMEM/F12 medium, 0.3% BSA (*see* **Note 1**), 1% ITS-X, 0.5% penicillin–streptomycin, 0.1 mM 2-mercaptoethanol, 50 ng/mL recombinant human EGF, 0.2% FBS, and 1.5 μg/mL L-ascorbic acid. The medium is stored at 4 °C.

2.4 hTSC Culture Medium

Combine hTSC basal medium, 0.1% inhibitor cocktail, 5 μM Y27632, and 0.0125% VPA. Prepare the medium just before use.

2.5 Directed Differentiation Basal Medium

Combine DMEM/F12 medium, 0.3% BSA (*see* **Note 1**), 1% ITS-X, 0.5% penicillin–streptomycin, and 0.1 mM 2-mercaptoethanol. The medium is stored at 4 °C.

2.6 STB Differentiation Medium

Combine directed differentiation basal medium, 2.5 μM Y27632, 4% KSR, and 2 μM Forskolin. Prepare the medium just before use.

2.7 EVT Differentiation Medium

EVTM1 (for the first 2 days of differentiation): Combine directed differentiation basal medium, 2.5 μM Y27632, 4% KSR, 7.5 μM A83-01, 100 ng/mL NRG1, and 2% Matrigel (#354234). Prepare the medium just before use.

EVTM2 (for day 3 and day 4 of differentiation): Combine directed differentiation basal medium, 2.5 μM Y27632, 4% KSR, 7.5 μM A83-01, and 0.25% Matrigel. Prepare the medium just before use.

EVTM3 (for day 5 and day 6 of differentiation): Combine directed differentiation basal medium, 2.5 μM Y27632, 7.5 μM A83-01, and 0.25% Matrigel (#354234). Prepare the medium just before use.

3 Methods

3.1 Maintenance of hTSC (Fig. 1)

3.1.1 Culture of hTSC

1. For 6-well plate, coat wells with 1 mL collagen (type I, final concentration of 5 μg/mL in PBS; *see* **Note 2**) per well for >30 min at 37 °C.
2. Aspirate collagen.
3. Wash wells with 2 mL PBS for three times (*see* **Note 4**).
4. Aspirate PBS.
5. Suspend 0.5–1 × 10^5 cells in 2 mL hTSC culture medium per well.
6. Replace the hTSC culture medium every day.

3.1.2 Passaging of hTSC

1. Aspirate hTSC culture medium.
2. Dissociate hTSC with TrypLE at 37 °C for 8 min, and halt the reaction by adding twice the volume of hTSC medium.
3. Pipet gently to break cell colonies into single cells.
4. Transfer the cell suspensions into 15 mL Falcon tubes and centrifuge at 1000 × rpm for 5 min.
5. Aspirate the supernatants.
6. Resuspend the cells in hTSC culture medium.
7. Seed into wells at a ratio of 1:4–1:6.

3.1.3 Cryopreservation of hTSC

1. Collect hTSCs following the enzymatic passaging protocol showed in Subheading 3.1.2.
2. Resuspend hTSC in freezing medium (Synth-a-Freeze CTS).
3. Mix gently and transfer to frozen pipe.
4. Place frozen pipe in cryofreezing container and store at –80 °C.
5. The next day, transfer frozen pipe to a liquid nitrogen tank.

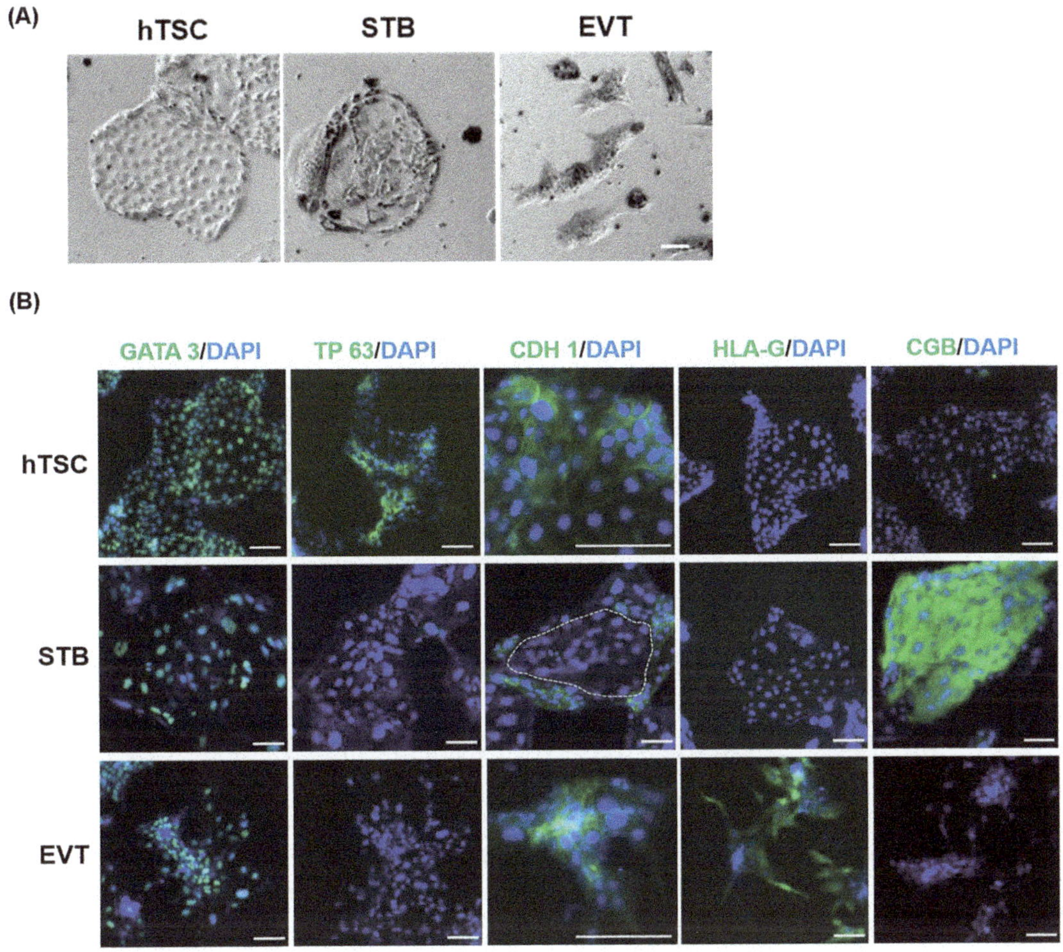

Fig. 1 Characterization of the morphology and gene expression of hTSC-derived trophoblast cells. (**a**) Bright-field images of hTSC and hTSC-derived STB and EVT. Scale bar: 100 μm. (**b**) Immunofluorescence images of GATA 3, TP 63, CDH 1, HLA-G, and CGB in hTSC and hTSC-derived STB and EVT. The white dashed line showing the outline of multinucleated STB. Scale bars: 100 μm

3.1.4 Thawing of hTSC

1. Remove vial from liquid nitrogen tank and thaw in 37 °C water bath until half of the vial is thawed.
2. Add 1 mL hTSC culture medium into frozen pipe and mix gently by pipetting up and down.
3. Transfer the liquid into 15 mL Falcon tubes and centrifuge at 1000 × rpm for 5 min.
4. Aspirate the supernatant.
5. Resuspend the cells in hTSC culture medium and seed into wells.

3.2 Directed Differentiation of hTSC (Fig. 1)

3.2.1 Differentiation of hTSC into STB (See Note 3)

1. For 6-well plate, coat wells with 1 mL collagen (type I, final concentration of 2.5 μg/mL in PBS; *see* **Note 2**) per well for >30 min at 37 °C.
2. Aspirate collagen.
3. Wash wells with 2 mL PBS for three times (*see* **Note 4**).
4. Aspirate PBS.
5. Suspend 1 × 10^5 hTSCs with 2 mL STB differentiation medium per well.
6. Replace the medium every other day.
7. Analyze the cells at day 6.

3.2.2 Differentiation of hTSC into EVT

1. For 6-well plate, coat wells with 1 mL collagen (type I, final concentration of 1 μg/mL in PBS; *see* **Note 2**) per well for >30 min at 37 °C.
2. Aspirate collagen.
3. Wash wells with 2 mL PBS for three times (*see* **Note 4**).
4. Aspirate PBS.
5. Suspend 1 × 10^5 hTSC in 2 mL EVTM1 per well.
6. At day 3, replace the medium with EVTM2.
7. At day 5, replace the medium with EVTM3.
8. Analyze the cells at day 6.

3.3 Generation and Differentiation of hTSC-Organoid (Fig. 2)

3.3.1 Generation of hTSC-Organoid (Fig. 2a–c)

1. Dissociate hTSCs with TryPLE at 37 °C for 8 min, and halt the reaction by adding twice the volume of hTSC medium.
2. Centrifuge the cells at 1000 × rpm for 5 min, and discard the supernatant.
3. Suspend 1 × 10^4 hTSCs in 100 μL Matrigel (#356231) (*see* **Notes 5** and **6**).
4. Drops (12.5 μL) were plated per well into an 8-well ibidi chamber (or other culture dish).
5. Set the chamber at 37 °C for 15 min.
6. Load 200 μL hTSC culture medium per well (in ibidi chamber).
7. Replace the medium every other day.
8. Analyze the hTSC-organoid at day 6 (*see* **Note 7**).

3.3.2 Passaging of hTSC-Organoid

1. Dissociate hTSC-organoids with TrypLE at 37 °C for 8 min.
2. Discard TrypLE medium.
3. Add 200 μL PBS and pipet gently to break cell aggregated into single cells.

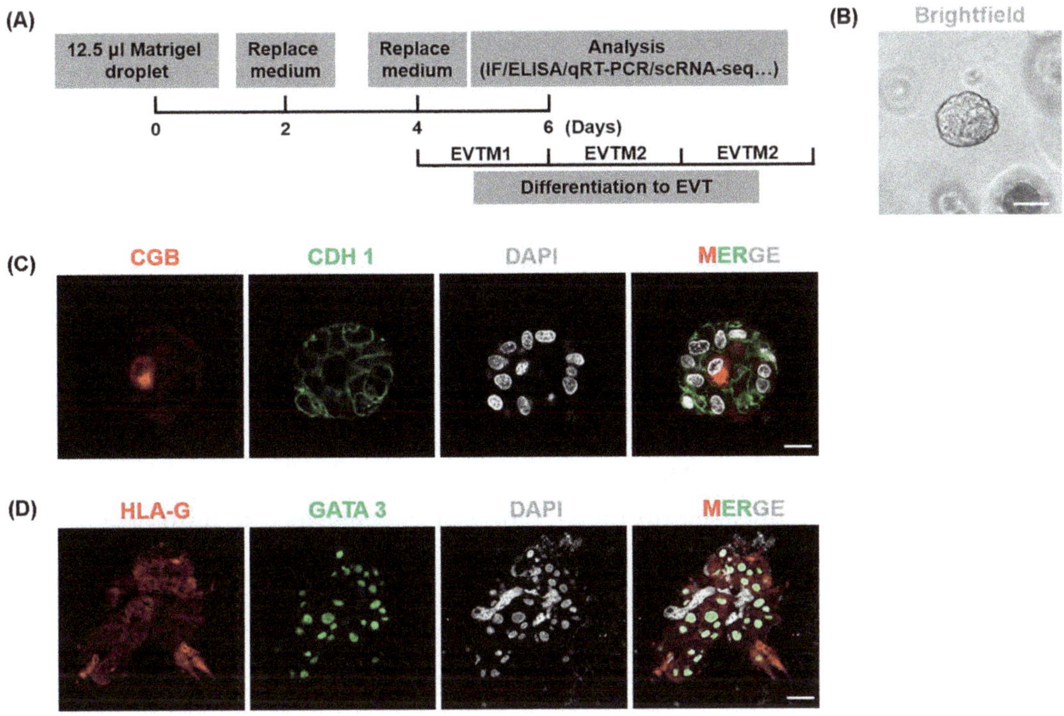

Fig. 2 Characterization of hTSC-organoids. (**a**) Flow diagram of the construction of hTSC-organoids and induction of differentiation to EVT. (**b**) Representative brightfield image of the day 6 hTSC-organoid. Scale bar: 50 μm. (**c**) Immunofluorescence staining for CGB and CDH 1 in day 6 hTSC-organoids. The results showing peripherally located mononucleated CTB and centrally located multinucleated STB. Scale bar: 20 μm. (**d**) Immunofluorescence staining for HLA-G and GATA 3 in the EVT differentiated from hTSC-organoids. Scale bar: 20 μm

4. Transfer the suspensions into a 1.5 mL Eppendorf tube, and centrifuge at 1000 × rpm for 5 min.
5. Wash the cells with PBS for 3 times.
6. Aspirate PBS.
7. Count the cell number.
8. Repeat the process of generation hTSC-organoids showed in Subheading 3.3.1.

3.3.3 Differentiation of hTSC-Organoids to Generate EVT Lineage (Fig. 2a, d)

1. Repeat the process of generation hTSC-organoids showed in Subheading 3.3.1.
2. At day 4, replace medium with EVTM1.
3. At day 6, replace the medium with EVT differentiation medium without NRG1 (EVTM2).
4. At day 8, repeat **step 3**.
5. Analyze the cells at day 10.

3.3.4 Validation of hTSC-Organoids

The hTSC-organoid is fully established by day 6 of culture. The supernatants, RNA extraction, and formalin fixation should be collected for characterization of the organoids.

1. Identify the cell identity and structure of hTSC-organoids by immunofluorescent staining (GATA 3 and CK 7 for trophoblast cells; CDH 1 and TP 63 for mononucleated trophoblast cells; CGB and SDC 1 for syncytiotrophoblast; HLA-G for extravillous trophoblast cells; *see* **Note 8**).
2. Confirm the expression of trophoblast-specific genes (GCM1, ERVW-1, ERVFRD-1, and ELF5) by qRT-PCR in hTSC-organoids (*see* **Note 9**).
3. Confirm the expression of microRNAs (miRNAs) from the chromosome 19 miRNA cluster (C19MC) in hTSC-organoids (*see* **Note 9**).
4. Confirm the secretion of placenta-specific hormones (hCG-β and GDF15) in the supernatants of hTSC-organoids by ELISA.

3.3.5 Collection of hTSC-Organoid

1. Release hTSC-organoid from Matrigel with Cell Recovery Solution at 4 °C for 20–30 min.
2. Aspirate Cell Recovery Solution gently.
3. Wash with PBS twice.
4. Aspirate PBS.
5. Suspend hTSC-organoids in PBS and break the aggregates into single hTSC-organoid by pipetting up and down gently.
6. Centrifuge the organoids at 1000 × rpm for 5 min, and discard the supernatants.
7. Wash with PBS for twice.
8. Aspirate the PBS, and the hTSC-organoid is ready for analysis.

4 Notes

1. BSA is important for hTSC culture; if replaced with a new production batch of BSA, several passages are required to adjust the cell status.
2. Prepare the diluted collagen just before use. Diluted collagen can be stored at 4 °C no more than 1 week.
3. Cells still have proliferation ability in the first 3 days of hTSC differentiation to STB; thus the initial cell density should be controlled; too much density will reduce fusion efficiency.

4. To avoid pre-coated dishes drying out, start washing the pre-coated dishes with PBS after resuspending the centrifuged cells.
5. To avoid Matrigel solidification, the process of building hTSC-organoids should to be performed on ice.
6. Matrigel should avoid repeated freezing and thawing and should not be stored at 4 °C for more than 1 week.
7. hTSC-organoid can be cultured for 12 days.
8. Recommendations of antibodies.

Antibodies	Source	Identifier
GATA 3	Cell Signaling Technology	#5852S
CK 7	ZSGB-BIO	#ZM-0071
CDH 1	Cell Signaling Technology	#3195S
TP 63	Cell Signaling Technology	#13109
CGB	ZSGB-BIO	#ZM-0134
SDC 1	Abcam	#ab128936
HLA-G	Santa Cruz Biotechnology	#sc-21799

9. When detecting the expression of placental feature genes and miRNAs in hTSC-organoids, the choriocarcinoma cell lines JEG-3 and JAR can be used as positive controls.

Acknowledgments

This work was supported by National Natural Science Foundation of China (82192870) and National Key R&D Program of China (2021YFC2700302, 2018YFE0201103).

References

1. Maltepe E, Fisher SJ (2015) Placenta: the forgotten organ. Annu Rev Cell Dev Biol 31:523–552
2. Delorme-Axford E, Sadovsky Y, Coyne CB (2014) The placenta as a barrier to viral infections. Annu Rev Virol 1(1):133–146
3. Bonnin A, Goeden N, Chen K, Wilson ML, King J, Shih JC, Blakely RD, Deneris ES, Levitt P (2011) A transient placental source of serotonin for the fetal forebrain. Nature 472(7343):347–350
4. Eriksson JG, Kajantie E, Thornburg KL, Osmond C, Barker DJP (2011) Mother's body size and placental size predict coronary heart disease in men. Eur Heart J 32(18): 2297–2303
5. Benirschke's pathology of the human placenta (2022)
6. Shahbazi MN, Zernicka-Goetz M (2018) Deconstructing and reconstructing the mouse and human early embryo. Nat Cell Biol 20(8): 878–887
7. Shahbazi MN (2020) Mechanisms of human embryo development: from cell fate to tissue shape and back. Development 147(14): dev190629

8. Okae H, Toh H, Sato T, Hiura H, Takahashi S, Shirane K, Kabayama Y, Suyama M, Sasaki H, Arima T (2018) Derivation of human trophoblast stem cells. Cell stem cell 22(1): 50–63. e56
9. Wang Y, Wu H, Jiang X, Jia L, Wang M, Rong Y, Chen S, Wang Y, Xiao Z, Liang X et al (2022) LMNA determines nuclear morphology during syncytialization of human trophoblast stem cells. Front Cell Dev Biol 10: 836390
10. Xiao Z, Yan L, Liang X, Wang H (2020) Progress in deciphering trophoblast cell differentiation during human placentation. Curr Opin Cell Biol 67:86–91
11. Liu Y, Fan X, Wang R, Lu X, Dang YL, Wang H, Lin HY, Zhu C, Ge H, Cross JC et al (2018) Single-cell RNA-seq reveals the diversity of trophoblast subtypes and patterns of differentiation in the human placenta. Cell Res 28(8):819–832
12. Lu X, Wang R, Zhu C, Wang H, Lin HY, Gu Y, Cross JC, Wang H (2018) Fine-tuned and cell-cycle-restricted expression of fusogenic protein syncytin-2 maintains functional placental syncytia. Cell Rep 23(13):3979
13. Chang WL, Liu YW, Dang YL, Jiang XX, Xu H, Huang X, Wang YL, Wang H, Zhu C, Xue LQ et al (2018) PLAC8, a new marker for human interstitial extravillous trophoblast cells, promotes their invasion and migration. Development 145(2):dev148932
14. Malassiné A, Cronier L (2002) Hormones and human trophoblast differentiation: a review. Endocrine 19(1):3–11
15. Genbacev O, Donne M, Kapidzic M, Gormley M, Lamb J, Gilmore J, Larocque N, Goldfien G, Zdravkovic T, McMaster MT et al (2011) Establishment of human trophoblast progenitor cell lines from the chorion. Stem Cells 29(9):1427–1436
16. Zdravkovic T, Nazor KL, Larocque N, Gormley M, Donne M, Hunkapillar N, Giritharan G, Bernstein HS, Wei G, Hebrok M et al (2015) Human stem cells from single blastomeres reveal pathways of embryonic or trophoblast fate specification. Development 142(23):4010–4025
17. Xu RH, Chen X, Li DS, Li R, Addicks GC, Glennon C, Zwaka TP, Thomson JA (2002) BMP4 initiates human embryonic stem cell differentiation to trophoblast. Nat Biotechnol 20(12):1261–1264
18. Wei Y, Wang T, Ma L, Zhang Y, Zhao Y, Lye K, Xiao L, Chen C, Wang Z, Ma Y et al (2021) Efficient derivation of human trophoblast stem cells from primed pluripotent stem cells. Sci Adv 7(33):eabf4416
19. Io S, Kabata M, Iemura Y, Semi K, Morone N, Minagawa A, Wang B, Okamoto I, Nakamura T, Kojima Y et al (2021) Capturing human trophoblast development with naive pluripotent stem cells in vitro. Cell Stem Cell 28(6):1023–1039.e1013
20. Pattillo RA, Gey GO (1968) The establishment of a cell line of human hormone-synthesizing trophoblastic cells in vitro. Cancer Res 28(7): 1231–1236
21. Kohler PO, Bridson WE (1971) Isolation of hormone-producing clonal lines of human choriocarcinoma. J Clin Endocrinol Metab 32(5):683–687
22. Barnea ER, Ashkenazy R, Tal Y, Kol S, Sarne Y (1991) Effect of beta-endorphin on human chorionic gonadotrophin secretion by placental explants. Hum Reprod 6(9):1327–1331
23. Turco MY, Gardner L, Kay RG, Hamilton RS, Prater M, Hollinshead MS, McWhinnie A, Esposito L, Fernando R, Skelton H et al (2018) Trophoblast organoids as a model for maternal-fetal interactions during human placentation. Nature 564(7735):263–267
24. Haider S, Meinhardt G, Saleh L, Kunihs V, Gamperl M, Kaindl U, Ellinger A, Burkard TR, Fiala C, Pollheimer J et al (2018) Self-renewing trophoblast organoids recapitulate the developmental program of the early human placenta. Stem Cell Rep 11(2): 537–551

Methods in Molecular Biology (2024) 2767: 53–62
DOI 10.1007/7651_2022_464

Published online: 15 December 2022

Workflow for Performing Genetic Manipulation in Human Trophoblast Stem Cells Using CRISPR/Cas9 Technology

Hao Wu, Yue Wang, and Hongmei Wang

Abstract

Clustered regularly interspaced short palindromic repeats (CRISPR)/Cas9-based tool kits greatly facilitate the investigation of gene function, especially gene editing and transcriptional regulation technology. Human trophoblast stem cell (hTSC) is a promising cell model for simulating the development and function of human placental trophoblast cells. Here, we provide an optimized workflow for efficient genome editing and transcriptional activation in hTSCs using a lentivirus-dependent method. We also present methods to evaluate the gene targeted mutations and transcriptional activation.

Key words Human trophoblast stem cell, CRISPR/Cas9, CRISPR/dCas9-VP64

1 Introduction

Human trophoblast stem cell (hTSC) has been derived from human blastocyst and the placental villi of early pregnancy [1]. hTSCs have the ability to differentiate into all types of trophoblast cells, such as syncytiotrophoblast (STB) and extravillous trophoblast cells (EVTs). Further, three-dimensional (3D) placental trophoblast organoids constructed using hTSC provide a power tool to study the effect of microenvironment on placental villous trophoblast and the secretory function of placental trophoblast (the methods of construction and culture of hTSC-organoids are described by our lab in Methods in Molecular Biology).

CRISPR/Cas is an RNA-mediated adaptive defense system that serves to remove invading viruses and plasmids in bacteria and archaea [2, 3]. Cas9 protein is a hallmark protein of Type II CRISPR/Cas system. The nuclease domains (HNH and RuvC) in the Cas9 protein can cleave target double-stranded DNA under the guidance of a CRISPR RNA (crRNA) that is complementary to the target DNA, a trans-activating (tracrRNA) that base-paired to crRNA, and a short protospacer-adjacent motif (PAM) [4]. Cas9-induced double-strand breaks will trigger cellular DNA repair processes including nonhomologous end-joining (NHEJ)-mediated error-prone DNA repair and homology-directed repair (HDR)-

mediated error-free DNA repair [5]. NHEJ-mediated DNA repair leads to rapid generation of small insertion and deletion (INDEL) mutations at target sites, which can be used to silence the function of target genes [6]. Furthermore, the development of technologies based on nuclease-deactivated Cas9 (dCas9) has been used for RNA-guided genome transcriptional regulation [7]. The fusion of transcriptional activation domain, such as VP64 and p65, to dCas9 can activate target endogenous genes [8–10]. Overall, RNA-guided CRISPR/Cas9-mediated genome editing and CRISPR/dCas9-VP64-mediated transcriptional activation provide fast, flexible, and efficient tools for loss-of-function and gain-of-function of target genes, respectively. This chapter described a step-by-step workflow for performing gene knockout and transcriptional activation in hTSC using CRISPR/Cas9-based technologies.

2 Materials

2.1 Reagents and Equipment

Reagents	Source	Identifier
DMEM/F12	Gibco	#11330032
DMEM high glucose	Hyclone	#SH30243.01
BSA	MP biomedicals	#180728
ITS-X	Gibco	#51500056
L-ascorbic acid	Sigma-Aldrich	#33034-100G
Valproic acid (VPA)	Wako	#227-01071
55mM 2-mercaptoethanol	Gibco	#M3148-250ML
FBS	Gibco	#10091148
Penicillin–streptomycin	Gibco	#15070063
DPBS	Hyclone	#SH30028.02
TrypLE	Gibco	#12605010
Collagen, Type I	Corning	#354236
Recombinant human EGF	MedChemExpress	#HY-P7109
A83-01	MedChemExpress	#HY-10432
CHIR-99021	MedChemExpress	#HY-10182
Y27632	MedChemExpress	#HY-10071
SB431542	MedChemExpress	#HY-10431
DMSO	Sigma-Aldrich	#V900090
PEI	Proteintech	#PR40001
One Shot™ TOP 10	Invitrogen	#C404010

(continued)

BsmBI	NEB	#R0739
T4 PNK	NEB	#M0201S
T4 Ligase	NEB	#M0202S
Reagent kits	**Source**	**Identifier**
DNA Clean & Concentrator	ZYMO	#D4006
Plasmid Mini Kit	QIAGEN	#12123
EndoFree Plasmid Maxi Kit	QIAGEN	#12326
Mouse Direct PCR Kit	Bimake	#B40013
Equipment	**Source**	**Identifier**
6-well plate	Corning	#3516
24-well plate	Corning	#3527
48-well plate	Corning	#3548
96-well plate	Corning	#3599
10 cm dish	Corning	#30167
Millex-HP Syringe Filter Unit	Millipore	#SLHP033RS

2.2 Inhibitor Cocktail

Components	Final concentration
SB431542	1 mM
A83-01	0.5 mM
CHIR-99021	2 mM

Store up to 3 months at −20 °C.

2.3 hTSC Basal Medium

Combine DMEM/F12 medium, 0.3% BSA, 1% ITS-X, 0.5% penicillin–streptomycin, 0.1 mM 2-mercaptoethanol, 50 ng/mL recombinant human EGF, 0.2% FBS, and 1.5 μg/mL L-ascorbic acid. The medium is stored at 4 °C.

2.4 hTSC Culture Medium

Combine hTSC basal medium, 0.1% inhibitor cocktail, 5 μM Y27632, and 0.0125% VPA. Prepare the medium just before use.

2.5 HEK293T Culture Medium

Combine DMEM high glucose medium, 10% FBS, and 1% penicillin–streptomycin.

3 Methods

3.1 CRISPR/Cas9-Dependent Gene Knockout in hTSCs (Fig. 1a, c, d)

3.1.1 sgRNA Design (See Notes 1–4)

1. Target sgRNA is designed using CRISPick (https://portals.broadinstitute.org/gppx/crispick/public).
2. Input NCBI gene ID of target gene or the target DNA sequence to design sgRNA.
3. Select 3–5 sgRNAs according to the on-target and off-target scores.

3.1.2 Construction of sgRNA Expression Vectors

1. To ensure lentiviral packaging efficiency, the sgRNA backbone is expressed separately in one lentiviral vector (Available vectors, such as addgene #52963 and addgene #62205). We will use addgene #52963 as an example for the following description: Cas9-NLS is expressed in another lentiviral vector with puromycin expression.
2. Digest 4 μg sgRNA vector with BsmB1 at 55 °C for 2 h.
3. Purify linearized sgRNA vector with DNA Clean & Concentrator kit (ZYMO) for use.
4. Synthesis of upstream and downstream sequences of sgRNA (add the sticky end of BsmB1 at the 5′ end).

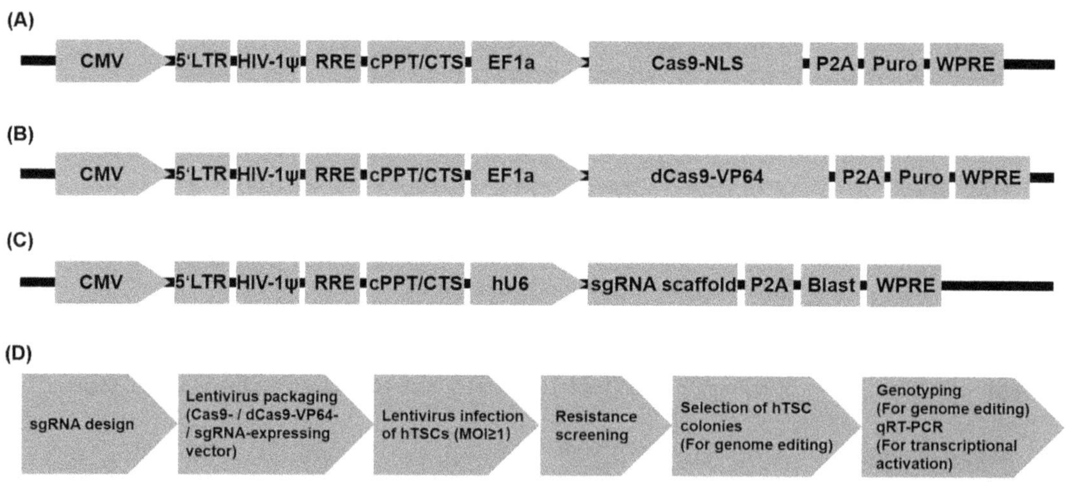

Fig. 1 Genetic manipulation in hTSCs. (**a**–**c**) The lentivirus vectors used for Cas9-NLS (**a**), dCas9-VP64 (**b**), and sgRNA (**c**) expression. CMV, human cytomegalovirus immediate early promoter; EF1a, elongation factor-1α core promoter; hU6, RNA polymerase II promoter for human U6 SnRNA; P2A, 2A self-cleaving peptide; Puro, puromycin selection marker; Blast, blasticidin selection marker; WPRE, posttranscriptional regulatory element. (**d**) Workflow for performing genome editing and transcriptional activation in hTSCs

5. Annealing synthesis of double-strand sgRNA.

Components	Volume
Forward oligo (100 μM)	1 μL
Reverse oligo (100 μM)	1 μL
T4 PNK	0.5 μL
T4 PNK buffer	1 μL
T4 ligase buffer	1 μL
ddH_2O	5.5 μL
Total	10 μL

6. Dilute annealing products at a ratio of 1:20–1:100.
7. Ligate the digested vector and 1 μL diluted annealing products using T4 ligase at 16 °C for 1 h.
8. Transformation of ligation products into TOP 10 chemically competent *Escherichia coli* (Invitrogen) cells using the heat shock method.
9. Amplification and resistance selection of bacteria in the Luria-Bertani (LB) agar plates with corresponding antibiotic.
10. Pick monoclonal colonies for expansion in LB liquid medium with corresponding antibiotic.
11. Extract plasmid using a Plasmid Mini Kit (QIAGEN).
12. Identify the sgRNA sequence in the plasmid using Sanger sequencing. (Primer for Sanger sequencing: hU6-F: 5′-GAG GGCCTATTTCCCATGATT-3′).

3.1.3 Lentivirus Production

1. Purify endotoxin-free plasmids for packaging lentivirus using a EndoFree Plasmid Maxi Kit (QIAGEN).
2. Plate HEK293T cells in a 10 cm dish at approximately 50% confluency with HEK293T culture medium the day before transfection. Cells should reach 70–80% confluency within 24 h.
3. Transfect a mixture of transfer plasmid and packaging vectors into HEK293T cells with polyethyleneimine (PEI). Set up the transfection as the following table.

Components	Amount (for 10 cm dish)
psPAX2	5.6 μg
pMD2.G	3.15 μg
sgRNA-expressing vector/Cas9-expressing vector	7.25 μg
PEI	48 μL

4. Change to the fresh HEK293T culture medium at 4–6 h after transfection.
5. Collect the culture supernatants containing lentivirus particles at 48 h and 72 h after transfection.
6. Filter supernatants through a 0.45 μm Millex-HP Syringe Filter Unit (Millipore).
7. Supernatants can be stored at 4 °C for several hours or stored at −80 °C for a long time after aliquoting (*see* **Note 5**).

3.1.4 Lentiviral Infection and Drug Resistance Selection

1. Infect hTSC with Cas9-expressing lentivirus, and establish a stable Cas9-expressing hTSC, followed by infection with an sgRNA-expressing lentivirus to establish a monoclonal gene mutant hTSC.
2. Seed hTSCs at 3×10^4 per well in a 24-well plate the day before infection.
3. Aspirate medium.
4. Add 1 mL hTSC culture medium containing 30 μL lentiviral particle solution and incubate overnight (*see* **Note 6**).
5. Replace with fresh hTSC culture medium at incubate for 12 h.
6. Aspirate medium.
7. Add hTSC culture medium containing antibiotic corresponding to vector information for screening (*see* **Notes 7** and **8**).
8. Replace the medium with hTSC culture medium containing antibiotic after 24 h.
9. After 48 h of resistance screening, aspirate the medium, and wash with fresh hTSC culture medium for twice.
10. Aspirate medium.
11. Add fresh medium and continue culture until 80–100% confluence.

3.1.5 Selection of Monoclonal hTSC by Gradient Dilution (See ***Note 9****)*

1. Dissociate the hTSCs after resistance screening with TrypLE at 37 °C for 8 min.
2. Halt the reaction by adding twice the volume of hTSC culture medium.
3. Pipet gently to break the cell colonies into single cells.
4. Count cell number.
5. Gradient dilution of 300 cells in 40 mL hTSC culture medium.
6. Mix cell suspension by pipetting up and down gently.
7. Seed 100 μL of cell suspension per well into four 96-well plates pre-coated with collagen.

8. Incubate the 96-well plates at 37 °C for 5 days without changing medium.
9. Scan and mark the monoclonal hTSC at day 6.
10. Change the medium every other day and passage the monoclonal hTSC to 48-well plate at day 7–10.

3.1.6 Design Primers for Genotyping

1. Primers used for genotyping is designed using Nucleotide BLAST (Nucleotide BLAST: search nucleotide databases using a nucleotide query (nih.gov)).
2. Input the DNA sequence of 500 bp before and after the target site.
3. Forward primer is designed between 1 and 300 bp of the input sequence, and the reverse primer is designed between the 700 and 1000 bp of the input sequence.

3.1.7 Genomic DNA Extraction and Genotyping

1. Dissociate hTSCs with TrypLE and harvest about 1×10^5 cells.
2. Extract the DNA for genotyping using Mouse Direct PCR Kit (Bimake).
3. Reaction system.

Components	Volume
ddH_2O	8 μL
Forward primer	0.5 μL
Reverse primer	0.5 μL
Template DNA	1 μL
2 × M-PCR OPTI™ Mix	10

4. Cycling condition.

PCR steps	Temperature (°C)	Time	Cycles
Pre-denaturation	94	5 min	1
Denaturation Annealing Extension	94 60 72	20 s 30 s 30 s	35
Extension	72	5 min	1
Holding	4	∞	

5. Load 5 uL PCR products into the agarose gel electrophoresis and check the size of DNA band.
6. Purify the PCR product using a DNA Clean & Concentrator Kits.
7. Identify the genotypes of the target site using Sanger sequencing.

3.2 CRISPR/dCas9-VP64-Dependent Transcriptional Activation in hTSCs (Fig. 1b–d)

3.2.1 sgRNA Design (See Notes 2 and 4)

1. Target sgRNA is designed using CRISPick (https://portals.broadinstitute.org/gppx/crispick/public).
2. Input NCBI gene ID of target gene to design sgRNA.
3. Select 3–5 sgRNAs according to the on-target and off-target scores.

3.2.2 Construction of sgRNA Expression Vectors

1. To ensure lentiviral packaging efficiency, the sgRNA backbone is expressed separately in one lentiviral vector. With the available vectors such as addgene #52963 and addgene #62205, we will use addgene #52963 as an example for the following description: dCas9-VP64 is expressed in another lentiviral vector with puromycin expression (*see* **Note 10**).
2. Refer to the procedures described in Subheading 3.1.2.

3.2.3 Lentiviral Infection and Drug Resistance Selection

1. Package the lentivirus expressing dCas9-VP64 and sgRNA, respectively, referring to the procedures described in Subheading 3.1.3.
2. Resistant enrichment of hTSCs co-expressing dCas9-VP64 and sgRNA refers to the procedures described in Subheading 3.1.3.

3.2.4 Identification of Target Gene Expression

1. Extract total RNA and protein.
2. Measure the expression level of mRNA and protein of targeted gene using quantitative real-time PCR (qRT-PCR) and Western blot, respectively.

4 Notes

1. The sgRNA used for genome editing is designed to target near the 5′ end of the coding region.
2. The sgRNA used for genome editing and transcriptional activation can be validated for efficiency by transient transfection in HEK293T cells, and the sgRNA with the highest efficiency can be used for the following application in hTSCs.
3. If the target gene has many transcripts, the sgRNA used for genome editing can be designed in the first exon at 5′ end that is involved in all transcripts.
4. U6 promoter prefers a guanine (G) to initiate transcription; thus a G can be added ahead of the sgRNA sequence whose first base is not G.
5. To increase the titer of lentivirus, the supernatant containing lentivirus particles can be concentrated by ultra-centrifugation at 25000 × rpm for over 2 h.

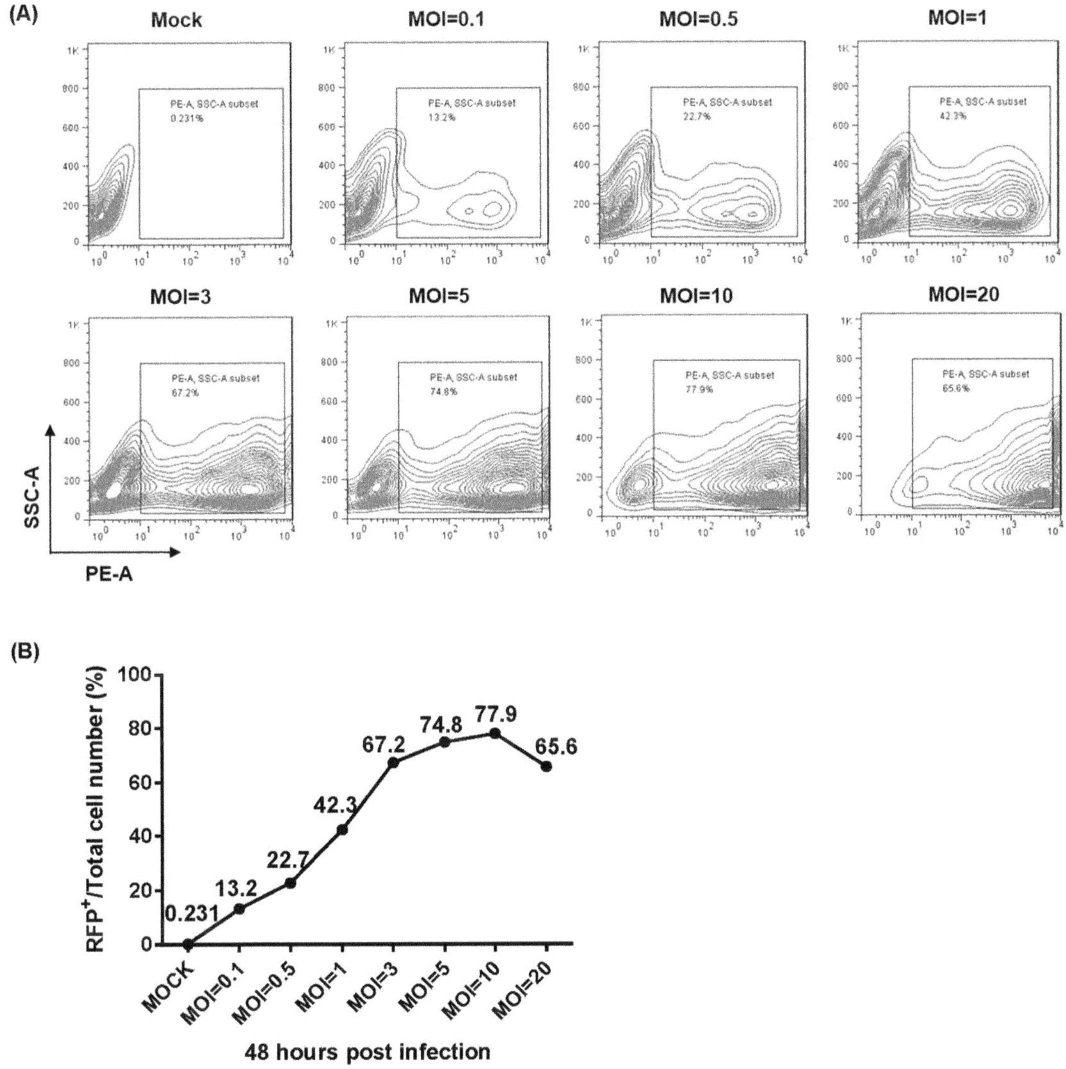

Fig. 2 Efficiency of lentivirus infection of hTSCs. (**a**) Analysis of RFP-expressing cells by flow cytometry after infection with RFP-expressing lentivirus at different titers. (**b**) Statistical analysis of the efficiency of lentivirus infection of hTSCs with different titers

6. Infection of hTSCs with lentivirus at multiplicity of infection (MOI) ≥ 1 will have a satisfying expression efficiency (Fig. 2).
7. Optimized concentrations for resistance screening in hTSCs: puromycin, 2 μg/mL; blasticidin, 5 μg/mL; hygromycin, 30 μg/mL.
8. During the resistance screening of lentivirus-infected hTSCs, uninfected hTSCs are used as negative controls, and all the cells of negative control died after 48 h of screening indicating the completion of screening.

9. The efficiency of obtaining the monoclonal hTSCs with mutations at target sites according to our described workflow is 40–70%.
10. The size of gene expression cassette (≤6.5 kb) carried by lentivirus vector limits lentiviral packaging efficiency; thus the dCas9-VP64 system is a better choice for transcriptional activation in hTSCs.

Acknowledgments

This work was supported by National Natural Science Foundation of China (82192870) and National Key R&D Program of China (2021YFC2700302, 2018YFE0201103).

References

1. Okae H, Toh H, Sato T, Hiura H, Takahashi S, Shirane K, Kabayama Y, Suyama M, Sasaki H, Arima T (2018) Derivation of human trophoblast stem cells. Cell Stem Cell 22(1): 50–63. e56
2. Mojica FJ, Díez-Villaseñor C, García-Martínez J, Soria E (2005) Intervening sequences of regularly spaced prokaryotic repeats derive from foreign genetic elements. J Mol Evol 60(2):174–182
3. Terns MP, Terns RM (2011) CRISPR-based adaptive immune systems. Curr Opin Microbiol 14(3):321–327
4. Cong L, Ran FA, Cox D, Lin S, Barretto R, Habib N, Hsu PD, Wu X, Jiang W, Marraffini LA et al (2013) Multiplex genome engineering using CRISPR/Cas systems. Science 339(6121):819–823
5. Zhang F, Wen Y, Guo X (2014) CRISPR/Cas9 for genome editing: progress, implications and challenges. Hum Mol Genet 23(R1):R40–R46
6. Jinek M, Chylinski K, Fonfara I, Hauer M, Doudna JA, Charpentier E (2012) A programmable dual-RNA-guided DNA endonuclease in adaptive bacterial immunity. Science 337(6096):816–821
7. Dominguez AA, Lim WA, Qi LS (2016) Beyond editing: repurposing CRISPR-Cas9 for precision genome regulation and interrogation. Nat Rev Mol Cell Biol 17(1):5–15
8. Maeder ML, Linder SJ, Cascio VM, Fu Y, Ho QH, Joung JK (2013) CRISPR RNA-guided activation of endogenous human genes. Nat Methods 10(10):977–979
9. Qi LS, Larson MH, Gilbert LA, Doudna JA, Weissman JS, Arkin AP, Lim WA (2013) Repurposing CRISPR as an RNA-guided platform for sequence-specific control of gene expression. Cell 152(5):1173–1183
10. Perez-Pinera P, Kocak DD, Vockley CM, Adler AF, Kabadi AM, Polstein LR, Thakore PI, Glass KA, Ousterout DG, Leong KW et al (2013) RNA-guided gene activation by CRISPR-Cas9-based transcription factors. Nat Methods 10(10):973–976

Methods in Molecular Biology (2024) 2767: 63–74
DOI 10.1007/7651_2023_495

Published online: 05 July 2023

Endometrial Assembloids to Model Human Embryo Implantation In Vitro

Thomas M. Rawlings, Maria Tryfonos, Komal Makwana, Deborah M. Taylor, Jan J. Brosens, and Emma S. Lucas

Abstract

Understanding the process of human embryo implantation is impeded by the inability to study this phenomenon in vivo, thus limiting opportunities to gain knowledge to in vitro modeling. Previous models have relied on monolayer co-cultures, which do not replicate the complexity of endometrial tissue. Here, we detail the establishment of three-dimensional endometrial assembloids, comprising gland-like epithelial organoids in a stromal matrix. Endometrial assembloids mimic endometrial tissue structure more faithfully and can be used to study human embryo-endometrial interactions. Co-cultures of human embryos and endometrial assembloids will enhance our fundamental understanding of these processes as well as allowing us to study the mechanisms of persistent reproductive failure.

Key words Assembloid, Decidualization, Embryo, Endometrium, Implantation, Primary culture

1 Introduction

Understanding the mechanisms underpinning human embryo implantation is impeded by the inability to study this process in vivo, due to ethical reasons and practical constraints. A small number of studies in the mid-twentieth century provided some insight into the early stages of human embryo implantation, but the ability to model this pivotal stage of development in vitro offers the opportunity to investigate mechanisms of implantation failure and early pregnancy loss with patient-specific samples [1]. The development of 3D in vitro models of the human endometrium represents a significant progression over 2D embryo-endometrium co-culture models [1]. Here we describe the establishment of endometrial assembloids, comprising epithelial gland-like organoids surrounded by a stromal matrix [2]. The endometrial assembloid model can be used to examine embryo-endometrial interactions using imaging as well as molecular and biochemical approaches, thus offering the prospect of furthering our understanding of this "black box" of human reproduction.

The assembloid model also represents a versatile tool for preclinical screening of pharmacological compounds aimed at preventing persistent reproductive failure [2].

2 Materials

Work in a Class II microbiological safety cabinet (MSC) using aseptic technique. Local waste disposal rules for clinical and chemical waste should be followed.

2.1 Digestion Medium

1. Stock solution of collagenase type IA (from *Clostridium histolyticum*; Sigma-Aldrich: C9891): add 10 mL Dulbecco's Modified Eagle Medium/Ham's F-12 (DMEM/F12) (1:1), HEPES with L-glutamine, phenol free (GibcoTM: 11039047), to a 500 mg vial and vortex-mix until dissolved. Aliquot and store at −20 °C.
2. Stock solution of deoxyribonuclease (DNase) I (Worthington Biochemical Corporation, Lorne Laboratories: LS002060): add 8.15 mL nuclease-free water to one vial and vortex-mix until dissolved. Aliquot and store at −20 °C.
3. Add 50 μL DNase I solution (0.1 mg/mL) and 50 μL collagenase type Ia solution (0.5 mg/mL) to 5 mL Dulbecco's Modified Eagle Medium/ Ham's F-12 (DMEM/F12) (1:1), HEPES with L-glutamine, phenol free (GibcoTM: 11039047).

2.2 Dextran-Coated Charcoal-Stripped Fetal Bovine Serum (DCC-FBS)

1. Thaw 500 mL heat-inactivated fetal bovine serum (FBS; Gibco:10500064) overnight at 4 °C (*see* **Note 1**).
2. Add 1.25 g activated charcoal (Merck: C-9157) and 125 mg dextran (Merck: D8821) to 500 mL FBS, mix thoroughly, and incubate for 2 h in a 56 °C water bath. Shake at regular intervals (~30 min).
3. Aliquot the mixture to 50 mL tubes and centrifuge for 30 min at 1800 × g.
4. Filter-sterilize using a 0.22 μm PES membrane vacuum filtration unit, e.g., Stericup S2GPU05RE.
5. Aliquot and store at −20 °C.

2.3 Growth Medium (Stromal Monolayers)

1. Thaw 50 mL DCC-FBS at RT or 4 °C.
2. Combine 50 mL DCC-FBS, 5 mL L-glutamine (Gibco: 25030), 5 mL antibiotic-antimycotic solution (Gibco: 15240), 100 μL insulin solution (10 mg/mL stock in acidified H_2O; Merck: 91077C), and 5 μL estradiol (100 μM stock in ethanol; Sigma-Aldrich: E2758). Add the mixture to 500 mL Dulbecco's Modified Eagle Medium/Ham's F-12

(DMEM/F12; 1:1), containing HEPES with L-glutamine with phenol red (GibcoTM: 31330038), by filtration through a 0.22 μm syringe filter. Store at 4 °C for up to 4 weeks.

2.4 Expansion Medium (Epithelial Organoids)

Expansion medium (ExM) recipe is as described by Turco et al. [3]. *See* **Note 2**.

1. Prepare stock solution aliquots for ExM according to Table 1.
2. Prepare 20 mL of medium by adding aliquoted growth factors, inhibitors, and supplements to advanced Dulbecco's Modified Eagle Medium/Ham's F-12 (DMEM/F12; Gibco: 12634010) according to Table 2.
3. Filter-sterilize the solution using a 0.22 μm syringe filter.
4. Store aliquots at 4 °C for up to 1 month.

2.5 Differentiation Medium

Differentiation medium (DM) recipe is as described by Rawlings et al. [2]. *See* **Note 2**.

1. Prepare the medium by adding growth factors, inhibitors, and supplements to advanced Dulbecco's Modified Eagle Medium/Ham's F-12 (DMEM/F12; Gibco: 12634010) according to Table 3.
2. Filter-sterilize the solution using a 0.22 μm syringe filter.
3. Store aliquots at 4 °C for up to 1 month.

2.6 Embryo Co-culture Medium

Embryo co-culture medium (EM) recipe is as described by Rawlings et al. [2]. *See* **Note 2**.

1. Prepare the medium by adding growth factors, inhibitors, and supplements to advanced Dulbecco's Modified Eagle Medium/Ham's F-12 (DMEM/F12; Gibco: 12634010) according to Table 4.
2. Filter sterilize the solution using a 0.22 μm syringe filter.
3. Store aliquots at 4 °C for up to 1 month.

3 Methods

Appropriate local ethical approvals for the collection and use of human tissues and embryos must be in place before work commences. Workers should be fully immunized against hepatitis B before commencing work with human samples. Work in a Class II microbiological safety cabinet (MSC) using aseptic technique. Local waste disposal rules for clinical and chemical waste should be followed.

Table 1
Expansion medium[a] components: preparation and storage

Material	Final concentration	Procedure	Aliquots	Stock temperature (°C)	Original form
Advanced DMEM/F12	1×		20 mL × 25	4	Liquid
N2 supplement 100x	1×		200 μL × 25	−20	Liquid
B27 supplement 50x	1×		400 μL × 25	−20	Liquid
N-acetyl –L-cysteine	1.25 mM	Reconsitute 102 mg in 5 mL advanced media	200 μL × 25	4	Powder
L-glutamine	1×	Use 5 mL of stock solution	200 μL × 25	−20	Liquid
Antibiotic/antimycotic	1×	Use 5 mL of stock solution	200 μL × 25	−20	Liquid
A83-01	500 nM	Dilute 5 mg in 2 mL of DMSO (5.9309 mM stock solution)	20 μL × 100	−20	Powder
Nicotinamide	10 mM	Reconsitute 610.6 mg in 5 mL advanced media	200 μL × 25	RT	Powder
EGF	50 ng/mL	Reconsitute one vial (100 μg) in 1 mL of sterile water. This makes a 100 μg/mL stock solution	10 μL × 100	−20	Powder
Noggin	100 ng/mL	Reconstitute one vial (100 μg) in 1 mL of sterile water. This makes a 100 μg/mL stock solution.	20 μL × 50	−20	Powder
R-spondin-1	500 ng/mL	Reconstitute each vial (250 μg) in 1 mL of sterile water. This makes a 250 μg/mL stock solution	40 μL × 25	−80	Powder
FGF10	100 ng/mL	Prepare a 0.5 M sodium phosphate solution, pH 7.4, mixing 0.5 M sodium phosphate monobasic and 0.5 M sodium phosphate dibasic. Put the dibasic (pH ~ 9) in the beaker and add the monobasic (pH ~ 4). Dilute solution to 5 mM by adding 10 μL of the 0.5 M sodium phosphate solution into 990 μL distilled water. Reconstitute FGF10 in 1 mL of 5 mM sodium phosphate (100 μg/mL solution)	20 μL × 50	−20	Powder
HGF	50 ng/mL	Reconstitute one vial (100 μg) in 1 mL of sterile water. This makes a 100 μg/mL stock solution	10 μL × 100	−20	Powder

[a]As described by Turco et al. [3]

Table 2
Expansion medium[a]

Component	Final concentration
Advanced DMEM/F12	1×
A83-01	500 nM
R-spondin-1	500 ng/mL
Nicotinamide	10 mM
N-acetyl-L-cysteine	1.25 mM
Noggin	100 ng/mL
FGF10	100 ng/mL
HGF	50 ng/mL
EGF	50 ng/mL
N2	1×
B27	1×
Antibiotic-antimycotic	1×
L-glutamine	2 mM

[a]As described by Turco et al. [3]

Table 3
Differentiation medium

Component	Final concentration
Advanced DMEM/F12	1×
N-acetyl-L-cysteine	1.25 mM
N2	1×
B27	1×
Antibiotic-antimycotic	1×
L-glutamine	2 mM
MPA	1 μM
8-Bromo-cAMP	0.5 mM
E2	1 μM

3.1 Mincing and Digestion of Human Endometrium Tissue

1. Endometrial tissue is collected by pipelle biopsy (*see* **Note 3**) and processed immediately for downstream handling, although other collection approaches can be used according to local procedures. Tissue is transferred immediately into DMEM/

Table 4
Embryo medium

Component	Final concentration
Advanced DMEM/F12	1×
N-acetyl-L-cysteine	1.25 mM
N2	1×
B27	1×
Antibiotic-antimycotic	1×
L-glutamine	2 mM
Non-essential amino acids	1×
Essential amino acids	1×
ITS-X	1×
MPA	1 μM
8-Bromo-cAMP	0.5 mM
E2	1 μM

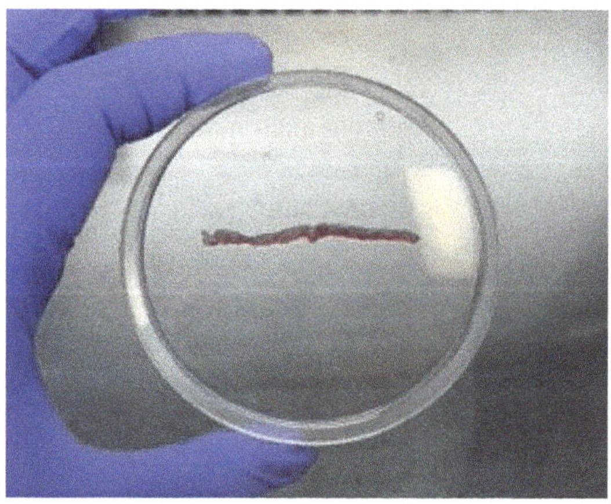

Fig. 1 Typical endometrial pipelle biopsy. A piece of biopsy tissue of a typical length when collected by pipelle. Tissue is placed on a 10 cm dish and excess transport medium has been removed. Tissue of this size would require ~5 min of chopping before enzymatic digestion

F12 in a 30 mL universal tube for washing and transport (*see* **Note 4**).

2. Tissue is removed from the transport media and transferred to a 10 cm non-coated dish. Aspirate any excess liquid by pipetting (Fig. 1).

3. Using a sterile scalpel, mince the tissue for ~5 min or until a uniform consistency is achieved (*see* **Note 5**).
4. Add 5–10 mL of sterile digestion media to the minced tissue and collect the suspension into a 50 mL tube. Incubate at 37 °C for 1 h with periodic shaking (15–20 min intervals).
5. Once digestion is complete, pass the suspension through a 40 μm cell sieve into a fresh tube and then wash the sieve membrane with 10 mL stromal growth medium. This tube now contains the stromal cell fraction, as well as immune and endothelial cells.
6. Invert the cell sieve over a new 50 mL tube and back-wash the membrane with 10 mL additive-free medium. This tube now contains undigested epithelial clumps.
7. Centrifuge both tubes at 280 × *g* for 5 min to pellet the cells, aspirate supernatant, and continue with Subheadings 3.2 and 3.3.

3.2 Separation, Plating, Expansion, and Passage of the Stromal Cell Fraction

1. Resuspend the stromal pellet in an appropriate volume of stromal growth medium (usually 5 mL or 10 mL) and plate the cells in either T25 or T75 culture flask, depending on the biopsy size (*see* **Note 6**). Change the media after 16–24 h in culture (*see* **Note** 7).
2. Allow cells to expand until ~90% confluence, with media changed every 48 h.
3. Once desired cell density is achieved, aspirate medium from cell culture flasks using a vacuum aspiration line and wash the cells in PBS.
4. Add 1 mL of 0.25% Trypsin-EDTA to the flask and ensure it covers the entire surface of the monolayer. Incubate for 5 min at 37 °C.
5. Remove from the incubator and dislodge remaining adherent cells by firmly tapping the side of the flask with the palm of your hand. Observe the flask under a light microscope to ensure all cells are detached from the surface.
6. Collect the cells by washing the flask with 9 mL growth medium and transfer the suspension to a 14 mL tube. Pellet the cells by centrifugation for 5 min at 300 × g. Meanwhile, pre-wet new flasks for continuing the culture. Cells can be passaged at ratios of 1:2 or 1:3, according to the rate of growth of each sample.
7. Aspirate the supernatant and resuspend the pellet in 10 mL growth medium. Distribute equally between the pre-wetted flasks. Transfer new flasks to the cell culture incubator, now labeled as Passage 1 (P1). Replace 10% DCC-FBS DMEM/F12 medium every 2–3 days.

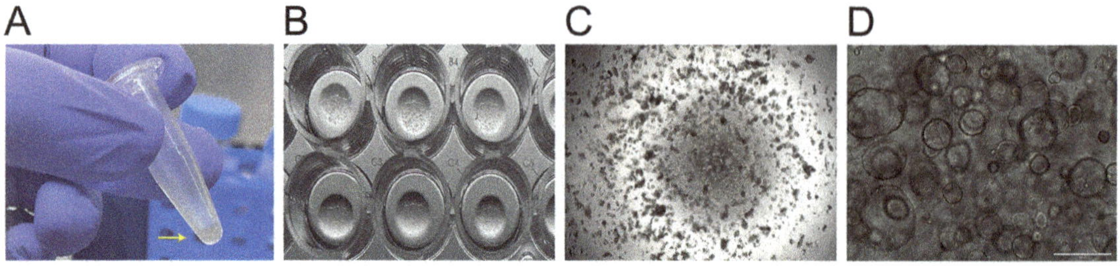

Fig. 2 Preparation of epithelial organoid culture. (**a**) The epithelial cell pellet within a 1.5 mL microcentrifuge tube. A pellet of this volume would be resuspended in 480 μL Matrigel. (**b**) Cured Matrigel/epithelial cell droplets (20 μL each) in a 48-well plate, prior to the addition of Expansion Medium. (**c**) Epithelial fragments in gel, viewed at ×4 magnification. (**d**) Gland-like structures in gel, viewed at ×10 magnification. Scale bar 400 μm

3.3 Plating, Expansion, and Passage of the Epithelial Cell Fraction

This protocol is based on the original published methods of Turco and colleagues [3].

1. Resuspend the epithelial cell pellet in 1 mL of additive-free DMEM/F12 medium with gentle pipetting. Transfer to a 1.5 mL microcentrifuge tube and centrifuge at 600 × *g* for 6 min.
2. Carefully aspirate the supernatant and estimate the volume of the cell pellet (Fig. 2a). Add ice-cold Matrigel (*see* **Note 8**) at a ratio of 1:20 (cells: Matrigel) and resuspend by gentle pipetting, taking care to avoid the introduction of bubbles. Immediately place the tube on wet ice to cool and prepare other samples if necessary.
3. Aliquot the epithelial cell/Matrigel suspension as 20 μL droplets into the center of wells in a 48-well plate. Transfer the plate to the 37 °C incubator for 20 min to allow the Matrigel to cure to the gel phase, resulting in 3D suspension of the epithelial cell clumps (Fig. 2b, c).
4. Gently add 200 μL organoid expansion medium to the wells (*see* **Note 9**) and return the plate to the incubator. Replace medium every 48–72 h. Using a light microscope, examine the droplets for the appearance of gland-like structures (Fig. 2d). Epithelial organoid cultures can be passaged at 7- to 10-day intervals, according to the growth profile of the sample as assessed by visual inspection of the cell density and the phenol red indicator in the ExM.
5. For routine passage, collect Matrigel droplets by dislodging with a 1000 μL pipette tip, and collect into a 1.5 mL microcentrifuge tube. Pool 6–8 gel droplets, plus a small volume of media, per tube, then centrifuge at 600 × g, 4 °C, for 6 min. Keep the tubes on wet ice while processing subsequent steps.

6. Carefully aspirate supernatant and resuspend the pellet in 200 μL ice-cold additive-free DMEM/F12 medium. Pipette each suspension 300× to release organoids from the Matrigel. Add a further 1 mL ice-cold additive-free DMEM/F12 medium to each tube, then centrifuge at 600 × g, 4 °C, for 6 min.
7. Repeat **step 6**, this time with vigorous pipetting 80× to break up the organoids. After centrifugation, remove the supernatant and estimate the volume of the cell pellet.
8. Add ice-cold Matrigel at a ratio of 1:20 (cells: Matrigel) and gently resuspend the cells to avoid introduction of bubbles and plate as described above in **steps 3** and **4**, now labeled as P1. Allow cells to expand for a further 7–10 days.

3.4 Establishment and Growth of Assembloid Cultures

1. Collect epithelial organoid droplets as per passage procedure (described above in Subheading 3.3, **steps 5–7**). Keep cells chilled on wet ice.
2. Collect stromal cells as per passage procedure (described above in Subheading 3.2, **steps 3–6**). Resuspend the cells in 1 mL additive-free DMEM/F12 medium and perform a cell count (*see* **Note 10**).
3. Stromal cells and gland organoid pellets are mixed at a ratio of approximately 1:1 (v/v). This is equivalent to 5×10^4 stromal cells and organoid fragments passaged at a ratio of 1:2 or 1:3 per assembloid culture. For example, 12 Matrigel droplets containing epithelial organoids obtained from the expansion phase plus 1.2×10^6 stromal cells (1–2 T75 flasks depending on the sample) would provide enough material to prepare 24 assembloid droplets.
4. Mix together the stromal and epithelial cells according to the ratios above and centrifuge at 600 × g, 4 °C, for 6 min (Fig. 3a).

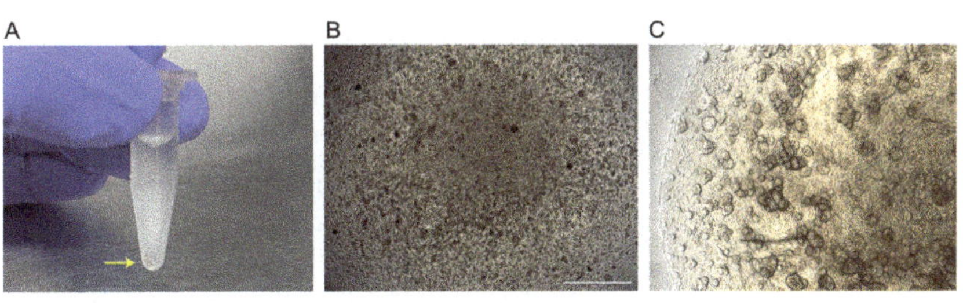

Fig. 3 Preparation of assembloid culture. (**a**) The stromal-epithelial cell pellet within a 1.5 ml microcentrifuge tube. A pellet of this volume would be resuspended in 300 μL collagen hydrogel. (**b**) Mixture of epithelial and stromal cells in gel at Day 0 of culture viewed at ×4 magnification. (**c**) Assembloid (gland-like structures surrounded by stromal cells) at Day 4 of culture viewed at ×4 magnification. Scale bar: 1000 μm

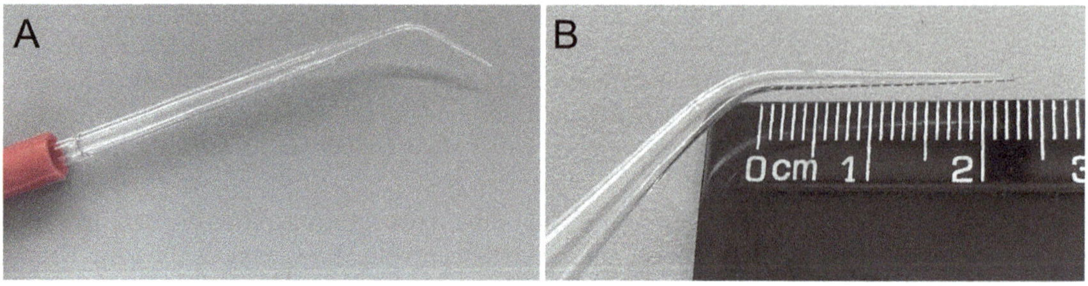

Fig. 4 Preparation of a pulled glass Pasteur pipette for embryo co-culture. (**a**) A glass Pasteur pipette pulled at an angle for embryo transfer for co-culture. (**b**) A pulled Pasteur pipette with ruler (cm/mm) for size reference

5. Carefully aspirate the supernatant and add ice-cold collagen hydrogel (e.g., PureCol® EZ Gel solution; Merck: 5074) at a volume of 20 μL per well required. Gently resuspend the cell pellet, taking care to avoid introduction of air bubbles.
6. Aliquot the suspension into 48-well plates to give a single 20 μL droplet per well. Allow the gel to cure for 45 min in the cell culture incubator (or as directed by the manufacturer if using an alternative matrix). Once cured, add 200 μL ExM + E2 (10 nM) to each well and return to the incubator. Refresh media every 48 h.

3.5 Differentiation of Assembloid Cultures

1. Following the required duration of growth (e.g., 4 days) (Fig. 3b), assembloid cultures can be differentiated according to preferred protocols (Fig. 3c). For example, treatment with minimal differentiation medium containing 8-bromo-cAMP and medroxyprogesterone acetate (MPA) for 4 days will induce a robust decidual response. Media should be changed every 48 h and spent media can be collected for analysis of secreted factors.
2. Media can be supplemented with additional factors to test the favored hypothesis of the experimenter. For example, for inhibition of tyrosine kinase activity, we supplement MDM with 250 nM dasatinib during either growth, or differentiation, or both.

3.6 Assembloid and Embryo Co-cultures

1. Prior to initiating a co-culture, assembloid wells should have media replaced with pre-warmed embryo media.
2. Prior to co-culture, warm vitrified blastocysts using the Kitazato vitrification warming media (Dibimed, Spain) and remove the zona pellucida using a Saturn 5 laser (CooperSurgical). Incubate blastocysts for 1 h under oil in 20 μL drops of culture media (ORIGIO Sequential Blast media, CooperSurgical) at 5% O_2, 6% CO_2, 89% N_2 at 37 °C to allow for re-expansion.

3. Using a pulled glass Pasteur pipette (Fig. 4), create a small well in the center of the assembloid droplet for the embryo to be transferred into.
4. Carefully collect a single blastocysts and transfer into the well created, Subheading 3.6, **step 3,** using a pulled glass Pasteur pipette, while visualizing under a microscope.
5. For timelapse microscopy, establish a suitable local protocol for imaging requirements. For example, we image wells every 20 min over 96 h. Restrictions to the duration of embryo culture must be observed and adhered to: in the UK, embryos must not be cultured beyond 14 days post-fertilization.

4 Notes

1. Thawing FBS at higher temperatures (e.g., 37 °C or above) can result in the formation of protein aggregates in the solution [4]. Ideally, FBS should always be thawed overnight at 4 °C, but incubation at RT for a shorter period is also suitable.
2. Stock solutions of growth factors, inhibitors, and supplements should be prepared according to manufacturers' instructions and aliquoted to minimize freeze-thaw cycles.
3. Biopsy tissue may be obtained according to local procedures and depending on the patient cohort, for example, surgical collection of tissue during [surgery] or collection of tissue from hysterectomy samples. Appropriate approvals and informed consent processes must be followed in any case.
4. Pipelle biopsies can normally be subdivided for several uses, according to requirements, for example, fixation in 10% neutral buffered formalin for immunohistochemistry; treatment with an RNA preservative, such as RNAlater; snap freezing for later isolation of proteins and/or DNA. Proportions will depend on individual requirements, but we find that 0.5–1.0 cm for each of the above is normally sufficient for most analyses, with the bulk of the sample being allocated to cell isolations and in vitro cultures.
5. Finely minced tissue will equate roughly to <0.5 mm pieces of tissue. Smaller sections of tissue (<1 cm in length) may only require 2–3 min of mincing, while larger pieces (>5 cm in length) may benefit from a longer process (i.e., 6 or 7 min).
6. Small biopsies (<2–3 cm in length) will benefit from plating in T25 flasks, while larger biopsies will have sufficient cell density to grow in T75 flasks.

7. If desired, immune cells in suspension can be isolated from the culture media collected at this point, either by magnetic-activated or flow cytometric cell sorting.
8. To remain in the liquid phase, Matrigel must be kept cold *at all times* until gelation (setting) is required. The use of prechilled pipette tips can assist with handling.
9. To avoid disturbing the gel droplets, add medium to the wells by directing the flow of medium from the pipette tip toward the wall of the wells, rather than into the center of the well.
10. Cell counts should be performed on a Neubauer Improved Hemocytometer. Automated cell counting may be used with a suitably optimized protocol.

References

1. Rawlings TM, Makwana K, Tryfonos M, Lucas ES (2021) Organoids to model the endometrium: implantation and beyond. Reproduction and Fertility:RAF-21-0023. https://doi.org/10.1530/raf-21-0023
2. Rawlings TM, Makwana K, Taylor DM, Mole MA, Fishwick KJ, Tryfonos M, Odendaal J, Hawkes A, Zernicka-Goetz M, Hartshorne GM, Brosens JJ, Lucas ES (2021) Modelling the impact of decidual senescence on embryo implantation in human endometrial assembloids. eLife 10. https://doi.org/10.7554/eLife.69603
3. Turco MY, Gardner L, Hughes J, Cindrova-Davies T, Gomez MJ, Farrell L, Hollinshead M, Marsh SGE, Brosens JJ, Critchley HO, Simons BD, Hemberger M, Koo BK, Moffett A, Burton GJ (2017) Long-term, hormone-responsive organoid cultures of human endometrium in a chemically defined medium. Nat Cell Biol 19(5):568–577. https://doi.org/10.1038/ncb3516
4. Fetal Bovine Serum (FBS) [Internet]. www.thermofisher.com, c2023. Available from: https://www.thermofisher.com/uk/en/home/life-science/cell-culture/mammalian-cell-culture/fbs.html

Methods in Molecular Biology (2024) 2767: 75–84
DOI 10.1007/7651_2022_470

Published online: 08 February 2023

Stem Cell-Derived Microfluidic Amniotic Sac Embryoid (μPASE)

Yi Zheng

Abstract

The microfluidic amniotic sac embryoid (μPASE) is a human pluripotent stem cell (hPSC)-derived multicellular human embryo-like structure with molecular and morphological features resembling the progressive development of the early post-implantation human embryonic sac. The microfluidic device is specifically designed to control the formation of hPSC clusters and expose the clusters to different morphogen environments, allowing the development of μPASEs in a highly controllable, reproducible, and scalable fashion. The μPASE model displays human embryonic developmental landmarks such as lumenogenesis of the epiblast, amniotic cavity formation, and the specification of primordial germ cells and gastrulating cells (or mesendoderm cells). Here, we provide detailed instructions needed to reproduce μPASEs, including the immunofluorescence staining and cell retrieval protocols for characterizing μPASEs obtained under different experimental conditions.

Key words Amnion, Embryonic development, Human embryoid, Human pluripotent stem cell, Microfluidics, Primitive streak, Primordial germ cell

1 Introduction

Study of human embryonic development is essential for prevention of pregnancy loss and birth defects, assisted reproductive technologies, and regenerative medicine [1–3]. So far, our knowledge of embryonic development has been derived from studying a few animal models that are convenient for experimentation. However, there is increasing evidence that significant genetic and morphological differences exist between humans and these animal models [4, 5]. To understand human embryonic development, ideally, we should examine and experiment on human embryonic tissue [6–9], but, due to ethical restrictions, this path is quite challenging [10–13]. Recent studies show that, under properly controlled conditions, hPSCs can self-organize into multicellular tissues in vitro, which resemble human embryo architecture and bare the molecular features of certain important developmental events [14–18]. Particularly, we have recently developed a hPSC-derived embryoid, termed post-implantation amniotic sac embryoid (PASE), that appears to closely recapitulate several landmarks during the development of the epiblast and amniotic ectoderm parts of the early post-implantation human embryo in a three-dimensional (3D), human-relevant tissue architecture [19, 20], including lumenogenesis of the epiblast and the resultant pro-amniotic cavity, symmetry breaking of the epiblast sac to form the bipolar embryonic sac, and specification of primordial germ cells (PGCs) and primitive streak/gastrulating cells. By using controllable microfluidic tools, we have further successfully developed a microfluidic PASE (μPASE) system allowing the development of PASEs in a highly controllable, reproducible, and scalable fashion [21–23] (Fig. 1a). We have successfully repeated the μPASE protocol using

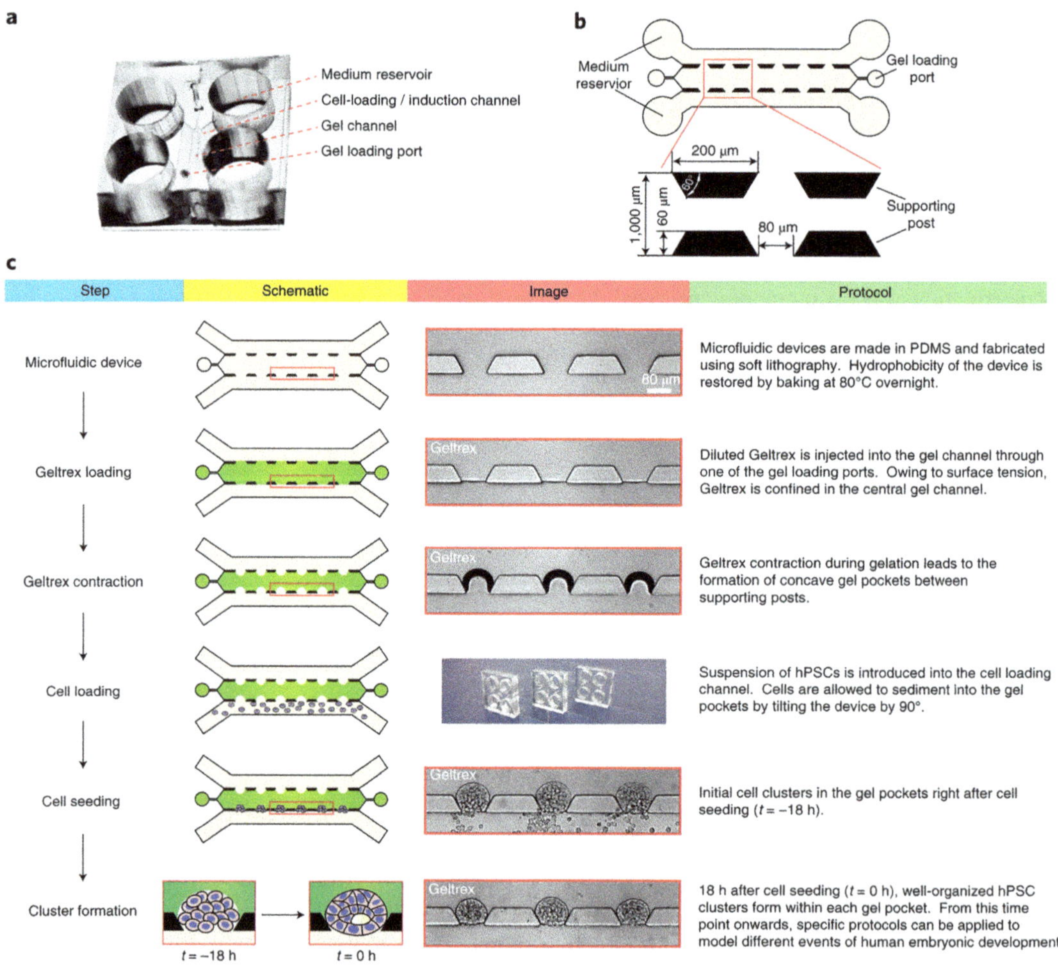

Fig. 1 Design of the microfluidic device and workflow for μPASE generation using hPSCs. (**a**) Photograph showing the microfluidic device. Medium reservoirs, gel-loading ports, and cell-loading and induction channels are indicated. (**b**) Design of the microfluidic device incorporating three parallel channels (80 μm in height) partitioned by trapezoid-shaped supporting posts spaced 80 μm apart. The central gel channel is preloaded with Geltrex. The other two side channels are used for cell loading (the cell-loading channel) and chemical induction (the induction channel), respectively. (**c**) Schematic diagrams and photographs showing experimental steps to establish initial clusters of hPSCs. (Reproduced with permission from ref. [23])

multiple human embryonic stem cell lines, human-induced pluripotent stem cell lines, and chimpanzee-induced pluripotent stem cell lines. This protocol has been independently reproduced by several laboratories to study different aspects of early human embryonic development.

2 Materials

2.1 Human Pluripotent Stem Cells

We have successfully repeated this protocol using multiple human embryonic stem cell lines, including H9, H1, ESI-017, and RUES2, and multiple human-induced pluripotent stem cell lines. To obtain optimal results, hPSCs should be maintained on Geltrex-coated tissue culture plates using mTeSR1. For hPSCs maintained using Matrigel or Essential 8 medium, μPASEs can still be generated but with suboptimal morphology and slightly different developmental timing.

2.2 Geltrex

Immediately, store Geltrex in a −20 or −80 °C freezer upon receiving. Thaw Geltrex by placing the tube on ice in a 4 °C fridge for at least 6 h. Before dividing into small aliquots (150–200 μL), mix the Geltrex tube embedded in ice using a rocking platform for 10 min. Thaw one new small aliquot before each experiment by embedding in ice in a fridge at 4 °C for at least 3 h (*see* **Note 1**).

3 Methods

3.1 Microfluidic Devices

1. Print a photomask with desired design using a transparency. See Fig. 1b for recommended device/photomask design (*see* **Note 2**).
2. Fabricate a positive silicon mold using the standard deep reactive-ion etching (DRIE) method. Consult technicians of specific microfabrication facilities for operational conditions of DRIE (*see* **Note 3**).
3. Expose the positive silicon mold to oxygen plasma for 1 min, and quickly place it in a vacuum desiccator together with 100 μL of trichlorosilane for 1 h to silanize the surface to prevent adhesion that may occur during the following casting procedure (*see* **Note 4**).
4. Mix PDMS pre-polymer with curing agent at 1:10 weight ratio and degas the mixture in a vacuum desiccator until no bubbles are visible (degassing) (*see* **Note 5**).
5. Pour PDMS mixture onto the silicon mold and bake in a convection oven at 110 °C for 60 min. Then cut and peal the PDMS layer off the silicon mold.

6. Punch medium reservoirs and gel-loading ports using biopsy punch tools (Fig. 1a). Cut the PDMS layer containing multiple devices into individual devices using a blade.
7. Firmly press adhesive tapes onto the PDMS layer to remove particles and debris.
8. Expose PDMS layers and coverslips to oxygen plasma for 3 s at a power of 40 W (*see* **Note 6**).
9. Quickly place the PDMS layer on top of the coverslip and then apply pressure using fingertips for 30 s on a 120 °C hotplate (*see* **Note 7**).
10. After all devices are bonded, move the devices into a convection oven at 80–100 °C and bake for at least 8 h (*see* **Note 8**).

3.2 Geltrex Injection

1. Transfer the microfluidic devices into a biosafety cabinet using a covered container and expose them to UV light while cooling down for 30 min. Use a marker to label one of the side channels as "cell-loading channel" and the other side channel as "induction channel" (*see* **Note 9**).
2. Prepare a humidified chamber that can fit several microfluidic devices, e.g., a 150-mm Petri dish with a 35-mm Petri dish filled with sterile DI water inside. Place the humidified chamber in a CO_2 incubator (37 °C, 5% CO_2) for at least 20 min before usage.
3. Place a 1.5 mL centrifuge tube on ice for 5 min. In the cold centrifuge tube, dilute Geltrex with cold Essential 6 medium at a ratio of 7:3. Then mix the diluted Geltrex by gently pipetting (*see* **Note 10**).
4. Use a 20 μL pipette tip to inject diluted Geltrex into one of the gel-loading ports to fill the central gel channel (Fig. 1c). The surface tension generated between PDMS supporting posts will confine the diluted Geltrex in the central channel (*see* **Note 11**).
5. After injection, place the microfluidic device in the prepared humidified chamber in a CO_2 incubator for 5 min to allow Geltrex to solidify. While solidifying, Geltrex contracts to form concave gel pockets between PDMS supporting posts in the microfluidic device, which will be used for stem cell seeding (*see* **Note 12**).
6. After 5 min of solidification in the incubator, visually check the gel pocket size every 2 min (*see* **Note 13**). When gel pockets reach a desired size (100–120 μm in diameter) (Fig. 1c), introduce 140 μL mTeSR1 medium immediately into one reservoir of the induction channel and apply vacuum to the other reservoir with a rubber dropper bulb to fill the induction channel. When a small droplet of medium is visible in the other

reservoir, then add 140 μL mTeSR1 medium to the other reservoir. Use the same method to fill the cell-loading channel with 180 μL medium in each cell-loading channel reservoir (*see* **Note 14**).

7. Place the filled microfluidic devices in the humidified chamber in a CO_2 incubator (37 °C, 5% CO_2) for 18–24 h to stabilize the Geltrex structure and remove trapped air bubbles.

3.3 μPASE Generation

1. When hPSCs in the culture reach 70–80% confluency, dissociate hPSCs into single cells using Accutase at 37 °C for 7–10 min and resuspend hPSCs in mTeSR1 containing 10 μM Y27632 at concentration of 8×10^6 cells mL^{-1}. Keep the cell suspension on ice while conducting the following steps (*see* **Note 15**).
2. Empty all four reservoirs of the microfluidic device using an aspirator by placing the tip of the aspirator at the bottom of the reservoirs (*see* **Note 16**).
3. After mixing thoroughly, pipette 10 μL hPSC suspension to the inlet of the cell-loading channel. Then place the device in the incubator in the position as shown in Fig. 1c. Allow the cells to settle into the gel pockets for 5 min. Then gently shake the device to agitate the cells and put the device back to the same position for another 5 min and visually inspect; hPSCs should have filled the gel pockets (Fig. 1c) (*see* **Note 17**).
4. Fill the reservoirs of the microfluidic device with mTeSR1 containing 10 μM Y27632: 200 μL per reservoir for the cell-loading channel and 50 μL per reservoir for the induction channel. Incubate in a CO_2 incubator for 1 h. Add 90 μL mTeSR1 containing 10 μM Y27632 to the medium reservoirs of the induction channel and incubate the device in the humidified camber in the incubator for 18–24 h (*see* **Note 18**).
5. 18–24 h after cell seeding, hPSC clusters should have initiated lumenogenesis (Fig. 1c). Aspirate out medium from all reservoirs. Add 160 μL Essential 6 medium supplemented with 20 ng mL^{-1} FGF2 and 50 ng mL^{-1} BMP4 into the reservoirs of the cell-loading channel; add 200 μL Essential 6 medium supplemented with 20 ng mL^{-1} FGF2 into the reservoirs of the induction channel (*see* **Note 19**). Place the devices back into the humidified chamber in the incubator. The hPSC clusters in the microfluidic device will start to go through morphogenesis and lineage diversification to form μPASEs. No medium change is needed if the experiment is terminated within 48 h.

3.4 Immuno-fluorescence Staining of μPASEs

1. μPASEs can be examined at desired timepoints using immuno-fluorescence staining. Replace culture medium in all medium reservoirs with 4% (wt/vol) paraformaldehyde (PFA) (150–200 μL each reservoir). Leave the microfluidic device at 4 °C in a fridge overnight for cell fixation (*see* **Note 20**).
2. Remove PFA in all medium reservoirs and wash by adding 200–250 μL PBS to all medium reservoirs for 10 min.
3. Add 150–200 μL 0.1% (vol/wt) sodium dodecyl sulfate (SDS) to each medium reservoir and incubate the microfluidic devices for 3 h at room temperature for permeabilization.
4. After washing with PBS, add 150–200 μL blocking solution (4% goat serum in PBS) to each reservoir and incubate for 24 h at 4 °C in a fridge.
5. Empty all reservoirs and add primary antibodies diluted using blocking solution at recommend concentrations (50 μL each). Incubate the microfluidic device at 4 °C in a fridge for 24 h.
6. Wash with PBS and add secondary antibodies diluted in blocking solution into each medium reservoir (50–60 μL), and then incubate the microfluidic device at 4 °C in a fridge for 24 h. Protect the microfluidic device from light.
7. Remove secondary antibody solution and wash the microfluidic device twice with PBS. Leave the devices filled with PBS. μPASEs can then be imaged using a fluorescent or confocal microscope by placing the device on a standard sample holder.

3.5 Live Cell Retrieval

For certain characterization purposes, e.g., single-cell RNA sequencing (scRNA-seq), cells of μPASEs in the microfluidic devices can be collected using the following protocol.

1. Remove cells that adhere to the bottom of the reservoirs using a pipette tip, then thoroughly wash twice with DMEM medium.
2. Add 80 μL Accutase to the reservoirs of the cell-loading channel and 20 μL Accutase to the reservoirs of the induction channel (*see* **Note 21**). Incubate the microfluidic device for 60 min in a CO_2 incubator (37 °C, 5% CO_2). Depending on the size and developmental stages of the μPASEs, the incubation can be prolonged to 120 min (*see* **Note 22**).
3. Visually confirm that the cells in μPASEs are mostly detached, and then gently agitate the Accutase solution until complete cell dissociation. Carefully collect the medium containing dissociated cells into a 1.5-mL microcentrifuge tube on ice.
4. Add filtered PBSA solution (2% BSA in PBS, wt/vol) into the cell suspension until the 1.5 mL microcentrifuge tube is full; then gently pipette.
5. Centrifuge at 300 g for 5 min at 4 °C and then resuspend the cells at desired concentration and solution.

4 Notes

1. Well-preserved Geltrex is critical for successful experiments; thus, Geltrex needs to be handled very carefully to prevent partial solidification before injecting into the microchannel. Embed Geltrex vials in ice whenever possible. Thoroughly mixing Geltrex before aliquoting is necessary to ensure consistency across all the aliquots. The aliquots should be stored in a −20 or −80 °C freezer. 4 °C fridge cannot be used for long-term storage of Geltrex.
2. The photomask design is related to the size of μPASEs. The gap between PDMS supporting posts should be 80–100 μm. We do not recommend major modifications of the provided design. If modifications are necessary for the experiments, the protocol needs to be adjusted accordingly.
3. DRIE is recommended for optimal geometry of the mold. However, when DRIE facility is not accessible, SU-8 photoresist can also be used for the mold fabrication. Please refer to related technical documents for detailed procedures and parameters.
4. Oxygen plasma time and power need to be selected for specific oxygen plasma instrument used in the experiments. After proper silanization, the mold is reusable for at least 100 times when handled properly.
5. The time for degassing depends on the volume of PDMS mixture and the vacuum pressure in the desiccator.
6. Optimal oxygen plasma time and power need to be selected for specific oxygen plasma instrument used in the experiments. Overtreatment with oxygen plasma will have negative effects on the bonding.
7. This step is important to ensure proper bonding. Applied pressure should not be too strong to cause microchannel collapse. The bonding quality can be visually inspected using a microscope.
8. If space is available in the oven, we suggest to leave the devices in the oven before usage. Long period of baking will not affect the devices.
9. The two-side channels are completely identical. But in the following steps, "cell-loading channel" and "induction channel" require different operational protocols. Thus, we need to tell our intended usage of the channels without using the microscope. A simple label will work.
10. It is critical to aspirate the needed volume of thawed Geltrex from the bottom of the aliquot vial into the cold centrifuge

tube. Then add corresponding volume of Essential 6 medium and mix thoroughly. Avoid bubbles. In the case that bubbles are visible after mixing, leave the Geltrex on the ice for 5–10 min to remove bubbles.

11. While injecting diluted Geltrex into the gel channel, the front of the Geltrex needs to be monitored visually. When Geltrex just fills the gel channel, stop filling immediately. Do not inject Geltrex to fill the other gel-loading port. This step is very important for proper gel contraction. Inject Geltrex as gently as possible to prevent leakage of Geltrex into the side channels.
12. Geltrex solidification time is strongly affected by concentration. If it takes significantly longer than the suggested time, increase the concentration of Geltrex. Geltrex solidification time is also influenced by environmental temperature and humidity. Thus, it may vary in different laboratory facilities.
13. After 5 min of solidification, the size of the gel pockets needs to be monitored carefully, since the pocket size will continuously increase until too larger for the following cell-seeding steps. Filling the channels with medium stops Geltrex contraction.
14. The medium level in the reservoirs of the cell-loading channel should be higher than the medium level in the reservoirs of the induction channel to help maintain the size of the gel pockets. Be gentle when applying vacuum to avoid damaging of the solidified Geltrex.
15. The concentration of cell suspension is optimized for H9 hPSCs cultured on Geltrex-coated tissue culture plates using mTeSR1. This concentration also works well for several other hPSC lines using the same culture system. But it should be adjusted if different culture systems are used for hPSC maintenance.
16. When aspirating medium from the reservoirs, keep the tip of the aspirator away from the inlets of channels. DO NOT empty the channels.
17. This step is critical for controlling the size of hPSC clusters. If hPSC cluster size is still smaller than desired after this step, fill the reservoirs of the cell-loading channel and the induction channel with 180 μL and 50 μL mTeSR1 containing 10 μM Y27632, respectively. Then place the devices back to the humidified chamber in the incubator for 30 min. Then add another 10 μL hPSC suspension and repeat this step until the desired size of hPSC cluster is obtained.
18. The medium level in the reservoirs of the cell-loading channel should be higher than the medium level in the reservoirs of the induction channel to help keep cells in the pockets.

19. The medium level in the induction channel should be higher than the cell-loading channel to press the hPSC clusters again the PDMS supporting posts.
20. No washing is needed before fixation. When emptying the medium from the reservoirs, DO NOT empty the channels.
21. To prevent loss of cells, small volume of Accutase is added into the devices. Extended Accutase treatment is required due to the low medium exchange rate in the microfluidic devices.
22. Accutase incubation time varies under different experimental conditions. Visual inspection to determine the completion of cell dissociation is required.

Acknowledgments

This work was supported by the National Institutes of Health (R21 HD105192 and R21 HD109635) and the Department of Biomedical and Chemical Engineering at Syracuse University.

Conflicts of Interest Two patents related to this work have been filed (US20190321415/WO2018106997 by J. Fu, Y. Zheng, and S.N. Esfahani; US2020049721/PCT/US20/49721 by J. Fu and Y. Zheng).

References

1. Bianco-Miotto T, Craig JM, Gasser YP, van Dijk SJ, Ozanne SE (2017) Epigenetics and DOHaD: from basics to birth and beyond. J Dev Orig Health Dis 8(5):513–519
2. Koot Y, Teklenburg G, Salker M, Brosens J, Macklon N (2012) Molecular aspects of implantation failure. Biochim Biophys Acta (BBA) 1822(12):1943–1950
3. Fu J, Warmflash A, Lutolf MP (2021) Stem-cell-based embryo models for fundamental research and translation. Nat Mater 20(2):132–144
4. Rossant J, Tam PP (2017) New insights into early human development: lessons for stem cell derivation and differentiation. Cell Stem Cell 20(1):18–28
5. Rossant J (2015) Mouse and human blastocyst-derived stem cells: vive les differences. Development 142(1):9–12
6. Tyser RC, Mahammadov E, Nakanoh S, Vallier L, Scialdone A, Srinivas S (2021) Single-cell transcriptomic characterization of a gastrulating human embryo. Nature 600(7888):285–289
7. Xiang L, Yin Y, Zheng Y, Ma Y, Li Y, Zhao Z, Guo J, Ai Z, Niu Y, Duan K (2020) A developmental landscape of 3D-cultured human pre-gastrulation embryos. Nature 577(7791):537–542
8. Shahbazi MN, Jedrusik A, Vuoristo S, Recher G, Hupalowska A, Bolton V, Fogarty NM, Campbell A, Devito LG, Ilic D (2016) Self-organization of the human embryo in the absence of maternal tissues. Nat Cell Biol 18(6):700–708
9. Deglincerti A, Croft GF, Pietila LN, Zernicka-Goetz M, Siggia ED, Brivanlou AH (2016) Self-organization of the in vitro attached human embryo. Nature 533(7602):251–254
10. Rossant J, Tam PP (2018) Exploring early human embryo development. Science 360(6393):1075–1076
11. Hyun I, Wilkerson A, Johnston J (2016) Embryology policy: revisit the 14-day rule. Nature 533(7602):169–171

12. Williams K, Johnson MH (2020) Adapting the 14-day rule for embryo research to encompass evolving technologies. Reprod Biomed Soc Online 10:1–9
13. Hurlbut JB, Hyun I, Levine AD, Lovell-Badge R, Lunshof JE, Matthews KR, Mills P, Murdoch A, Pera MF, Scott CT (2017) Revisiting the Warnock rule. Nat Biotechnol 35(11):1029–1042
14. Zheng Y, Xue X, Resto-Irizarry A, Li Z, Shao Y, Zheng Y, Zhao G, Fu J (2019) Dorsal-ventral patterned neural cyst from human pluripotent stem cells in a neurogenic niche. Sci Adv 5(12):eaax5933
15. Haremaki T, Metzger JJ, Rito T, Ozair MZ, Etoc F, Brivanlou AH (2019) Self-organizing neuruloids model developmental aspects of Huntington's disease in the ectodermal compartment. Nat Biotechnol 37(10):1198–1208
16. Xue X, Sun Y, Resto-Irizarry AM, Yuan Y, Aw Yong KM, Zheng Y, Weng S, Shao Y, Chai Y, Studer L (2018) Mechanics-guided embryonic patterning of neuroectoderm tissue from human pluripotent stem cells. Nat Mater 17(7):633–641
17. Warmflash A, Sorre B, Etoc F, Siggia ED, Brivanlou AH (2014) A method to recapitulate early embryonic spatial patterning in human embryonic stem cells. Nat Methods 11(8): 847–854
18. Moris N, Anlas K, van den Brink SC, Alemany A, Schröder J, Ghimire S, Balayo T, van Oudenaarden A, Martinez Arias A (2020) An in vitro model of early anteroposterior organization during human development. Nature 582(7812):410–415
19. Shao Y, Taniguchi K, Townshend RF, Miki T, Gumucio DL, Fu J (2017) A pluripotent stem cell-based model for post-implantation human amniotic sac development. Nat Commun 8(1): 1–15
20. Shao Y, Taniguchi K, Gurdziel K, Townshend RF, Xue X, Yong KMA, Sang J, Spence JR, Gumucio DL, Fu J (2017) Self-organized amniogenesis by human pluripotent stem cells in a biomimetic implantation-like niche. Nat Mater 16(4):419–425
21. Zheng Y, Yan RZ, Sun S, Kobayashi M, Xiang L, Yang R, Goedel A, Kang Y, Xue X, Esfahani SN (2022) Single-cell analysis of embryoids reveals lineage diversification roadmaps of early human development. Cell Stem Cell 29 (9):1402–1419.e1408
22. Zheng Y, Xue X, Shao Y, Wang S, Esfahani SN, Li Z, Muncie JM, Lakins JN, Weaver VM, Gumucio DL (2019) Controlled modelling of human epiblast and amnion development using stem cells. Nature 573(7774):421–425
23. Zheng Y, Shao Y, Fu J (2021) A microfluidics-based stem cell model of early post-implantation human development. Nat Protoc 16(1):309–326

Methods in Molecular Biology (2024) 2767: 85–103
DOI 10.1007/7651_2023_496

Published online: 05 July 2023

Generation of 3D Trophoblast Organoids from Human Naïve Pluripotent Stem Cells

Rowan M. Karvas and Thorold W. Theunissen

Abstract

The human placenta is a transient organ that functions to support the needs of the fetus throughout gestation. Trophoblasts are the major epithelial cells found within the placenta and comprise a variety of distinct cell types with specialized roles in fetal-maternal communication. Our understanding of human trophoblast development remains limited due to ethical and legal restrictions on accessing first-trimester placental tissues, as well as the inability of common animal models to replicate primate placental development. It is therefore important to advance in vitro models of human trophoblast development as a basis for studying pregnancy-associated complications and diseases. In this chapter, we describe a protocol for generating 3D trophoblast organoids from naïve human pluripotent stem cells (hPSCs). The resulting stem-cell-derived trophoblast organoids (SC-TOs) contain distinct cytotrophoblast (CTB), syncytiotrophoblast (STB), and extravillous trophoblast (EVT) cell types, which closely correspond to trophoblast identities in the human post-implantation embryo. We discuss methods for characterizing SC-TOs by immunofluorescence, flow cytometry, mRNA and microRNA expression profiling, and placental hormone secretion. Furthermore, SC-TOs can undergo differentiation into specialized 3D EVT organoids, which display robust invasion when co-cultured with human endometrial cells. Thus, the protocol described herein offers an accessible 3D model system of human placental development and trophoblast invasion.

Key word Cytotrophoblast, Extravillous trophoblast, Human trophoblast stem cells, Naïve pluripotent stem cells, Syncytiotrophoblast, Trophoblast organoids

1 Introduction

The placenta is a major conduit for the exchange of nutrients, gases, and waste products between the mother and the developing fetus. Since access to first-trimester placental tissues is limited and animal models do not fully recapitulate human placental development [1, 2], there has been significant interest in creating faithful in vitro model systems of human placentation. In recent years, both 2D and 3D model systems of the human trophoblast lineage have been established from primary tissues. Human trophoblast stem cells (hTSCs) that can undergo lineage-directed differentiation into specialized EVT and STBs have been derived from blastocysts and first-trimester placental tissues [3]. In addition, trophoblast organoids that contain a variety of trophoblast cell types within an integrated 3D structure have been established from first-trimester CTBs [4, 5]. We and others demonstrated that naïve hPSCs, which resemble pluripotent cells in the pre-implantation embryo, can directly differentiate into self-renewing and bipotent hTSCs [6–10]. In addition, we recently reported that hTSCs obtained from various sources (naïve hPSCs, blastocysts, and first-trimester placental tissues) can efficiently self-organize into 3D trophoblast organoids that acquire a similar villous architecture as primary trophoblast organoids [11]. Single cell transcriptome profiling revealed that these stem-cell-derived trophoblast organoids (SC-TOs) comprise discrete CTB, STB, and EVT subpopulations, which closely correspond to trophoblast identities in the human post-implantation embryo [12]. Furthermore, we have shown that SC-TOs model selective placental susceptibility to emerging viral pathogens (SARS-CoV-2 and ZIKV) and can undergo differentiation into specialized EVT organoids (SC-EVTOs), which display robust invasion in a 3D matrix coated with human endometrial cells [11].

In this chapter, we present a detailed protocol for generating 3D trophoblast organoids from naïve hPSCs. We recently outlined protocols in the *Methods in Molecular Biology* series for inducing naïve pluripotency in hPSCs [13] and generating hTSCs from naïve hPSCs [14]. Therefore, this chapter will focus on specialized techniques for generating SC-TOs from naïve hPSC-derived hTSCs, which we refer to as "naïve hTSCs." However, this protocol for generating SC-TOs works equally well on primary hTSCs derived from blastocysts or first-trimester placental tissues [3]. We will also discuss methods for characterizing SC-TOs by immunofluorescence, flow cytometry, mRNA and microRNA (miRNA) expression, placental hormone secretion, and single cell RNA-sequencing (scRNA-seq). Finally, we will outline methods for differentiating SC-TOs into specialized SC-EVTOs and functionally testing their invasive properties in the presence of human endometrial cells. The methods described in this protocol enable the generation of 3D trophoblast organoids from a broad range of hPSC lines, facilitating studies into human placental development and diseases.

2 Materials

2.1 Maintenance of Naïve and Primary hTSCs

1. Fibroblast medium: supplement DMEM with 10% FBS, 1× GlutaMAX, and 1% penicillin-streptomycin. It can be stored at 4 °C for up to 4 weeks.
2. hTSC medium: supplement DMEM/F12 with 0.1 mM 2-mercaptoethanol, 0.2% FBS, 0.5% penicillin-streptomycin, 0.3% BSA, 1% ITS-X, 1.5 μg/mL L-ascorbic acid, 50 ng/mL EGF, 2 μM CHIR99021, 0.5 μM A83-01, 1 μM SB431542,

0.8 mM VPA, and 5 μM Y-27632. It can be stored at 4 °C for up to 2 weeks.
3. Collagen IV, stock concentration at 1 mg/mL.
4. Dulbecco's phosphate buffered saline (DPBS).
5. 0.4% Trypan blue.
6. TrypLE Express.

2.2 Generation and Maintenance of SC-TOs

1. Trophoblast organoid medium (TOM): Advanced DMEM/F12 (Life Technologies #12634010), 1× N2 Supplement (Life Technologies #17502048), 1× B27 Supplement minus vitamin A (Life Technologies 12587010), 100 μg/mL Primocin (Invivogen #ant-pm-1), 1.25 mM N-Acetyl-L-Cysteine (Sigma #A9165), 2 mM L-Glutamine (Life Technologies #25030-024), 500 nM A83-01 (Peprotech #9094360), 1.5 μM CHIR99021 (Stemgent #04-0004), 2 μM Y-27632 (Stemgent #04-0012), 50 ng/mL rhEGF (Peprotech #AF-100-15), 50 ng/mL rhHGF (Peprotech #100-39), 80 ng/mL rhR-Spondin1 (Peprotech #120-38), 100 ng/mL rhFGF2 (Peprotech #100-18B), and 2.5 μM Prostaglandin E2 (Millipore Sigma #P0409).
2. TrypLE express.
3. 0.4% Trypan blue.
4. Growth factor-reduced hESC-qualified Matrigel.
5. Cell Recovery Solution (Corning).

2.3 Differentiation of SC-TOs into Specialized SC-EVTOs

1. EVT medium-1 (EVTm-1): Advanced DMEM/F12, 0.1 mM 2-mercaptoethanol, 0.5% penicillin/streptomycin, 0.3% bovine serum albumin, 1% ITS-X, 100 ng/mL NRG1, 7.5uM A83-01, and 4% knockout serum replacement.
2. EVT medium-2 (EVTm-2): Advanced DMEM/F12, 0.1 mM 2-mercaptoethanol, 0.5% penicillin/streptomycin, 0.3% bovine serum albumin, 1% ITS-X, 7.5 μM A83-01, and 4% knockout serum replacement.

2.4 Characterization of Cellular Identity

1. FACS buffer: PBS + 5% FBS.
2. Blocking and permeabilization solution: PBS + 5% FBS + 4% BSA + 0.5% Triton-X.
3. Staining solution: PBS + 5% FBS + 4% BSA + 0.1% Tween-20.
4. Clearing solution: 60% glycerol and 2.5 M fructose in water.
5. BD LSRFortessa X-20 (BD Biosciences).
6. FlowJo software package.
7. 4% Paraformaldehyde.

8. anti-HLA-A, B, C antibody (W6/32-488) (BioLegend),
9. Antibodies:

Target	Manufacturer	Product number
KRT7	Dako/Agilent	M7018
TP63	Abcam	Ab124762
SDC1	Abcam	Ab34164
CDH1 (rabbit)	Cell Signaling	3195S
CDH1 (mouse)	Cell Signaling	14472S
hCG beta	Abcam	Ab53087
hCG alpha	R&D	MAB4169
MMP2	Cell Signaling	40994S
HLAG	Santa Cruz	sc-21799
Ki67	Cell Signaling	9449S
ITGA5B1	Abcam	Ab275977
Secondary ab—Donkey anti-mouse 488	Thermo Fisher Scientific	A21202
Secondary ab—Donkey anti-rabbit 488	Thermo Fisher Scientific	A21206
Secondary ab- Donkey anti-mouse 555	Thermo Fisher Scientific	A31570
Secondary ab—Donkey anti-rabbit 555	Thermo Fisher Scientific	A31572

10. Phalloidin-670 (Cytoskeleton PHDN1).
11. DAPI.
12. Human pregnancy tests (Wondflo).
13. miRNeasy Mini Kit (Qiagen: 1038703),
14. TaqMan Advanced miRNA cDNA Synthesis Kit (Applied Biosystems: A28007).
15. TaqMan Fast Advanced Master Mix (Applied Biosystems: 4444556).
16. High capacity cDNA reverse transcription kit.
17. PowerUp SYBR green master mix.

18. StepOnePlus Real-Time PCR System (Applied Biosystems).
19. Primer sequences:

Target	Forward sequence	Reverse sequence
RPLP0	GCTTCCTGGAGGGTG TCC	GGACTCGTTTGTACCCG TTG
KLF17	CTGCCTGAGCGTGGTA TGAG	TCATCCGGGAAGGAG TGAGA
DNMT3L	TTCTGGATGTTCG TGGACAA	ACATCTGGGATGGTGAC TGG
ZIC2	CCCTTCAAGGCCAAA TACAA	TGCATGTGCTTCTTCCTG TC
VIM	TGTCCAAATCGATGTGGA TGTTTC	TTGTACCATTCTTCTGCC TCCTG
TFAP2C	TCTTGGAGGACGAAA TGAGATGG	GGGCTTCTTTGATGTAG TTCTGC
KRT7	AGGATGTGGATGCTGCC TAC	CACCACAGATGTG TCGGAGA
GATA3	TGCAGGAGCAGTATCA TGAAGCCT	GCATCAAACAACTG TGGCCAGTGA
HLAG	CTCTCAGGCTGCAATG TGAA	CATGAGGAAGAGGGTCA TGG
ASCL2	AACTTGAGCTGC TGGAGGGACA	TCTTGGCCAGCA TGGAAAACTC
ITGA2	TCACCAGGAACA TGGGAACT	GTCAGAACACACACCCG TTG
CGB	GTCAACACCACCATGTG TGC	GGTAGTTGCACACCACC TGA
SDC1	GCTGACCTTCACAC TCCCCA	CAAAGGTGAAGTCCTGC TCCC
ERVW1	CTACCCCAACTGCGG TTAAA	GGTTCCTTTGGCAGTA TCCA
ELF5	AGTCTGCACTGACA TTTTCTCATC	CAGAAGTCC TAGGGGCAGTC
TEAD4	CAGGTGGTGGAGAAAG TTGAGA	GTGCTTGAGCTTGTGGA TGAAG
MMP2	TGGCACCCATTTACACC TACAC	ATG TCAGGAGAGGCCCCA TAGA
FN1	TGGTATTCAGCTTCC TGGCA	CGGGTATGGTCTTGGCC TAT

20. TaqMan miRNA Advanced Assays:
    (a) 477860_mir Assay Name: hsa-miR-16-5p (Applied Biosystems: A25576)
    (b) 478703_mir Assay Name: miR-1323 (Applied Biosystems: A25576)
    (c) 479396_mir Assay Name: miR-525-5p (Applied Biosystems: A25576)
    (d) 479485_mir Assay Name: miR-517a-3p (Applied Biosystems: A25576)
    (e) 478148_mir Assay Name: miR-518b (Applied Biosystems: A25576)
    (f) 478986_mir Assay Name: miR-519d-3p (Applied Biosystems: A25576).
21. ELISA kits:
    (a) hCG ELISA—CalBiotech (HC251F).
    (b) MMP2 ELISA—Abcam (ab100606).

2.5 In vitro Human Endometrium Invasion Assay

1. Matrigel.
2. Ibidi microwell 8 chamber slides.
3. Human endometrial stromal cells (immortalized): ATCC CRL-4003.
4. Human endometrial epithelial cells—EM-TERTs (immortalized).
5. Human trophoblast stem cells labeled with a constitutive FUW-GFP lentivirus.
6. Microscope equipped with GFP fluorescence.
7. Immortalized human endometrial cell medium: phenol-red free DMEM/F12 supplemented with 7.5% charcoal-stripped fetal bovine serum (Gibco), 1× non-essential amino acids (Gibco), 1× Insulin/Transferrin/Selenium (Gibco), and 1× Antibiotic-Antimicotic (Thermo Fisher).

3 Methods

Unless otherwise specified, perform all experiments at room temperature and all cell culture procedures in a sterile laminar flow hood. Warm all cell culture media to room temperature before use.

3.1 Derivation of hTSCs from Naïve hPSCs (See Note 1)

1. Coat cell culture plates with 5 μg/mL Collagen IV in DPBS at 37 °C for at least 3 h. Coated plates can be sealed with parafilm and stored at 4 °C for at least 1 to 2 weeks.

2. Aspirate cell culture media from a confluent well of naïve hPSCs cultured in 5i/L/A or PXGL media (*see* **Notes 1–4**).
3. Add 1 mL TrypLE Express per well of a 6-well plate and incubate at 37 °C for 5 min.
4. Completely single-cell dissociate the naïve hPSCs by pipetting up and down, and then transfer the cells to a 15 mL Falcon tube with 5 mL fibroblast medium.
5. Centrifuge the cells at 250 rcf for 3 min.
6. Aspirate the media.
7. Resuspend the cell pellet in 1 mL hTSC medium.
8. Count the number of naïve hPSCs (*see* **Note 5**).
9. Seed 0.5–1.0 × 10^6 naïve hPSCs per Collagen IV coated well of a 6-well plate. Each well should eventually contain 2 mL hTSC medium. Culture the cells in 5% CO_2 and 20% O_2.
10. Replace media every 2 days with 2 mL fresh hTSC medium until the cells reach 80–100% confluence (*see* **Note 6**).
11. At this point, passage the cells by first aspirating cell culture media from the well.
12. Add 1 mL TrypLE Express per well of a 6-well plate and incubate at 37 °C for 5 min (*see* **Note 7**).
13. Single-cell dissociate the cells by pipetting up and down. Transfer the cells to a 15 mL Falcon tube with 5 mL fibroblast medium.
14. Centrifuge the cells at 200 rcf for 3 min.
15. Aspirate the media.
16. Resuspend the cell pellet in hTSC medium and passage it at a ratio of 1:2. Each well should eventually contain 2 mL hTSC medium. Culture the cells in 5% CO_2 and 20% (atmospheric) O_2.
17. Repeat **steps 10–16**. Allow 5–10 passages for highly proliferative naïve hTSCs to become homogeneous in the culture (*see* **Note 8**).
18. Naïve hTSCs can be passaged every 3 days at a ratio of 1:4.

3.2 Generation of SC-TOs from Naïve and Primary hTSCs

1. Aspirate cell culture media from an ~80% confluent well of naïve or primary hTSCs.
2. Add 1 mL TrypLE Express per well of a 6-well plate and incubate at 37 °C for 5 min.
3. Completely single-cell dissociate the cells by pipetting up and down, and then transfer the cells to a 15 mL Falcon tube with 10 mL hTSC medium (SAVECY).
4. Centrifuge the cells at 300 rcf for 3 min.

5. Aspirate the media.
6. Resuspend in 1 mL of SAVECY medium. Pipette suspension through a 40 μm strainer to remove large clumps.
7. Count suspension. Resuspend 3000–5000 cells per Matrigel droplet. Generate 72% Matrigel droplets in cold Advanced DMEM/F12. Matrigel should always be manipulated on ice until solidification.
8. Place 30 μL droplets in the center of the wells in a 24-well plate. Allow these to sit for 2 min at room temperature. Flip the plate over gently, and finish solidifying Matrigel droplets in the 37 °C incubator for 30 min.
9. Flip plates right-side-up and add 500 μL of TOM medium. Replace in the incubator. Change media every 1–2 days.
10. Allow 8–12 days of growth for the development of multilineage organoids containing CTBs, STBs, and a small population of EVTs (Fig. 1a–c).
11. To passage SC-TOs, aspirate TOM medium and wash once in PBS. Replace PBS with 1 mL TrypLE and disrupt Matrigel droplet by criss-crossing a P1000 pipette tip across the bottom of the well (*see* **Note 9**).
12. Transfer the TrypLE suspension with cells and Matrigel into a 1.5 mL microcentrifuge tube. Place this tube at 37 °C for 15 min, inverting the tube every 5 min.
13. Pipette the suspension up and down until organoids are broken into approximately single cells.
14. Centrifuge the tube at 300 rcf for 5 min.
15. Remove TrypLE supernatant, resuspend the pellet in 1 mL Advanced DMEM/F12.
16. Centrifuge at 300 rcf for 5 min.
17. Resuspend cells in TOM and pass suspension through a 40 μm strainer.
18. Count the resulting cell suspension and replate 3000–5000 cells per droplet. Solidify the Matrigel droplets and maintain cultures as **steps 7–9** above.

3.3 Detection of Secreted hCG with hCG Pregnancy Test Strips

1. Add 50 μL of medium from SC-TO cultures carefully with a pipette into the test side of the pregnancy test. Turn test on its side and pipette medium into the test through to the "MAX" line.
2. Place test strip flat on the table surface, medium will move through the test strip by capillary action. Allow the test to develop for 5 min. *See* **Note 10** regarding assessing quality of the test.

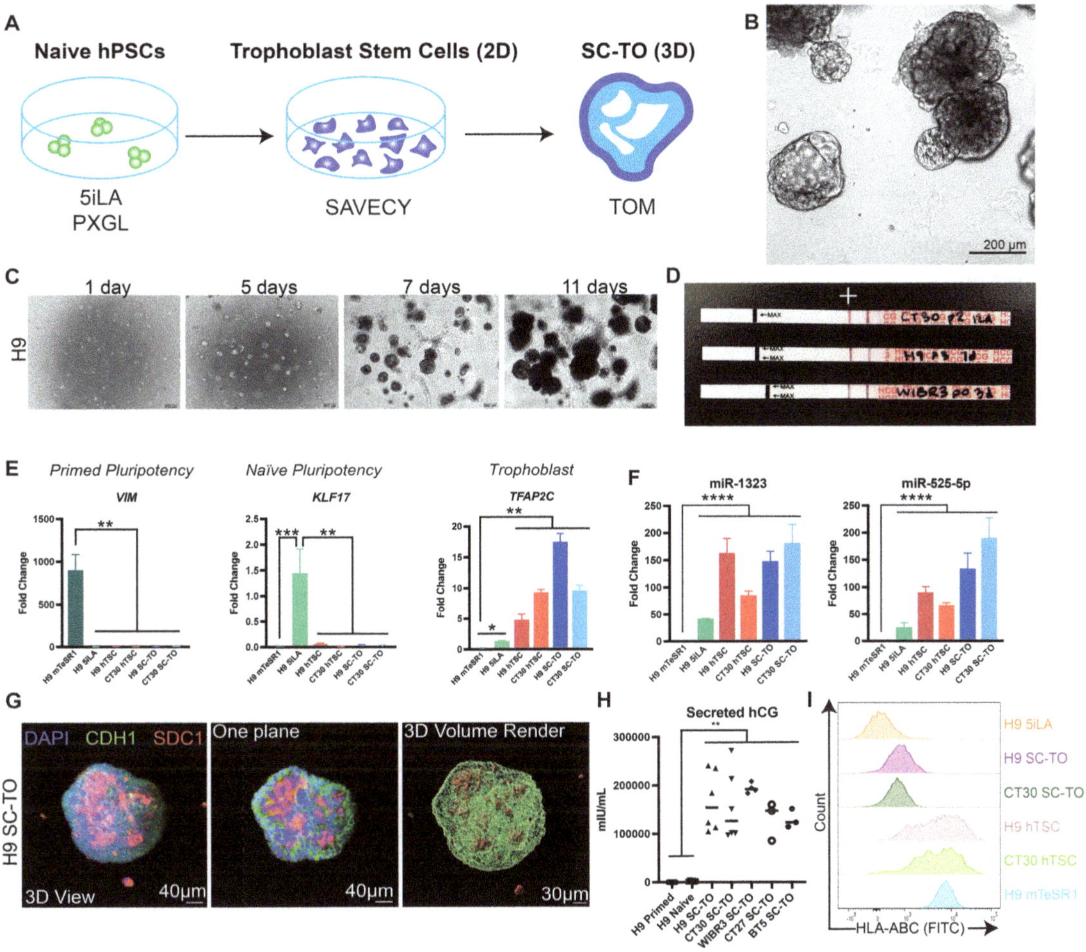

Fig. 1 Generation and characterization of naïve stem-cell-derived trophoblast organoids (SC-TOs). Figure is partially adapted from Karvas et al., 2022 [11]. (**a**) Schematic describing the derivation of naïve stem-cell-derived trophoblast organoids (SC-TOs). Naïve human pluripotent stem cells (hPSCs) are maintained in 5iLA or PXGL media. Trophoblast identity is achieved by culturing cells in SAVECY medium in 2D culture conditions. Subsequently, hTSCs self-organize into 3D SC-TOs upon transfer into 3D Matrigel droplets in the presence of trophoblast organoid medium (TOM). (**b**) Brightfield image of SC-TOs derived from H9 naïve hTSCs. Scale bar indicates 200 μm. (**c**) Time course of H9 SC-TO growth from days 1–11 of TOM treatment. Scale bars indicate 200 μm. (**d**) hCG pregnancy test strips show positive bands for hCG in SC-TO cultures. CT30 (top) is derived from the primary hTSC line CT30. H9 (middle) and WIBR3 (bottom) are both naïve hTSC-derived SC-TOs. (**e**) qRT-PCR analysis demonstrates mRNA transcripts upregulated at each stage of SC-TO derivation. *VIM* is highly upregulated in primed hPSCs maintained in mTeSR1+ compared to all other conditions. *KLF17* is highly upregulated in naïve hPSCs maintained in 5iLA. The trophoblast marker *TFAP2C* is highly upregulated in hTSCs and SC-TOs (CT30 and H9) maintained in SAVECY and TOM conditions, respectively. All experiments represent 3 technical replicates. $*p < 0.05$, $**p < 0.01$, $***p < 0.001$. (**f**) Chromosome 19 miRNA cluster expression is induced in naïve hPSCs and trophoblast cells compared to primed hPSCs. The miRNAs miR-1323 and miR-525-5p are highly expressed in hTSCs and SC-TOs. All experiments represent 3 technical replicates. $****p < 0.0001$. (**g**) Immunofluorescence and light sheet microscopy of H9 SC-TOs demonstrates inside-out morphology based on marker expression of CDH1 (CTB) and SDC1 (STB). (**h**) ELISA analysis for hCG in SC-TO cultures derived from both primary hTSCs (CT27, CT30, BT5) and naïve hTSCs (H9) compared to primed and naïve hPSCs. All cell lines tested represent 3–6 technical replicates. $**p < 0.01$. (**i**) Flow cytometry analysis for HLA-A,B,C surface marker shows a reduction in SC-TOs. Primed hPSCs (mTeSR1) highly express HLA-A,B,C, while naïve hPSCs (5iLA) are negative for these HLAs. hTSCs grown in SAVECY medium in 2D show variable expression, while SC-TOs in 3D are largely negative

3. The control line closest to the side labeled hCG is a positive control for serum albumin. Ensure this line is highly visible. hCG is positive if there are two lines observed (Fig. 1d) and negative if there is only one line (the positive control) visible.

3.4 Expression Analysis of Trophoblast-Specific mRNAs and miRNAs

1. Harvest SC-TOs grown for 10 days in TOM with Cell Recovery Solution. Add 1 mL of Cell Recovery Solution to the well and break apart the Matrigel droplet with perpendicular strokes across the bottom of the well. Pipette this solution into a 1.5 microcentrifuge tube.
2. Remove Cell Recovery Solution with two washes of Advanced DMEM/F12 + 0.1% BSA. Centrifuge tube at 600 rcf for 3 min. Remove supernatant and wash again with Advanced DMEM/F12 + 0.1% BSA.
3. Following the last wash, centrifuge the tube at 600 rcf for 3 min and remove as much supernatant as possible. Snap-freeze the pellet of organoids in liquid nitrogen, being careful not to completely submerge the microcentrifuge tube. Liquid nitrogen leaking into the tube can cause the tube to burst open.
4. For mRNA expression analysis, isolate total RNA from the cells using the E.Z.N.A. total RNA kit I (*see* **Note 11**), or flash-freeze the cell pellets in liquid nitrogen and store at −80 °C to isolate the RNA at a later time. The RNA should be stored at −80 °C. Follow **steps 5–7** for cDNA synthesis and qRT-PCR analysis. For miRNA expression analysis, follow **steps 8–11**.
5. Perform cDNA synthesis from the isolated total RNA using the high-capacity cDNA reverse transcription kit (*see* **Note 11**). The cDNA should be stored at −20 °C.
6. Perform qRT-PCR using the PowerUp SYBR green master mix on the StepOnePlus Real-Time PCR System on cDNA samples from primed hPSCs, naïve hPSCs, hTSCs, and SC-TOs (*see* **Note 11**). *RPLP0* is used as the housekeeping control. *ELF5*, *KRT7*, *GATA3*, and *TEAD4* are used as trophoblast markers, *VIM* and *ZIC2* are used as markers of primed pluripotency, and *DNMT3L* and *KLF17* are used as markers of naïve pluripotency.
7. As shown in Fig. 1e, *VIM* expression should be ~800× higher in primed hESCs compared to SC-TOs; *KLF17* should be ~60× higher in naïve hESCs compared to SC-TOs; *TFAP2C* expression should be ~17 times higher in SC-TOs than in naïve hPSCs.
8. For miRNA expression analyses, isolate RNA using a Qiagen miRNeasy mini kit. Follow manufacturer's instructions.
9. Synthesize cDNA using the TaqMan Advanced microRNA cDNA Synthesis Kit. Follow manufacturer's instructions.

10. The following miRNAs are included in the chromosome 19 microRNA cluster: miR-519d-3p, miR-518b, miR-517a-3p, miR-525-5p, miR-1323. hsa-mir-16-5p is used as a ubiquitously expressed control miRNA.
11. Perform qRT-PCR, follow manufacturer's instructions included in the TaqMan Fast Advanced Master Mix. Thermocycler conditions are as follows: 95 °C for 20 s, 40 cycles of 95 °C for 1 s, and 60 °C for 20 s.
12. As shown in Fig. 1f, miR-1323 should be ~145× higher in SC-TOs compared to primed hESCs, while miR-525-5p should be ~131× higher in SC-TOs compared to primed hESCs.

3.5 Immunofluorescence Analysis and Whole Organoid Imaging

1. Mature SC-TOs with CTBs, STBs, and a small EVT compartment arise after approximately 8–12 days of growth in TOM media. The information below pertains to immunofluorescence (IF) analysis of SC-TOs, but can also be applied to characterize stem-cell-derived EVT organoids (SC-EVTOs) (see Subheading 3.8).
2. Aspirate medium and add 1 mL of Cell Recovery Solution to the well. Disrupt the Matrigel droplet by passing a 1 mL pipette tip side-to-side and top-to-bottom across the bottom of the well to dislodge the droplet containing organoids from the well. Draw up the Cell Recovery Solution and broken-up Matrigel with organoids and deposit into a 1.5 mL Eppendorf tube.
3. Place the 1.5 mL tube with organoids and Cell Recovery Solution on ice and rotate the tube so that it is on its side and the organoids do not collect together at the bottom of the tube. This will prevent the SC-TOs from excessively sticking to each other. Place this solution on ice for 20–30 min.
4. Pulse organoids down to the bottom of the tube with a table top centrifuge. Aspirate the Cell Recovery Solution. Resuspend SC-TOs in Advanced DMEM/F12 + 0.1% BSA. Gently triturate to remove extra Matrigel.
5. Pulse organoids to the bottom of the tube with a table top centrifuge. Remove supernatant and repeat wash in Advanced DMEM/F12 + 0.01% BSA. Pulse SC-TOs to the bottom of the tube again and remove supernatant.
6. Fix organoids in 4% paraformaldehyde (PFA) on ice (place Eppendorf tube on its side) for 30 min.
7. Pulse organoids to the bottom of the Eppendorf tube with a table top centrifuge. Remove PFA and dispose in a PFA waste container.

8. Wash organoids with PBS + 0.01% BSA for 10 min, repeat two more times. Allow organoids to settle to the bottom of the tube by gravity for 10 min or using short table top centrifuge pulses. Aspirate supernatant. Repeat wash 2 more times. Fixed SC-TOs can be stored in PBS + 0.01% at 4 °C for approximately 1 month, but antibody staining is best performed within 1–2 weeks.
9. Apply permeabilization and blocking solution (see Subheading 2.4) overnight at 4 °C.
10. Pulse organoids to the bottom of the tube. Remove supernatant and resuspend organoids in primary antibody solution, which consists of staining solution (see Subheading 2.4) supplemented with the following primary antibodies:
    (a) CTB-specific: E-cadherin (CDH1) 1:100, TP63 1:50
    (b) STB-specific: CGB 1:300, SDC1 1:100
    (c) Both CTB and STB: KRT7 1:200, CGA 1:300, GATA3 1:300
    (d) EVT-specific: HLAG 1:100, ITGA5B1 1:100, MMP2 1:100
11. Incubate overnight at 4 °C.
12. Remove staining solution and wash SC-TOs three times with PBS + 0.01% BSA. Washes should be 10 min, allow SC-TOs to settle by gravity or short tabletop centrifuge pulses.
13. Resuspend SC-TOs in staining solution containing fluorescent secondary antibodies (see Subheading 2.4). Secondary antibodies are diluted 1:300. If nuclear localization or actin structure visualization is desired, add DAPI (1:1000) and appropriate dilution of fluorescent phalloidin during this step. Incubate overnight at 4 °C.
14. Remove secondary antibody staining solution. Wash SC-TOs three times in PBS + 0.01% BSA for 10 min each.
15. Following the last wash, resuspend SC-TOs in 50 μL clearing solution (see Subheading 2.4).
16. Prepare the microscope slide:
    (a) Use a 25 mm × 50 mm coverslip and add a Grace biolab chamber to the center of this coverslip (Catalog number: 645501).
    (b) Carefully remove 40 μL of SC-TOs in clearing solution with a wide-bore pipette or preferably a glass pipette. Add SC-TOs to the center of the chamber slide. Cover the chamber with a 22 mm × 22 mm glass coverslip.

17. Image with a single or multiphoton confocal microscope equipped with appropriate lasers for fluorophore excitation. Store slides flat at 4 °C.
18. If preparing the sample for light sheet microscopy, do not mount within a microscope chamber slide and instead resuspend the sample in an agarose + clearing solution capillary tube:
    (a) Cleared organoids were resuspended and embedded in 50% clearing solution and 50% water +2.5% low melting point agarose. SC-TOs in agarose were drawn into a light sheet microscope capillary tube and allowed to solidify overnight at 4 °C before imaging the next day on a Zeiss Lattice Lightsheet 7 Microscope.
    (b) Following acquisition of samples with light sheet microscopy, images can be rendered in 3D using Amira software (Fig. 1g).

3.6 Preparing SC-TOs for Single Cell RNA-Sequencing

1. Prepare the following solution

Activated papain solution	Stock concentration	Per sample
Papain		9 μL
HEPES in calcium-, magnesium-free HBSS	100 mM HEPES in HBSS	351 μL
Cysteine	50 mM	20 μL
EDTA	10 mM	20 μL

2. Incubate papain solution in 37 °C bath for 15 min. When active, the solution should become clear.
3. Aspirate medium and add 1 mL ice-cold HBSS. Dislodge Matrigel droplet with SC-TOs by dragging pipette in perpendicular strokes across the bottom of the well. Coat the walls of the pipette and the tube with 0.1% BSA to prevent organoids from sticking to plastic.
4. Wash organoids three times in ice-cold HBSS.
5. When the papain solution is active, centrifuge SC-TOs at 1500 rcf for 30 s and remove excess HBSS.
6. Add 400 μL of active papain solution to each sample.
7. Incubate at 37 °C for 10 min, gently vortexing the tube every 2 min.
8. Add 600 μL of DMEM/F12 + 10% FBS with 5 mM magnesium to each sample.
9. Gently triturate the sample 5 times.

10. Add 10 μL DNase I to each tube. Invert gently to mix.
11. Incubate samples at 37 °C for 5 min.
12. Gently triturate the sample 5 times with a P1000 pipette.
13. Centrifuge samples at 1500 rcf for 30 s.
14. Add 600 μL of PBS + 0.04% BSA and triturate 3–5 times.
15. Filter sample through a 40 μm cell strainer.
16. Count cells by using a cell counting device. Ensure cells are dissociated to single cells and that viability is greater than 90% for single-cell RNA sequencing.
17. Submit single-cell suspension for gene expression library preparation, as recommended by 10× Genomics.

3.7 Analysis of HLA Surface Antigen Expression by Flow Cytometry

1. Dissociate organoids using Papain solution as described above in Subheading 3.6.
2. Resuspend dissociated cells in FACS buffer (HBSS +5% FBS). Block cells in FACS buffer on ice for 30 min.
3. Add anti-HLA-A,B,C antibody (W6/32-488) 1:50, incubate on ice for 1 h.
4. Wash cells 3 times in PBS + 0.1% BSA.
5. In the last wash, pass solution through a 40 μm cell strainer and analyze cells on a flow cytometer for FITC-labeled HLA-A,B,C surface expression. Primed hPSCs and hTSCs should have increased HLA-A,B,C expression, while naïve hPSCs and SC-TOs should have reduced HLA-A,B,C expression (Fig. 1i). *See* **Note 12** for the description of positive and negative controls.

3.8 Differentiation of SC-TOs into Specialized 3D EVT Organoids (SC-EVTOs)

1. Passage organoids from 2D hTSC cultures into Matrigel droplets in TOM medium.
2. Change TOM medium daily for 3 days.
3. Switch medium to EVTm-1 for 5 days, change medium daily.
4. Switch medium to EVT-m2 for 2–3 days. At this stage, migratory EVTs should be visible (Fig. 2a, b).
5. To validate successful differentiation into SC-EVTOs, perform total RNA isolation and cDNA synthesis as described above in Subheading 3.4 and perform qRT-PCR analysis using the CTB markers *ELF5* and *TEAD4*, the STB marker *ERVW-1*, and EVT markers *FN-1*, *HLA-G*, and *MMP2*.
6. As shown in Fig. 2c, *ELF5* should be ~65× higher in SC-TOs compared to SC-EVTOs, *ERVW-1* should be ~6× higher in SC-TOs compared to SC-EVTOs, *HLA-G* should be ~13× higher in SC-EVTOs compared to SC-TOs, and *MMP2* should be ~250× higher in SC-EVTOs compared to SC-TOs.

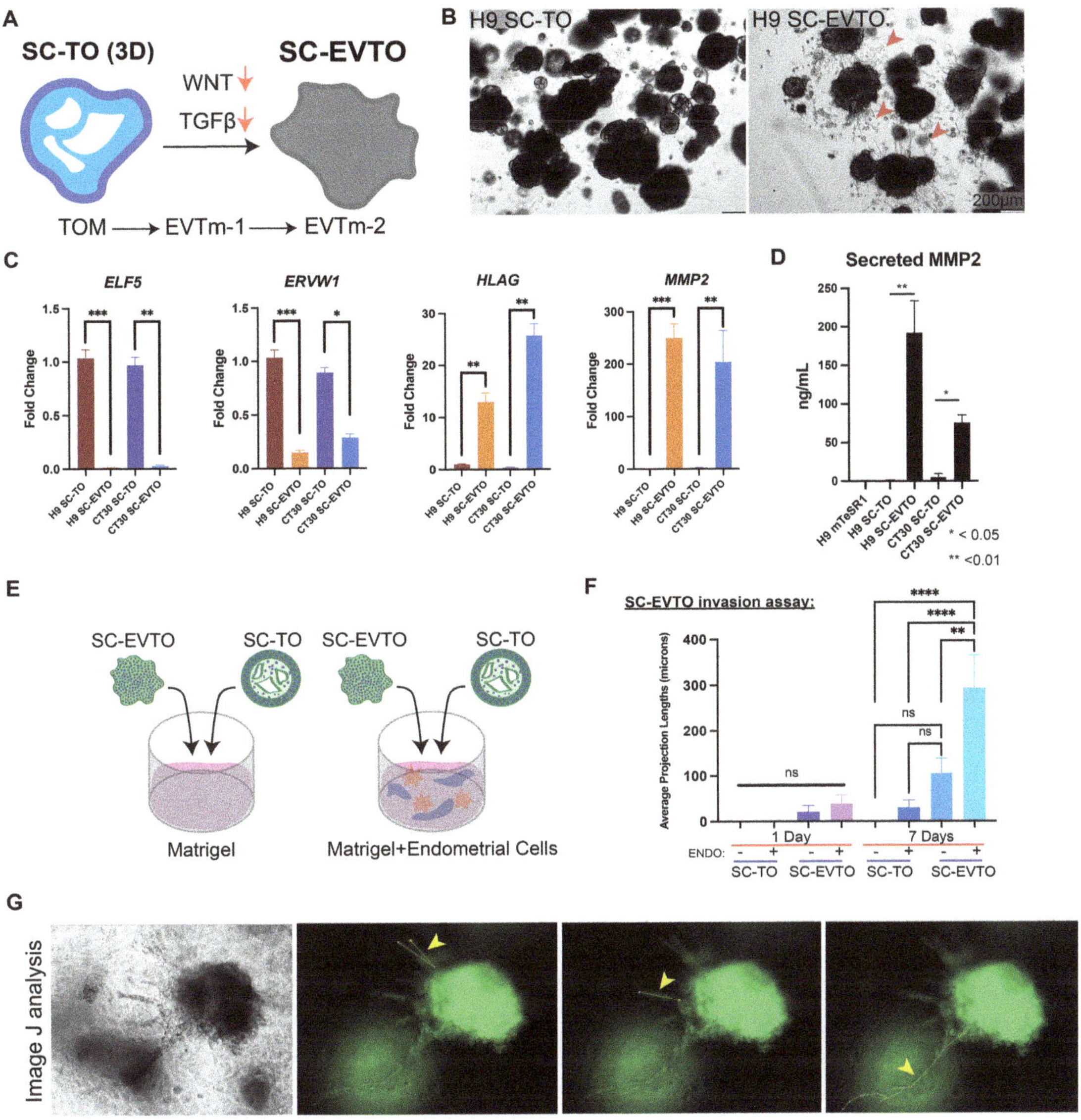

Fig. 2 Differentiation and invasive potential of SC-EVTOs. Figure is partially adapted from Karvas et al. [11]. (**a**) Schematic of SC-TO to SC-EVTO differentiation. EVT identity is established by the removal of WNT-promoting factors in the medium and reinforced inhibition of TGFβ signaling. (**b**) Brightfield imaging reveals morphological changes associated with EVT identity. Red arrows indicate invasive projections characteristic of SC-EVTOs. Scale bar represents 200 μm. (**c**) qRT-PCR analysis reveals that markers of CTB and STB identity (*ELF5, ERVW1*) are downregulated in SC-EVTOs. Genes associated with EVT differentiation (*HLA-G, MMP2*) are highly upregulated in SC-EVTOs compared to SC-TOs in both H9 (naïve hTSC) and CT30 (primary hTSC) cell lines. Experiments were performed in technical triplicates. $^*p < 0.05$, $^{**}p < 0.01$, $^{***}p < 0.001$. (**d**) The levels of secreted MMP2 are significantly increased in SC-EVTOs compared to SC-TOs and primed hPSCs. Samples were collected from 3 technical replicates. $^*p < 0.05$, $^{**}p < 0.01$. (**e**) Schematic diagram describing in vitro invasion assays. GFP-labeled H9 hTSCs were used to generate SC-TOs and SC-EVTOs. 3D Matrigel matrices were generated and embedded with human endometrial cells (stromal fibroblasts and glandular epithelium). (**f**) SC-EVTOs formed longer invasive projections when in contact with endometrial cells than Matrigel alone. SC-TOs maintained in TOM did not form significant invasive projects in any condition. Experiments were performed in triplicate wells. ns not significant, $^{**}p < 0.01$, $^{***}p < 0.001$, $^{****}p < 0.0001$. Statistical analysis—2-way ANOVA with multiple comparisons. (**g**) Schematic of invasive projection measurement methods. Image J was used to measure invasive projection lengths (in μm). Yellow arrows indicate tracked GFP-positive invasive EVTs

3.9 Quantifying Proteins Secreted from SC-TO and SC-EVTO Culture by ELISA

1. For accurate ELISA quantification of hCG secretion in SC-TOs (Fig. 1h) and MMP secretion in SC-EVTOs (Fig. 2d), plate the same starting numbers of cells and collect medium on the same day of organoid growth (e.g., day 10). Collect spent medium from SC-TO and SC-EVTO cultures 24 h after final medium change.
2. Store samples at −80 °C until all samples for comparison are acquired.
3. Minimize the number of freeze-thaws between samples as this will degrade your proteins of interest and may introduce variability. *See* **Note 13** to determine the dilution factor when analyzing hCG, given its high expression levels in SC-TOs. For MMP2 ELISAs, it is not necessary to dilute the spent medium.
4. Follow manufacturer's instructions for ELISA analysis.

3.10 Invasion Assay in the Presence of Human Endometrial Cells (Fig. 2e–g)

1. Maintain immortalized human endometrial stromal cells (HESC) and immortalized human endometrial epithelial cells (EM-TERT) in endometrial cell maintenance medium as described in Subheading 2.5.
2. In an ibidi microwell slide (8 chambers), plate 40,000 stromal cells suspended in 100 μL of ice-cold Matrigel. Place chamber slide in the incubator and solidify the Matrigel with fibroblasts for 30 min.
3. Add 25,000 EM-TERT cells in endometrial medium. Allow these cells to attach overnight in a 37 °C, 5% CO_2 incubator.
4. The following day, replace medium with endometrial medium with added hormones—36 μM estradiol, 1 μM medroxy progesterone 17-acetate, and 100 μM c-AMP. Allow hormone treatment to decidualize the combined endometrial cells for 24 h in the incubator.
5. Prepare GFP-labeled SC-EVTOs: extract SC-EVTOs from Matrigel droplets on the day when they are switched from EVTm-1 to EVTm-2. Add 1 mL of Cell Recovery Solution, break up the Matrigel droplet, and pipette the solution into a 1.5 microcentrifuge. Place tube on its side on ice for 20 min.
6. Centrifuge SC-EVTOs to the bottom of the tube by pulsing them to bottom with a table top centrifuge. Remove supernatant containing Cell Recovery Solution. Wash with Advanced DMEM/F12 + 0.1% BSA, pipette SC-EVTOs to remove Matrigel. Repeat centrifugation and wash one more time.
7. Resuspend SC-EVTOs in endometrial medium containing 7.5 μM A83-01, 36 nM estradiol, and 1 μM medroxy progesterone 17-acetate.

8. Remove medium from ibidi microwell chambers containing endometrial cells.
9. Add approximately 10 SC-EVTOs per well in 200 μL total medium.
10. Image organoids on the first day and periodically throughout 7 days of culture. Change medium daily.
11. Track EVT projections emerging from the center of the SC-EVTOs by using image J software (Fig. 2f, g).

4 Notes

1. For detailed protocols for inducing naïve pluripotency and deriving hTSCs from naïve hPSCs, please see our recent *Methods in Molecular Biology* book chapters [6, 13, 14]. Subheading 3.1 and the following **Notes 2–8** were reproduced from our prior book chapter describing hTSC derivation from naïve hPSCs.
2. Most of our hTSC differentiation experiments were performed using naïve cells derived and maintained in 5i/L/A [15], but we have also been able to derive hTSCs from naïve hPSCs maintained in the alternative media formulation PXGL [16].
3. It is recommended that naïve hPSCs within 10 passages be used for hTSC differentiation to reduce the possibility of the cells acquiring karyotypic abnormalities as a result of prolonged culture in naïve media.
4. In our experience, the most ideal stage to harvest naïve hPSCs is when they reach ~80% confluence. That is usually ~6 to 8 days after passage.
5. Any cell counter/hemocytometer will be sufficient for this step. We prefer to exclude dead cells using trypan blue.
6. The cells will maintain round, naïve-like morphology for the first ~3 days in hTSC medium, but will then adopt a more flat, cobblestone-like morphology. It takes approximately 1 week for the cells to reach 80–100% confluence.
7. We found that TrypLE Express is the best dissociation reagent for hTSCs. Other dissociation reagents such as Accutase will result in low viability after passaging.
8. For the first ~3 passages, the cells will grow slowly and appear flat. This is normal and not a cause for concern. Highly proliferative hTSC-like cells will then emerge and become homogeneous within the next several passages.

9. SC-TOs can be passaged but, in order to preserve the cellular heterogeneity of the organoids, we recommend generating fresh SC-TOs from 2D hTSC (SAVECY) cultures.
10. Occasionally there are defective pregnancy test strips. If medium stops moving through the test strip by capillary action, discard test strip and repeat on a new test strip. An uninterpretable result may occur when the hCG line appears positive, but the control line does not appear. This is also defective and should be rerun on a new strip. It is not possible to determine the concentration of hCG with test strips. To quantify this result, it is required to perform an ELISA.
11. Most commercial total RNA isolation and cDNA reverse transcription kits and qRT-PCR systems will work for this step.
12. As positive and negative controls for flow cytometry, primed hESCs grown in mTeSR1 medium express HLA-A,B,C, while naïve hESCs do not. Commonly grown choriocarcinoma cell lines also serve as positive and negative controls: JEG-3 is a positive control and JAR is a negative control [17].
13. hCG is highly expressed from SC-TOs and if analyzed without first diluting the sample, the points will fall outside of the range of the standard curve. It is important to run an initial pilot experiment with one sample of medium testing a number of dilution factors before running your experimental samples. Begin with 1×, 1:100, 1:250, 1:500, 1:1000, etc. The sample that best fits within the final standard curve is the dilution factor you will use, then recalculate your final values based on this dilution.

Acknowledgments

Work in our laboratory is supported by the NIH Director's New Innovator Award (DP2 GM137418) and grants from the Shipley Foundation Program for Innovation in Stem Cell Science, the Edward Mallinckrodt, Jr. Foundation, and the Washington University Children's Discovery Institute. R.M.K. is supported by a Training in Regenerative Medicine training grant (T32 EB028092) from the NIH. Federal NIH/NIGMS funds were not used to develop integrated 3D models of human embryonic development.

Conflicts of Interest R.M.K. and T.W.T. are co-inventors on a patent application related to stem-cell-derived trophoblast organoids.

References

1. Carter AM (2007) Animal models of human placentation--a review. Placenta 28 Suppl A: S41–S47. https://doi.org/10.1016/j.placenta.2006.11.002
2. Guttmacher AE, Maddox YT, Spong CY (2014) The Human Placenta Project: placental structure, development, and function in real time. Placenta 35(5):303–304. https://doi.org/10.1016/j.placenta.2014.02.012
3. Okae H, Toh H, Sato T, Hiura H, Takahashi S, Shirane K et al (2018) Derivation of human trophoblast stem cells. Cell Stem Cell 22(1): 50–63 e6. https://doi.org/10.1016/j.stem.2017.11.004
4. Turco MY, Gardner L, Kay RG, Hamilton RS, Prater M, Hollinshead MS et al (2018) Trophoblast organoids as a model for maternal-fetal interactions during human placentation. Nature 564(7735):263–267. https://doi.org/10.1038/s41586-018-0753-3
5. Haider S, Meinhardt G, Saleh L, Kunihs V, Gamperl M, Kaindl U et al (2018) Self-renewing trophoblast organoids recapitulate the developmental program of the early human placenta. Stem Cell Reports 11(2): 537–551. https://doi.org/10.1016/j.stemcr.2018.07.004
6. Dong C, Beltcheva M, Gontarz P, Zhang B, Popli P, Fischer LA et al (2020) Derivation of trophoblast stem cells from naive human pluripotent stem cells. elife 9. https://doi.org/10.7554/eLife.52504
7. Cinkornpumin JK, Kwon SY, Guo Y, Hossain I, Sirois J, Russett CS et al (2020) Naive human embryonic stem cells can give rise to cells with a trophoblast-like transcriptome and methylome. Stem Cell Reports 15(1):198 213. https://doi.org/10.1016/j.stemcr.2020.06.003
8. Guo G, Stirparo GG, Strawbridge SE, Spindlow D, Yang J, Clarke J et al (2021) Human naive epiblast cells possess unrestricted lineage potential. Cell Stem Cell 28(6): 1040–56 e6. https://doi.org/10.1016/j.stem.2021.02.025
9. Castel G, Meistermann D, Bretin B, Firmin J, Blin J, Loubersac S et al (2020) Induction of human trophoblast stem cells from somatic cells and pluripotent stem cells. Cell Rep 33(8):108419. https://doi.org/10.1016/j.celrep.2020.108419
10. Io S, Kabata M, Iemura Y, Semi K, Morone N, Minagawa A et al (2021) Capturing human trophoblast development with naive pluripotent stem cells in vitro. Cell Stem Cell 28(6): 1023–39 e13. https://doi.org/10.1016/j.stem.2021.03.013
11. Karvas RM, Khan SA, Verma S, Yin Y, Kulkarni D, Dong C et al (2022) Stem-cell-derived trophoblast organoids model human placental development and susceptibility to emerging pathogens. Cell Stem Cell 29(5): 810–25 e8. https://doi.org/10.1016/j.stem.2022.04.004
12. Xiang L, Yin Y, Zheng Y, Ma Y, Li Y, Zhao Z et al (2020) A developmental landscape of 3D-cultured human pre-gastrulation embryos. Nature 577(7791):537–542. https://doi.org/10.1038/s41586-019-1875-y
13. Fischer LA, Khan SA, Theunissen TW (2022) Induction of human naive pluripotency using 5i/L/A medium. Methods Mol Biol 2416:13–28. https://doi.org/10.1007/978-1-0716-1908-7_2
14. Dong C, Theunissen TW (2022) Generating trophoblast stem cells from human naive pluripotent stem cells. Methods Mol Biol 2416: 91–104. https://doi.org/10.1007/978-1-0716-1908-7_7
15. Theunissen Thorold W, Powell Benjamin E, Wang H, Mitalipova M, Faddah Dina A, Reddy J et al (2014) Systematic identification of culture conditions for induction and maintenance of naive human pluripotency. Cell Stem Cell 15(4):471–487. https://doi.org/10.1016/j.stem.2014.07.002
16. Bredenkamp N, Stirparo GG, Nichols J, Smith A, Guo G (2019) The cell-surface marker Sushi Containing Domain 2 facilitates establishment of human naive pluripotent stem cells. Stem Cell Reports. https://doi.org/10.1016/j.stemcr.2019.03.014
17. Sheridan MA, Fernando RC, Gardner L, Hollinshead MS, Burton GJ, Moffett A et al (2020) Establishment and differentiation of long-term trophoblast organoid cultures from the human placenta. Nat Protoc 15(10): 3441–3463. https://doi.org/10.1038/s41596-020-0381-x

Methods in Molecular Biology (2024) 2767: 105–113
DOI 10.1007/7651_2023_483

Published online: 28 May 2023

Induction of Human Extraembryonic Mesoderm Cells from Naive Pluripotent Stem Cells

Amitesh Panda, Thi Xuan Ai Pham, Sherif Khodeer, and Vincent Pasque

Abstract

The human extraembryonic mesoderm (EXM) is an important tissue in the postimplantation embryo which is specified before gastrulation in primates but not in rodents. EXM is mesenchymal and plays an important role in embryogenesis, including early erythropoiesis, and provides mechanical support to the developing embryo. Recently, it has been shown that self-renewing extraembryonic mesoderm cells (EXMCs) can be modeled in vitro by using human naive pluripotent stem cells. Here, we present a detailed step-by-step protocol to induce EXMCs from naive pluripotent stem cells in vitro.

Key words Extraembryonic mesoderm, Human embryo, Naive pluripotent stem cells

1 Introduction

Human extraembryonic mesoderm (EXM) is a key cell type in the postimplantation embryo. The timing of EXM specification differs significantly between species. In primates, the EXM appears before gastrulation, whereas in rodents it is specified only after gastrulation [1–9]. EXM is known to play several important roles during embryogenesis such as in primitive erythropoiesis, formation of extracellular matrix and primitive umbilical cord [4, 5, 7, 10–12]. However, despite its importance, our current understanding of human EXM development is limited, in part due to the unavailability of an in vitro model system.

Great efforts have been made to culture human embryos *ex utero* for up to 14 days [13–17]. However, due to extensive ethical and legal constraints, obtaining human embryos remains a challenge [18]. Thus, our knowledge of human postimplantation development remains limited. Naive human pluripotent stem cells (hPSCs) which represent the preimplantation epiblast have served as a great system to study early embryonic development and have the capacity to differentiate into specific cell lineages such as the trophoblast and the primitive endoderm lineages [19–26]. In

Amitesh Panda and Thi Xuan Ai Pham contributed equally with all other contributors.

addition, naive hPSCs have the ability to efficiently self-organize into blastocyst-like structures termed "blastoid" which can be used as attractive models to better understand preimplantation human embryogenesis [27–30]. In addition, a recent study has shown that naive hPSCs can give rise by differentiation to human extraembryonic mesoderm cells (EXMCs) in vitro [31]. This serves as a new model system to understand human EXM development in vitro as well as to model postimplantation human embryo development.

In this chapter, we describe a detailed step-by-step protocol for the derivation and validation of human EXMCs in vitro from naive hPSCs.

2 Materials

2.1 General Culture Reagents

1. 0.1% Gelatin solution: dissolve 1 g gelatine in 1 L of distilled water. Sterilize the solution by autoclaving and store it at room temperature.
2. Geltrex.
3. 1× PBS
4. Collagen IV.
5. TryplE express.
6. Mycostrip.

2.2 General Culture Media

1. Mouse embryonic fibroblast (MEF) medium: 90% of DMEM supplemented with 10% FBS, 1% Glutamax, 1× Penicillin-streptomycin, 1× non-essential amino acid, and 0.1 mM β-mercaptoethanol. Filter sterilize the medium using a 0.22 μm filter and store at 4 °C.
2. N2B27 base medium: 50% DMEM/F12, 50% neurobasal, 0.5% N2 supplement, 1% B27 supplement, 2 mM L-glutamine, 0.1 mM β-mercaptoethanol, and 1× penicillin-streptomycin. Filter sterilize the medium using a 0.22 μm filter and store at 4 °C.
3. Naive hPSC culture (PXGL): N2B27 base medium with 1 μM PD0325901, 2 μM XAV939, 2 μM Gö6983, 20 pg/mL human LIF, and 10 μM Y-27632 (ROCK inhibitor).
4. ASECRiAV medium: DMEM/F12 supplemented with 0.3% BSA, 0.2% FBS, 1% penicillin-streptomycin, 1% insulin-transferrin-selenium-ethanolamine-X 100 supplement, 1.5 μg/mL L-ascorbic acid, 0.5 μM A83-01, 1 μM SB431542, 50 ng/mL hEGF, 2 uM CHIR99021, 0.8 mM valproic acid, 0.1 mM β-mercapto-EtOH, and 5 μM Y-27632. Filter sterilize the medium using a 0.22 μm filter and store at 4 °C.

5. Washing medium: DMEM/F12 with 0.1% BSA.
6. Freezing medium: 90% FBS + 10% DMSO.

2.3 Characterization of Cellular Identity

1. FACS buffer: PBS with 0.1%BSA.
2. Blocking buffer: PBST with 5% normal donkey serum and 0.2% fish skin gelatin.
3. FlowJo software package.
4. 16% paraformaldehyde.
5. Rat monoclonal IgG2a anti-human GATA4.
6. Rabbit monoclonal IgG anti-human Anti-Periostin (POSTN).
7. Mouse monoclonal IgG1 anti-human Anti-Vimentin (VIM).
8. Rabbit monoclonal IgG anti-human BST2.
9. Donkey IgG anti-rabbit (H + L) polyclonal secondary antibody, Alexa Fluor 647 conjugated.
10. Donkey IgG anti-mouse (H + L) polyclonal secondary antibody, Alexa Fluor 488 conjugated.
11. Donkey IgG anti-rat (H + L) polyclonal secondary antibody, Alexa Fluor 488 conjugated.
12. DAPI.
13. Prolong Gold antifade reagent with DAPI.
14. ImageJ.
15. Zeiss Axio Imager A1 inverted microscope.

3 Methods

3.1 Preparation of Tissue Culture Plate Coated with MMC-MEF (Day 0)

1. Coat 6-well plate by adding 1 mL of 0.1% gelatin per well and incubate at 37 °C at least for 30 min.
2. Take 1 cryovial containing 2 million cells of mitomycin-treated MEF (MMC-MEF) from liquid nitrogen and thaw it in a water bath for 3–4 min and then transfer the cryovial content to a 15 ml tube containing 5 mL of MEF medium.
3. Centrifuge at 1200 rpm for 3 min and then aspirate the supernatant, and resuspend the pellet in 12 mL of MEF culture medium.
4. Dispense 2 mL of medium with cells per well of a gelatin-coated 6-well plate, and incubate the plate in a 37 °C incubator overnight with 5% CO_2.

3.2 Seeding Naive Human Pluripotent Stem Cells (Day 1)

1. Take 1 cryovial of naive hPSCs containing 300,000 cells from the liquid nitrogen storage and thaw it in a water bath for 3–4 min and then transfer the cryovial content to 5 mL of naive hPSCs medium (PXGL) containing 10 μM of ROCK inhibitor (*see* **Notes 1** and **2**).
2. Centrifuge at 1200 rpm for 5 min and then aspirate the supernatant, and resuspend the pellet in 2 mL of PXGL medium containing 10 μM ROCK inhibitor.
3. Take out the MMC-MEF 6-well plate from 37 °C incubator, aspirate the MEF medium, and wash each well of the MMC-MEF 6-well plate with 2 mL of 1× PBS.
4. Aspirate the 1× PBS and add the 2 mL of PXGL medium containing cells with 10 μM ROCK inhibitor.
5. Incubate the plate at 37 °C hypoxia incubator with 5% CO_2 and 5% O_2. Presence of small dome-shaped naive colonies can be found the next day (Fig. 1b).

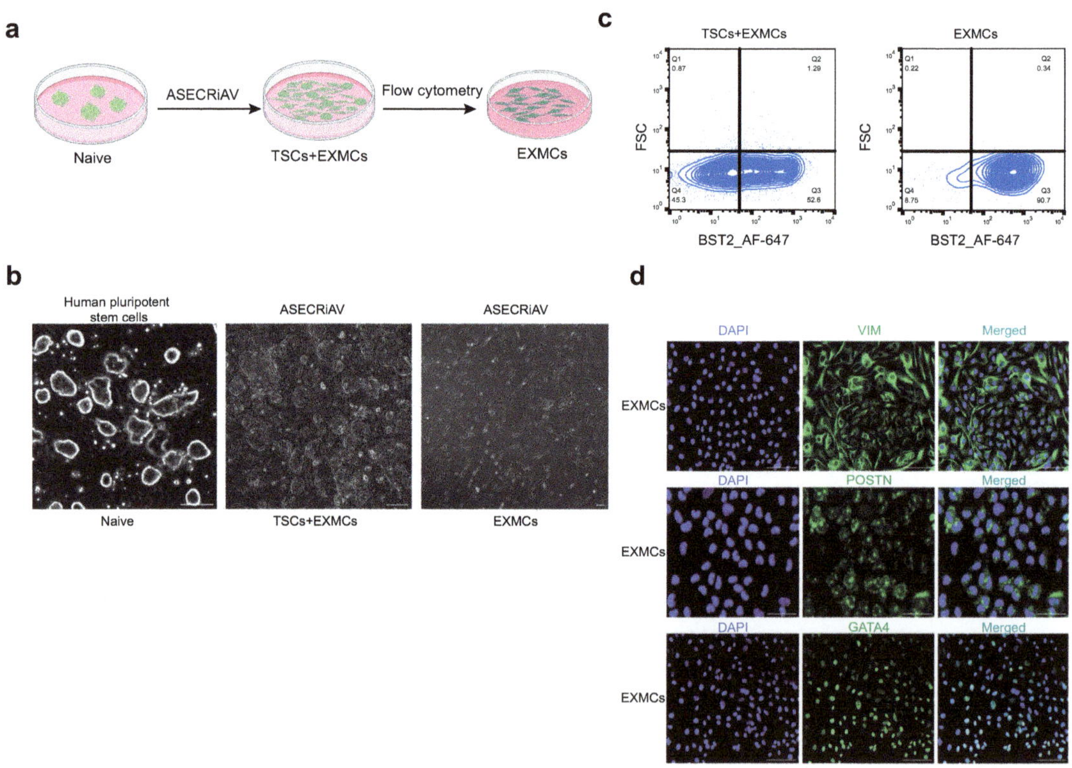

Fig. 1 Induction and characterization of EXMCs from naive hPSCs. (**a**) Schematic of conversion and purification of EXMCs from naive hPSCs. (**b**) Phase contrast images of naive hPSCs, mixed population of TSCs-EXMCs, and sorted EXMCs in the ASECRiAV culture condition. Scale bar 200 μm. (**c**) Flow cytometry analysis of mixed population of TSCs-EXMCs and sorted EXMCs for BST2 marker. (**d**) Immunofluorescent images of EXMCs for a unique combination of markers GATA4, VIM, and POSTN. Scale bar 100 μm

3.3 Conversion of Naive Pluripotent Stem Cells to Extraembryonic Mesoderm Cells (Day 2–5)

1. Aspirate the old PXGL medium and wash the cells with 1× PBS.
2. Add 2 mL of ASECRiAV into one well of 6-well plate with 5 μM ROCK inhibitor (Fig. 1a).
3. Change the medium daily with fresh ASECRiAV medium with 5 μM ROCK inhibitor.
4. After 5 days of culturing the cells on MMC-MEF, passage the cells onto a plate coated with 5 μg/mL Collagen IV. Elongated mesenchymal-like EXMCs should appear along with epithelial trophoblast stem cells (Fig. 1b).
5. Passage the cells every 5 days on Collagen IV.

3.4 Purification and Validation of EXMCs by Flow Cytometry (Day 17–18)

3.4.1 Purification Using Fluorescence-Activated Cell Sorting (FACS)

1. Aspirate the ASECRiAV medium and wash the cells with 1× PBS.
2. Add 1 mL TrypLE Express and incubate at 37 °C for 10–15 min.
3. Thoroughly dissociate the cells into single cells by pipetting up and down, and then transfer the cells to a 15 mL tube containing 5–10 mL washing medium.
4. Centrifuge the cells at 1200 rpm for 5 min. Aspirate the supernatant.
5. Wash the cell pellet by resuspending in 1 mL of FACS buffer and pipetting up and down gently.
6. Centrifuge the cells at 1200 rpm for 3 min. Aspirate the FACS buffer.
7. Resuspend cell pellet with approximately one million cells in 49 μL of FACS buffer with 1 μL of primary antibody BST2.
8. Incubate the cells with the antibody on ice and in the dark for 30 min.
9. Following incubation, wash the cells 2 times by adding 1 mL of FACS buffer.
10. Centrifuge the cells at 1200 rpm for 3 min. Aspirate the supernatant.
11. Resuspend the cell pellet with approximately one million cells in 500 μL of FACS buffer with 1 μL of secondary antibody Donkey Anti-Rabbit IgG Alexa Fluor 647.
12. Incubate the cells with the antibody on ice and in the dark for 30 min.
13. Following incubation, wash the cells 2 times by adding 1 mL of FACS buffer.
14. Centrifuge the cells at 1200 rpm for 3 min. Aspirate the supernatant.

15. Resuspend cell pellet in 300–500 μL FACS buffer and pass through a 40 μM cell strainer into a polystyrene round-bottom tube.
16. Sort the cells for surface marker BST2 on APC filter using BD Influx after adjusting the gate using an unstained sample as a negative control (Fig. 1a, c).
17. After the sorting, check the purity and accuracy of sorting (*see* **Note 3**).
18. Centrifuge the cells at 1200 rpm for 3 min and remove the supernatant.
19. Resuspend the pellet in 2 mL ASECRiAV medium with 5 μM ROCK inhibitor and seed them on Collagen IV pre-coated 6-well plate at the density of 300,000 cells per well. A homogeneous population of EXMCs should be present without any traces of trophoblast stem cells (Fig. 1b).

3.4.2 Validation by Immunofluorescence (IF)

1. To prepare EXMCs samples for IF, plate EXMCs on collagen IV pre-coated coverslips at 37 °C hypoxia incubator with 5% CO_2 and 5% O_2.
2. After 2 days, fix the cells in 4% paraformaldehyde-PBS for 10 min at room temperature in the dark.
3. Permeabilize the cells with 0.5% Triton X-100 in PBS for 10 min at room temperature.
4. Wash twice with 0.2% Tween-20 in PBS (PBST) for 5 min each.
5. Incubate the cells with a unique combination of primary antibodies diluted in the blocking buffer at 4 °C overnight (*see* **Note 4**).
6. Wash the coverslip twice with 0.2% PBST for 5 min each.
7. Incubate the cells with diluted secondary antibodies at RT for 1 h in the dark (*see* **Note 5**).
8. Wash the coverslip twice with 0.2% PBST for 5 min each.
9. Wash the coverslip again with 0.2% Tween20 in PBS + DAPI (1:50,000) for 5 min.
10. Mount the coverslip on the glass slide using Prolong Gold antifade reagent with DAPI.
11. Store the mounted coverslips at 4 °C in the dark.
12. Evaluate the identity of the cells by examining the expression of the desired proteins by using any standard fluorescence microscope (Fig. 1d).

3.5 Routine Culture and Maintenance of EXMCs

1. Coat a 6-well plate with 5 μg/mL Collagen IV in 1 mL of 1× PBS and incubate the plate at 37 °C incubator with 5% CO_2 for at least 30 min.
2. Take out EXMCs from the hypoxia incubator and aspirate the ASECRiAV medium.
3. Wash the cells with 1× PBS, add 1 mL of TrypLE, and incubate at 37 °C hypoxia incubator for 10–15 min.
4. Dissociate the cells into single cells and collect the cells into a 15 mL tube in 4 mL washing medium.
5. Centrifuge at 1200 rpm for 5 min and then aspirate the supernatant.
6. Resuspend the pellet in 2 mL of ASECRiAV medium containing 5 μM ROCK inhibitor.
7. Incubate the plate at 37 °C hypoxia incubator with 5% CO_2 and 5% O_2.
8. Change the medium every other day with fresh ASECRiAV medium with 5 μM ROCK inhibitor.
9. Split the cells every 5 days at ratio 1:6.
10. For freezing EXMCs, transfer 500,000 cells into 1 cryovial containing 1 mL freezing medium, store the cryovial in −80 °C, and then next day transfer the cryovials to liquid nitrogen.

4 Notes

1. ROCK inhibitor should be added at a final concentration 10 μM only during thawing and splitting naive hPSCs to increase survival rate.
2. The inhibitors for naive hPSCs medium (PXGL) should be added freshly everyday.
3. Always check the purity after sorting EXMCs using BST2 cell surface marker to avoid having any residual of TSCs in your culture.
4. All the primary antibodies for GATA4, POSTN, and VIM immunostaining were optimized and used at the dilution of 1:100 in the blocking buffer.
5. The secondary antibodies were used at the dilution of 1:500 in the blocking buffer.

Acknowledgments

Research in the Pasque laboratory was supported by the Research Foundation-Fladers (FWO; FWO; Odysseus Return Grant G0F7716N to V.P.; FWO grants G0C9320N and G0B4420N to V.P.), the KU Leuven Research Fund (C1 GRANT C14/21/19 to V.P.), and FWO PhD fellowships to T.X.A.P. (11N3122N). We are also grateful to Susan Schlenner and the KU Leuven FACS core team for running the facility and especially Reena Chinaraj who helped us during flow cytometry experiments.

References

1. Arnold SJ, Robertson EJ (2009) Making a commitment: cell lineage allocation and axis patterning in the early mouse embryo. Nat Rev Mol Cell Biol 10:91–103. https://doi.org/10.1038/nrm2618
2. Saykali B, Mathiah N, Nahaboo W et al (2019) Distinct mesoderm migration phenotypes in extra-embryonic and embryonic regions of the early mouse embryo. elife 8:10.7554/eLife.42434
3. Tam PP, Beddington RS (1987) The formation of mesodermal tissues in the mouse embryo during gastrulation and early organogenesis. Development 99:109–126
4. Luckett WP (1978) Origin and differentiation of the yolk sac and extraembryonic mesoderm in presomite human and rhesus monkey embryos. Am J Anat 152:59–97. https://doi.org/10.1002/aja.1001520106
5. Ross C, Boroviak TE (2020) Origin and function of the yolk sac in primate embryogenesis. Nat Commun 11:3760. https://doi.org/10.1038/s41467-020-17575-w
6. Cui G, Feng S, Yan Y et al (2022) Spatial molecular anatomy of germ layers in the gastrulating cynomolgus monkey embryo. Cell Rep 40:111285. https://doi.org/10.1016/j.celrep.2022.111285
7. Enders AC, King BF (1988) Formation and differentiation of extraembryonic mesoderm in the rhesus monkey. Am J Anat 181:327–340. https://doi.org/10.1002/aja.1001810402
8. Gasser RF (1975) Atlas of human embryos. HarperCollins Publishers
9. Kinder SJ, Tsang TE, Quinlan GA et al (1999) The orderly allocation of mesodermal cells to the extraembryonic structures and the anteroposterior axis during gastrulation of the mouse embryo. Development 126:4691–4701
10. Spencer Chapman M, Ranzoni AM, Myers B et al (2021) Lineage tracing of human development through somatic mutations. Nature 595:85–90. https://doi.org/10.1038/s41586-021-03548-6
11. Shepard TH (1989) Developmental stages in human embryos. R O'Rahilly, F Müller (eds), Carnegie Institution of Washington, Washington, DC, 1987, 306 pp., $52. Teratology 40:85–85
12. Sadler TW (1988) Langman's Medical Embryology. Plast Reconstr Surg 81:131
13. Zhou F, Wang R, Yuan P et al (2019) Reconstituting the transcriptome and DNA methylome landscapes of human implantation. Nature 572:660–664. https://doi.org/10.1038/s41586-019-1500-0
14. Xiang L, Yin Y, Zheng Y et al (2020) A developmental landscape of 3D-cultured human pre-gastrulation embryos. Nature 577:537–542. https://doi.org/10.1038/s41586-019-1875-y
15. Deglincerti A, Croft GF, Pietila LN et al (2016) Self-organization of the in vitro attached human embryo. Nature 533:251–254. https://doi.org/10.1038/nature17948
16. Shahbazi MN, Zernicka-Goetz M (2018) Deconstructing and reconstructing the mouse and human early embryo. Nat Cell Biol 20:878–887. https://doi.org/10.1038/s41556-018-0144-x
17. Shahbazi MN, Jedrusik A, Vuoristo S et al (2016) Self-organization of the human embryo in the absence of maternal tissues. Nat Cell Biol 18:700–708. https://doi.org/10.1038/ncb3347
18. Lovell-Badge R, Anthony E, Barker RA et al (2021) ISSCR guidelines for stem cell research and clinical translation: the 2021 update. Stem Cell Reports 16:1398–1408. https://doi.org/10.1016/j.stemcr.2021.05.012

19. Dong C, Beltcheva M, Gontarz P et al (2020) Derivation of trophoblast stem cells from naïve human pluripotent stem cells. elife 9:10.7554/eLife.52504
20. Io S, Kabata M, Iemura Y et al (2021) Capturing human trophoblast development with naive pluripotent stem cells in vitro. Cell Stem Cell 28:1023–1039.e13. https://doi.org/10.1016/j.stem.2021.03.013
21. Guo G, Stirparo GG, Strawbridge SE et al (2021) Human naive epiblast cells possess unrestricted lineage potential. Cell Stem Cell 28:1040–1056.e6. https://doi.org/10.1016/j.stem.2021.02.025
22. Castel G, Meistermann D, Bretin B et al (2020) Induction of human trophoblast stem cells from somatic cells and pluripotent stem cells. Cell Rep 33:108419. https://doi.org/10.1016/j.celrep.2020.108419
23. Cinkornpumin JK, Kwon SY, Guo Y et al (2020) Naive human embryonic stem cells can give rise to cells with a trophoblast-like transcriptome and methylome. Stem Cell Reports 15:198–213. https://doi.org/10.1016/j.stemcr.2020.06.003
24. Linneberg-Agerholm M, Wong YF, Romero Herrera JA et al (2019) Naïve human pluripotent stem cells respond to Wnt, Nodal and LIF signalling to produce expandable naïve extra-embryonic endoderm. Development 146:10.1242/dev.180620
25. Rostovskaya M, Andrews S, Reik W, Rugg-Gunn PJ (2022) Amniogenesis occurs in two independent waves in primates. Cell Stem Cell 29:744–759.e6. https://doi.org/10.1016/j.stem.2022.03.014
26. Karvas RM, Khan SA, Verma S et al (2022) Stem-cell-derived trophoblast organoids model human placental development and susceptibility to emerging pathogens. Cell Stem Cell 29:810–825.e8. https://doi.org/10.1016/j.stem.2022.04.004
27. Yu L, Wei Y, Duan J et al (2021) Author correction: blastocyst-like structures generated from human pluripotent stem cells. Nature 596:E5. https://doi.org/10.1038/s41586-021-03635-8
28. Yanagida A, Spindlow D, Nichols J et al (2021) Naive stem cell blastocyst model captures human embryo lineage segregation. Cell Stem Cell 28:1016–1022.e4. https://doi.org/10.1016/j.stem.2021.04.031
29. Kagawa H, Javali A, Khoei HH et al (2022) Human blastoids model blastocyst development and implantation. Nature 601:600–605. https://doi.org/10.1038/s41586-021-04267-8
30. Liu X, Tan JP, Schröder J et al (2021) Modelling human blastocysts by reprogramming fibroblasts into iBlastoids. Nature 591:627–632. https://doi.org/10.1038/s41586-021-03372-y
31. Pham TXA, Panda A, Kagawa H et al (2022) Modeling human extraembryonic mesoderm cells using naive pluripotent stem cells. Cell Stem Cell 29:1346–1365.e10. https://doi.org/10.1016/j.stem.2022.08.001

Methods in Molecular Biology (2024) 2767: 115–122
DOI 10.1007/7651_2023_507

Published online: 17 October 2023

Modeling Human Paraxial Mesoderm Development with Pluripotent Stem Cells

Yuchuan Miao, Margarete Diaz-Cuadros, and Olivier Pourquié

Abstract

Paraxial mesoderm in the early embryo is segmented into epithelial blocks called somites that establish the metameric organization of the vertebrate body plan. Somites are sequentially formed from head to tail in a rhythmic manner controlled by an oscillating gene regulatory network known as the segmentation clock. We know very little about this important process during human development due to limited access to human embryos and ethical concerns. To bypass these difficulties, model systems derived from human pluripotent stem cells have been established. Here, we detail three protocols modeling different aspects of human paraxial mesoderm development in vitro: a 2D cell monolayer system recapitulating dynamics of the human segmentation clock, a 3D organoid system called "somitoid" supporting the simultaneous formation of somite-like structures, and another organoid system called "segmentoid" reconstituting in vivo–like hallmarks of somitogenesis. Together, these complementary model systems provide an excellent platform to decode somitogenesis and advance human developmental biology.

Key words Directed differentiation, Human development, Organoid, Paraxial mesoderm, Pluripotent stem cell, Somite, Somitogenesis, The segmentation clock

1 Introduction

In vertebrate embryos, the bilateral paraxial mesoderm flanking the neural tube is segmented into repeated epithelial blocks called somites [1]. The periodically arranged somites provide the blueprint to build the segmentally organized musculoskeletal axis, vascular system, and peripheral nervous system of the adult body. They give rise to our vertebrae, skeletal muscle, brown adipocyte tissue, dermis on the back, etc., playing an instrumental role for the normal development of our body plan. Conversely, defects associated with paraxial mesoderm development in humans lead to a range of diseases such as congenital scoliosis [2]. The process of somite formation, or somitogenesis, is a dynamic, complex developmental process. Somites are periodically formed from the anterior-most part of the unsegmented region called presomitic mesoderm (PSM). The formation occurs sequentially in an anterior to posterior direction, in coordination with body axis elongation. These elegant features have attracted the focus of generations of scientists. Animal models such as mice, chicken, and zebrafish have been used to study the mechanisms of somitogenesis. An oscillating gene regulatory network called the segmentation clock controls the rhythm of somite formation in all vertebrates [1, 3]. In each cycle, the clock activity is initiated in the posterior PSM and propagates anteriorly as a wave of gene expression. In the anterior region of the PSM, clock activity together with posterior to anterior signaling gradients specifies a segmental domain where a program of epithelialization is initiated to form a somite. In humans, this process occurs between 3 and 4 weeks after fertilization [4]. Embryos at this stage are rarely accessible, impeding the understanding of human-specific mechanisms.

A variety of in vitro model systems has recently allowed studying somitogenesis in the context of human development. To date, both 2D and 3D systems derived from human pluripotent stem cells (PSC) have been established to recapitulate paraxial mesoderm development [5–11]. They open the door to investigate human-specific traits and elucidate causes of congenital diseases. In this chapter, we detail the experimental procedures of three systems that use human-induced pluripotent stem cells (iPSC) to generate almost exclusively paraxial mesoderm (Fig. 1). Using signaling factors to activate Wnt and inhibit BMP signaling, a 2D monolayer system promotes the efficient differentiation of epiblast staged PSC to neuromesodermal progenitors (NMP) and then to posterior PSM [6]. Live imaging of the induced PSM cells demonstrated oscillatory expressions of the core clock gene HES7, identifying the human segmentation clock. This tractable system offers unlimited materials to elucidate signaling dynamics of paraxial mesoderm development. Next, when the directed differentiation method was adapted to iPSC aggregates, a 3D organoid system called "somitoid" was established that can generate a field of somite-like structures [9]. Somitoids recapitulate the temporal sequence of somitogenesis uncoupled from tissue elongation, with all cells undergoing differentiation and morphogenesis in a synchronous manner. This simplified system is amenable to high-resolution live imaging and will allow exploring the patterning and morphogenetic processes at an unprecedented level of detail. Furthermore, another organoid model termed "segmentoid" reconstitutes the spatio-temporal features of somitogenesis, including gene expression dynamics, tissue elongation, sequential somite morphogenesis, and somite anteroposterior polarity patterning [9]. It therefore provides an excellent proxy to study human somitogenesis. Altogether, these three systems model different aspects of human somitogenesis and provide complementary experimental strategies, representing a valuable platform to advance knowledge of human development.

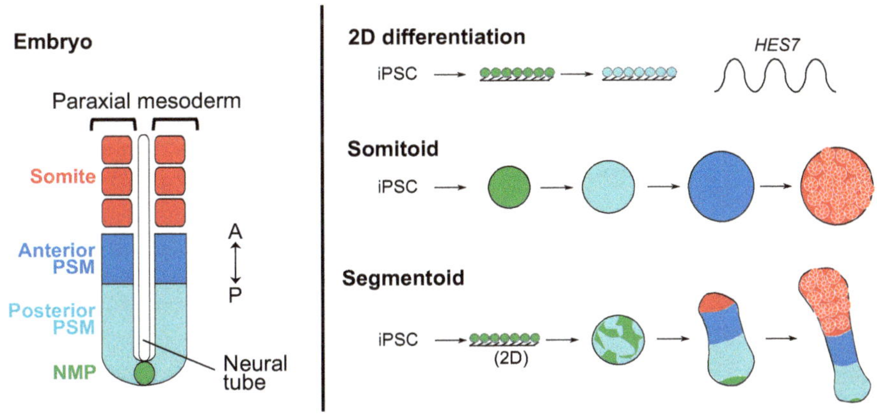

Fig. 1 Illustration of the embryo (left) and differentiation protocols of human paraxial mesoderm (right). The paraxial mesoderm is derived from neuromesodermal progenitors (NMP) and form repeated somites flanking the neural tube. The 2D differentiation system produces posterior PSM cells and recapitulates oscillations of the human segmentation clock. The somitoid system supports synchronized differentiation in 3D and generates somite-like structures simultaneously. The segmentoid system reconstitutes embryo-like dynamics and can serve as a proxy of human somitogenesis. The same cell type is represented as the same color, with NMP as green, posterior PSM as cyan, anterior PSM as blue, and somite as red. *HES7* is a core gene of the human segmentation clock. A anterior; P posterior

2 Materials

All reagents should be sterile and handled in cell culture hoods at room temperature (unless indicated otherwise). Culture medium supplemented with signaling factors should be freshly prepared before experiments.

2.1 Mediums and Solutions

mTeSR1 (StemCell Technologies 05851), DMEM/F12 GlutaMAX (Gibco10565042), Neurobasal medium (Gibco 21103049), Accutase (Corning 25058CI), 0.1% Gelatin solution (Stemcell Technologies 07903), DPBS (Gibco 14190-250).

2.2 Factors and Supplements

Y-27362 dihydrochloride (Tocris Bioscience1254), insulin-transferrin-selenium (Gibco 41400045), CHIR99021 (Tocris 4423), LDN193189 (Stemgent 04-0074), mFgf4 (R&D Systems 5846-F4-025), heparin (Sigma Aldrich H3393-100KU), BMS493 (Sigma Aldrich B6688-5MG), fetal bovine serum (Sigma-Aldrich EmbryoMax ES-009-B), N2 Supplement-B (Stemcell Technologies 07156), B-27 Supplement (Gibco 17504044), 2-mercaptoethanol (Gibco 21985023), Matrigel (Corning 35277), Matrigel-B (Corning CB-40234A) (*see* **Note 1**).

2.3 Culture Vessels and Equipment

Six-well TC cell culture plate (Falcon 353046), 96-well U-bottom non-treated plate (Greiner bio-one 650185 or Falcon 351177), 96-well cell-repellent U-bottom plate (Greiner bio-one 650970), automatic cell counter (Nexcelom Cellometer), cell counting chamber (Nexcelom CHT4-SD100-002), stereo microscope in the cell culture hood, incubator (37 °C, 5.5% CO_2, and humidity).

2.4 DICL Medium

DMEM/F12, 1% insulin-transferrin-selenium (ITS), 3 μM CHIR99021, 0.5 μM LDN193189 (*see* **Note 2**).

2.5 DICLFBR Medium

DMEM/F12, 1% insulin-transferrin-selenium (ITS), 3 μM CHIR99021, 0.5 μM LDN193189, 50 ng/mL mFgf4, 1 μg/mL heparin, 2.5 μM BMS493, and 10 μM Y-27362 dihydrochloride.

2.6 SIM Medium

DMEM/F12, 1% ITS, 1.5 μM CHIR99021, 0.5 μM LDN193189, 5% fetal bovine serum (FBS) (*see* **Note 3**).

2.7 N2B27 Medium

1:1 mix of DMEM/F12 and neurobasal medium, 1% N_2 supplement-B, 2% B-27 supplement, 0.1% 2-mercaptoethanol. Store at 4 °C for up to a couple of weeks.

2.8 Matrigel-Coated Plates

Dilute Matrigel (Corning 35277) with ice-cold DMEM/F12 (dilution fold as suggested by the manufacturer). On ice, add 1 mL ice-cold Matrigel solution into one well of a six-well cell culture plate. Store at 4 °C for up to a week. Incubate at 37 °C for 20 min before use.

3 Methods

Carry out all procedures with open containers in cell culture hoods at room temperature (unless indicated otherwise).

3.1 Dissociation and Passage of Human iPSC

1. We designate the day of cell dissociation as Day −1. Remove the medium of iPSC cultured in one well of the six-well plate, wash twice with 1 mL DPBS, add 1 mL Accutase, and incubate at 37 °C for 5 min. Then gently pipette, mix, and transfer into a conical tube with 4 mL DMEM/F12. Centrifuge at 300 rcf for 4 min.
2. Remove supernatant and gently resuspend cell pellet using 2 mL mTeSR1 with 10 μM Y-27362 (ROCKi). Load 20 μL cell mix into the cell counting chamber and determine cell density using the automatic cell counter. The suspended cell solution can be used to start protocols of differentiation as detailed in Subheadings 3.2, 3.3, and 3.4.
3. For iPSC maintenance, calculate the volume needed for 500,000 cells. Remove all medium in the Matrigel-coated plate and add 2 mL mTeSR1 supplemented with 10 μM Y-27362. Add 500,000 cells into the well, shake the plate well to mix evenly, and put it into the 37 °C incubator with

5% CO_2 and humidity (Day 0). For the next 3 days, change fresh mTesR1 medium daily before cell passage on Day 4 again starting from **step 1**.

3.2 2D Differentiation

1. On Day −1, take the cell solution from Subheading 3.1 to seed iPSCs at a density of 3 × 10^4 cells/cm^2 on Matrigel-coated dishes using mTeSR1 with 10 μM Y-27362. Shake the plate to distribute cells evenly and incubate at the 37 °C, 5% CO_2 overnight.
2. The next day (Day 0), replace the media by mTeSR1 without Y-27362. Incubate for 5 h to allow cells to come together in small clusters. Confluency should be 15–20% at this point (*see* **Note 4**). Replace media by DICL medium to initiate the differentiation process.
3. On Day 1, cells will remain in tight epithelial clusters but will have already transitioned to the NMP state. Refresh the DICL medium to continue the differentiation.
4. On Day 2, cells will have undergone an epithelial to mesenchymal transition and adopted posterior PSM fate. Confluency should be 80–90%. To sustain the posterior PSM fate for up to two additional days, replace media by DICLFBR medium.

3.3 Somitoid

1. On Day −1, dilute the suspended cell solution from Subheading 3.1 using mTeSR1 with 10 μM Y-27362, so that 100 μL diluted cell mix contains 3000 cells (*see* **Note 5**). Dispense 100 μL into each well of the 96-well U-bottom non-treated plates (Greiner bio-one 650185 or Falcon 351177). Centrifuge the plate at 300 rcf for 2 min and put it into the incubator without disturbing the pellet.
2. On Day 0, remove all medium from each well under a stereo dissection microscope in the cell culture hood (*see* **Note 6**). Add 150 μL DICL medium. Gently pipette the aggregate under a stereo dissection microscope to make sure it is unattached from the well. Put the plate back into the incubator.
3. On Day 1, the aggregate should appear as a sphere resting on the bottom of the well. Remove 130 μL medium from each well and add in 150 μL fresh DICL medium (*see* **Note 7**).
4. On Day 2, add 1 mL 0.1% gelatin solution or other coating solution to a well of six-well TC plate and incubate at 37 °C for 1 h. Remove all solution from the gelatin-coated well and add in 2 mL SIM medium. Transfer the organoids one by one from the U-bottom wells into the flat TC plate well with little carryover medium, using the 200 μL pipette (*see* **Note 8**). Up to 20 organoids can be put into one well of the six-well plate. Shake the plate to mix well the organoids and incubate without further medium change (*see* **Note 9**).

3.4 Segmentoid

1. On Day −1, remove all medium in the Matrigel-coated plate and add 2 mL mTeSR1 with 10 μM Y-27362. Add 220,000 cells from the suspended cell solution in Subheading 3.1 into the well (*see* **Note 10**), shake the plate to mix evenly, and put it into the 37 °C incubator with 5.5% CO_2 and humidity.
2. On Day 0, prepare DICL medium. Remove all medium from each well and add 2 mL fresh DICL medium.
3. On Day 1, 24 h after the previous step, remove all DICL medium and wash once with PBS (*see* **Note 11**). Add 1 mL Accutase and incubate at 37 °C for 5 min. Gently pipette, mix, and transfer into a conical tube with 4 mL DMEM/F12. Centrifuge at 300 rcf for 4 min. Remove supernatant and gently resuspend cell pellet using 300 μL N2B27 medium with 10 μM Y-27362. Load 20 μL cell mix into the cell counting chamber, and determine cell density using the automatic cell counter. Add 150 μL N2B27 containing 10 μM Y-27362 into each well of the 96-well cell-repellent U-bottom plate (Greiner bio-one, 650970) (*see* **Note 12**). Calculate the volume needed and add 6000–8000 cells into each well. Centrifuge the plate at 300 rcf for 2 min and put it into the incubator without disturbing the pellet.
4. On Day 2, a single spheroid should be formed in each well. Thaw Matrigel-B aliquot (Corning CB-40234A) on ice and prepare ice-cold 10% Matrigel solution using N2B27 medium. Remove medium under a dissection microscope and add 30 μL ice-cold 10% Matrigel-B solution. Gently tap the plate to let each organoid sink to the bottom. Incubate the plate at 37 °C for one hour. Carefully add 150 μL N2B27 medium to each well. Lean the tip against the inside wall to avoid breaking the solidified Matrigel. Incubate the plate at 37 °C and no further medium change is needed (*see* **Note 13**).

4 Notes

1. Different kinds of Matrigel are used for plate coating and embedding. Matrigel used for organoid embedding is designated as Matrigel-B in this chapter.
2. We found that CHIR99021 from different companies might have varied qualities, thus the concentration might need to be adjusted accordingly.
3. Note that the concentration of CHIR99021 is lowered by half from what is used in DICL medium.
4. Initial cell density is crucial, as sparse cultures will crash and overcrowded cultures will be heterogeneous.

5. The aggregation size is important. Aggregates of <2000 cells often lead to cell death, while aggregates of >4000 cells often lead to poor differentiation in the center. Also, 100% confluent iPSC culture for aggregation often results in unhealthy somitoids, and for this protocol, we dissociate cells when the cell culture is ~80% confluent.
6. With the 96-well U-bottom non-treated plates (Greiner bio-one 650185 or Falcon 351177), the aggregate often appears as a flat disk attached at the bottom, which makes it easy to remove as much medium as possible. Sometimes, several small aggregates are formed but they will fuse in subsequent culture.
7. This step is optional as we did not observe any difference when no medium change is performed on Day 1.
8. Under the stereo microscope, position the organoid at the very tip of the pipette tip, so very gentle release with little carry-over medium is sufficient for the transfer.
9. The organoid should spread to the fullest by the end of Day 3, appearing as a flat disk attached on the surface. "Cracks" should be observed on Day 4 and a field of somite-like structures should be visible under a bright-field microscope on Day 5.
10. The seeding cell number is the single most important parameter throughout the protocol. Higher cell number during the 2D differentiation step will lead to poorly elongated and poorly patterned structures with accelerated epithelization process. We want to highlight that the exact cell number listed in our protocol is only meaningful in specific contexts. Many routines and habits of cell culture can introduce variations. For example, different labs grow iPSC in different ways. In our cell culture method, dissociation of iPSC at different confluency will lead to different cell survival rates next day. Thus, seeding the same number of cells does not necessarily mean the same actual cell density next day. In addition, different counting methods are adopted among labs. We suggest that to start with, a range of cell numbers (e.g., 180K, 210K, 240K, 270K) should be tried on Day −1. Once a working number is calibrated, keep cell culture routine consistent to lower variability: dissociating cell culture at the same confluency, counting cells in the same way, paying attention to the time between media change, etc.
11. The DICL medium treatment should be precisely around 24 h. Avoid significantly shorter or longer (>1 h) treatment which might jeopardize the outcomes of subsequent steps.
12. The plate for aggregation is important; note that it is different from the U-bottom plate used in the somitoid protocol. The

96-well cell-repellent U-bottom plate (Greiner bio-one, 650970) is the ideal plate for segmentoid.

13. The organoid should break symmetry on Day 3 and start to elongate; on Day 4, the organoid should contain half segmented and half unsegmented regions, while on Day 5, the organoid should appear as a segmented rod with somite-like structures clearly visible.

References

1. Hubaud A, Pourquié O (2014) Signalling dynamics in vertebrate segmentation. Nat Rev Mol Cell Biol 15:709–721
2. Eckalbar WL, Fisher RE, Rawls A, Kusumi K (2012) Scoliosis and segmentation defects of the vertebrae. Wiley Interdiscip Rev Dev Biol 1:401–423
3. Oates AC, Morelli LG, Ares S (2012) Patterning embryos with oscillations: structure, function and dynamics of the vertebrate segmentation clock. Development 139:625–639
4. Schoenwolf GC, Bleyl SB, Brauer PR, Francis-West PH (2020) Larsen's human embryology E-book. Elsevier Health Sciences
5. Chu L-F, Mamott D, Ni Z, Bacher R, Liu C, Swanson S, Kendziorski C, Stewart R, Thomson JA (2019) An in vitro human segmentation clock model derived from embryonic stem cells. Cell Rep 28:2247–2255.e5
6. Diaz-Cuadros M, Wagner DE, Budjan C, Hubaud A, Tarazona OA, Donelly S, Michaut A, Al Tanoury Z, Yoshioka-Kobayashi K, Niino Y, Kageyama R, Miyawaki A, Touboul J, Pourquié O (2020) In vitro characterization of the human segmentation clock. Nature. https://doi.org/10.1038/s41586-019-1885-9
7. Matsuda M, Yamanaka Y, Uemura M, Osawa M, Saito MK, Nagahashi A, Nishio M, Guo L, Ikegawa S, Sakurai S, Kihara S, Maurissen TL, Nakamura M, Matsumoto T, Yoshitomi H, Ikeya M, Kawakami N, Yamamoto T, Woltjen K, Ebisuya M, Toguchida J, Alev C (2020) Recapitulating the human segmentation clock with pluripotent stem cells. Nature 580:124–129
8. Sanaki-Matsumiya M, Matsuda M, Gritti N, Nakaki F, Sharpe J, Trivedi V, Ebisuya M (2022) Periodic formation of epithelial somites from human pluripotent stem cells. Nat Commun 13:1–14
9. Miao Y, Djeffal Y, De Simone A, Zhu K, Lee JG, Lu Z, Silberfeld A, Rao J, Tarazona OA, Mongera A, Rigoni P, Diaz-Cuadros M, Song LMS, Di Talia S, Pourquié O (2022) Reconstruction and deconstruction of human somitogenesis in vitro. Nature. https://doi.org/10.1038/s41586-022-05655-4
10. Yamanaka Y, Hamidi S, Yoshioka-Kobayashi K, Munira S, Sunadome K, Zhang Y, Kurokawa Y, Ericsson R, Mieda A, Thompson JL, Kerwin J, Lisgo S, Yamamoto T, Moris N, Martinez-Arias A, Tsujimura T, Alev C (2022) Reconstituting human somitogenesis in vitro. Nature. https://doi.org/10.1038/s41586-022-05649-2
11. Yaman YI, Ramanathan S (2023) Controlling human organoid symmetry breaking reveals signaling gradients drive segmentation clock waves. Cell 186:513–527.e19

Methods in Molecular Biology (2024) 2767 123–134
DOI 10.1007/7651_2023_493

Published online: 08 June 2023

Generation of Stem Cell-Based Mouse Embryo-Like Structures

Charlotte E. Handford, Pallavi Panda, Ikbal B. Mohammad Choudhury, Gianluca Amadei, and Magdalena Zernicka-Goetz

Abstract

In this chapter, we detail the experimental protocol leading to the generation of stem cell-based mouse embryo-like structures termed "ETiX-embryoids." ETiX-embryoids are formed from combined embryonic stem cells, trophoblast stem cells, and embryonic stem cells transiently induced to express Gata4. Cells are seeded into AggreWell dishes where they form aggregates that develop to resemble post-implantation mouse embryos following 4 days of culture. ETiX-embryoids establish an anterior signaling center and undergo gastrulation over the following 2 days. By day 7, ETiX-embryoids undergo neurulation and form an anterior-posterior axis with head folds at one end and a tail bud on the other. On day 8, they develop a brain and form a heart-like structure and a gut tube.

Key words Embryo-like structures, ETiX-embryoids, Neurulation, Organogenesis, Stem cells

1 Introduction

In addition to the study of the natural embryo, the use of stem cell-based culture models has become increasingly prevalent in recent years [1–6]. Critically, they allow for the characterization of developmental processes without the requirement for large numbers of natural embryos. Moreover, they are amenable to genetic manipulations as well as live imaging techniques and therefore offer opportunities for molecular studies to be performed without the need for generation of knockout animals. In this chapter, we describe a procedure to generate ETiX-embryoids that over 8 days of culture are capable of recapitulating the development of the post-implantation mouse embryo up to early organogenesis.

2 Materials

2.1 Reagents

1. 2-β-mercaptoethanol (Gibco 31350010).
2. Advanced DMEM/F12 (Gibco, 21331-020).
3. AggreWell rinsing solution (StemCell Technology 07010).

4. B27 (Gibco 10889038).
5. β-estradiol (Tocris 50-28-2, stock: 10 uM in DMSO).
6. CHIR99021 (StemCell Technology 72052).
7. DMEM (Gibco 41966052).
8. DMEM, bicarbonate-buffered, without glutamine (Gibco 11054).
9. DMEM low-glucose, with no bicarbonate buffer (Gibco 11880).
10. DMEM/F-12 (Gibco 21331020).
11. Doxycycline (Sigma-Aldrich D9891-5G, stock: 1 mg/mL in water).
12. Fetal bovine serum (Gibco 10270106).
13. FGF4 (R&D Systems 7486-F4-025, stock: 100 ug/mL).
14. Gelatin type A (Sigma-Aldrich G1890).
15. D-(+)-Glucose (G8644-100ML).
16. GlutaMAX (Gibco 35050038).
17. Heparin (Sigma-Aldrich H3149-25KU, stock: 1 mg/mL in PBS).
18. HEPES (Thermo Fisher Scientific 15630056).
19. Human cord serum (Cambridge Blood and Stem cell Biobank, UK).
20. Human adult blood serum (BIOIVT, USA #HP1022).
21. ITS-X (Thermo Fisher Scientific 51500-056).
22. Leukemia inhibitory factor LIF (StemCell Technologies 78056).
23. MEM non-essential amino acids (Gibco 11140035).
24. N-acetyl-L-cysteine (Sigma-Aldrich A7250, stock: 50 mM in water).
25. N2 Supplement (Life Technologies 17502048).
26. Neurobasal-A (Gibco 10888022).
27. PBS (Life Technologies 10010056).
28. PD0325901 (Cambridge Stem Cell Institute).
29. Penicillin/streptomycin (Gibco 15140122).
30. Progesterone (Sigma-Aldrich P0130-25G, stock: 1 mg/mL in DMSO).
31. Rat serum (adult rats, Charles River/Janvier Labs/BIOIVT).

32. RPMI 1640, no Glutamine (Thermo Fisher Scientific 21870076).
33. Sodium pyruvate (Gibco 11360039).
34. Trypsin-EDTA 0.05% (Life Technologies 25300054).

2.2 Experimental Plates

1. Tissue culture treated 6-well plates (e.g., Thermo Fisher Scientific 140675).
2. Suspension culture 6-well plate (CELLSTAR, Greiner Bio-One 657185).
3. Suspension culture 48-well plate (CELLSTAR, Greiner Bio-One 677102).
4. AggreWells (24-well format, StemCell Technology 34415).

2.3 Cell Culture Media

2.3.1 N2B27 Medium (for 50 mL)

1. 25 mL of Neurobasal-A.
2. 25 mL of DMEM/F12.
3. 500 uL of P/S.
4. 500 uL of GlutaMAX.
5. 100 uL of 2-β-mercaptoethanol.
6. 500 uL of B27.
7. 250 uL of N2.
8. Supplement with LIF, Chiron, and PD03 to make N2B27/2iLIF.

2.3.2 Feeder Cell Medium (FC, for 1 Full Bottle)

1. 1 bottle of DMEM (Gibco 41966052).
2. 90 mL of heat-inactivated FBS.
3. 6 mL of P/S.
4. 6 mL of GlutaMAX.
5. 6 mL of MEM non-essential amino acids.
6. 6 mL of Sodium pyruvate.
7. 1.2 mL of 2-β-mercaptoethanol.

2.3.3 Trophoblast Stem Cell Medium (TS, for 50 mL)

1. 50 mL of RPMI 1640.
2. 10.4 mL of FBS.
3. 520 uL of P/S.
4. 520 uL of GlutaMAX.
5. 520 uL of Sodium pyruvate.
6. 104 uL of 2-β-mercaptoethanol.
7. For 10 mL of TSF4H medium, add 10 uL of Heparin stock and 2.5 uL of FGF4 stock.

2.4 In Vitro Culture (IVC) Medium [7]

2.4.1 Base for IVC (ADF+++)

1. 50 mL of Advanced DMEM/F12.
2. 500 uL of GlutaMAX.
3. 500 uL of ITS-X.
4. 125 uL of P/S.

2.4.2 IVC-20% FBS (10 mL)

1. 8 mL of ADF+++.
2. 2 mL of FBS.
3. 8 uL of β-estradiol.
4. 2 uL of Progesterone.
5. 5 uL of N-acetyl-L-cysteine.

2.4.3 IVC-30% FBS (10 mL)

1. 7 mL of ADF+++.
2. 3 mL of FBS.
3. 8 uL of β-estradiol.
4. 2 uL of Progesterone.
5. 5 uL of N-acetyl-L-cysteine.

2.4.4 DMEM/Rat Serum Medium/Human Cord Serum (DRH, for 1 mL) [8]

1. 250 uL of DMEM, bicarbonate-buffered, without glutamine (Gibco 11054).
2. 500 uL of rat serum.
3. 250 uL of human cord serum.
4. 10 uL of GlutaMAX.
5. 10 uL of P/S.

2.4.5 Human and Rat Serum Preparation

Store at −80 °C. Thaw slowly and heat-inactivate for 30 min at 56 °C and filter-sterilize prior to use.

2.5 Roller System Components

1. Rotary bottle system.
2. Gas mixer.
3. Gas cylinders (O_2, N_2, and CO_2).

The gas mixer has input from all three gas cylinders (O_2, N_2, and CO_2). The input pressure for each gases is maintained at 10 psi. The gases are mixed in the desired ratio (in this case 21% O_2 and 5% CO_2) and fed into the rotary bottle system. The final gas mixture is humidified before feeding into the rotatory bottles. The pressure of the gas (mixed + humidified) before at the entrance of the rotary bottle is 0.5 psi. The flow rate at the exit of the rotatory bottle system is 8–10 bubbles/s. The bottles are rolled at 3 rpm.

3 Methods

3.1 Culture of Mouse Embryonic Stem Cells (ESCs)

Embryonic stem cells may be cultured in N2B27/2iLIF medium on gelatinized, tissue culture-grade plates in an incubator set to 37 °C with 5% CO_2 and 21% O_2. Cells should be routinely passaged every 3–4 days or when 70–80% confluent, with regular media change every day or every second day, in between passages (*see* **Notes 1** and **2**).

1. To passage, begin by preparing a tissue culture dish of the appropriate size for replating. To this end, add the appropriate volume of 0.1% gelatin solution to a well (i.e., add 1 mL gelatin to 1 well of a 6-well plate) and set plate aside until needed. Pre-warm required media and solutions (N2B27/2iLIF, FC, Trypsin) in 37 °C water bath for no more than 10 min.
2. Aspirate the media from the well of cells and add appropriate volume of Trypsin (i.e., for 1 well of a 6-well plate add 1 mL Trypsin).
3. Incubate the plate in the incubator for 3–4 min.
4. Gently pipette cells 2–3 times to produce a single-cell suspension and add 1–2 volumes of FC media to quench the Trypsin activity (*see* **Note 3**).
5. Transfer the cell suspension to a 15 mL Falcon tube and centrifuge at 0.2 × g (RCF) for 4 min.
6. Aspirate supernatant leaving a tiny amount above the cell pellet so as not to disturb it. Wash the cells with PBS (1 mL for 1 well of a 6-well plate) and repeat centrifugation step as above (**step 5**).
7. Aspirate supernatant as above and resuspend cells in 1 mL N2B27/2iLIF medium.
8. Aspirate the gelatin solution from the prepared plate (**step 1**) and add appropriate volume of N2B27/2iLIF (2 mL for 1 well of a 6-well plate).
9. Plate the cells at desired dilution (i.e., 1:5, 1:10, 1:20).
10. Distribute the cells evenly by gently moving the plate in a figure of 8 motion and place in the incubator (*see* **Note 4**).

3.2 Culture of Mouse Trophoblast Stem Cells (TSCs)

Trophoblast stem cells may be cultured in TSF4H on inactivated MEFs in an incubator kept at 37 °C with 5% CO_2 and 21% O_2. As with ESCs, cells should be routinely passaged every 3–4 days or when 70/80% confluent, with regular media top-up in between passages (*see* **Note 5**).

Passaging of TSCs is similar to that of ESCs with the following distinctions:

1. Medium conditioning for ~5–6 h is advisable and helps prevent differentiation of TSCs. To condition media, aspirate the FC medium from a well coated with iMEFs and add the appropriate volume of TSF4H.
2. After 6 h, begin passage by aspirating the media from the well of TSCs, and washing with PBS.
3. Add the appropriate volume of Trypsin and incubate the plate for 4 min.
4. Quench the Trypsin activity with 2 mL of TS base medium.
5. Gently dissociate cells by pipetting and centrifuge in 15 mL Falcon tube at 0.2 × g (RCF) for 4 min.
6. Aspirate supernatant and resuspend cells with 1 mL of TSF4H.
7. Dilute at 1:10 or 1:20 and plate in the well of iMEFs pre-conditioned with TSF4H (**step 1**).

3.3 Culture and Inactivation of Mouse Embryonic Fibroblasts (MEFs)

Culture and expand active MEFs only up until P5. Cell harvests decrease with the passage number. Cells should not be passaged more often than at 3-day intervals. Should the cell number at confluency be lower or equal to the number of cells plated, the cells need to be inactivated immediately, regardless of passage number. Overcrowding promotes the outgrowth of neural cell types.

To passage active MEFs:

1. Aspirate the media and add PBS to wash once.
2. Aspirate the PBS and add TryplE for 2–7 min or until cells are visibly detached.
3. Add at least 2 volumes of FC media and wash the surface to ensure all the cells are removed. Add cells to a 50 mL Falcon tube.
4. If passaging multiple plates, pool all the cells and quantify cell number using a hemocytometer.
5. Seed appropriate number of cells to dishes:

Size	Seeding number
15 cm dish/T150	1.3×10^6
T75	6.5×10^5
T25	2.2×10^5

MEFs may be inactivated through irradiation or with mitomycin-C. For mitomycin-C inactivation proceed as follows:

1. Aspirate FC media from plates and add appropriate volume of fresh FC (enough volume to cover the surface, i.e., for 15 cm plate use 10 mL).
2. Add the appropriate volume of mitomycin-C for a final concentration of 10 ug/mL dropwise to the cells taking care to distribute the drops across the entire dish. Mix by swirling the dish very gently.
3. Incubate treated cells for 1,5 h at 37 °C with 5% CO_2 and 21% O_2.
4. Remove media and dispose of all mitomycin-C containing media appropriately.
5. Wash cells with PBS 5 times prior to the addition of TryplE.
6. Incubate for 2–7 min or until the cells begin to detach.
7. Add at least 2 volumes of FC media and wash the surface to ensure all the cells are removed. Add cells to a 50 mL Falcon tube.
8. Quantify cell number and plate/freeze down in appropriate numbers as needed (i.e., for TSC culture 0.5×10^6 iMEFs frozen down can be subsequently thawed and used for 3 wells of a 6-well dish).

3.4 Preparation and Seeding of Cells for ETiX-Embryoid Formation

1. One or two days prior to the experiment, passage tetO-Gata4 ESCs at a dilution of roughly 1:5. A well with many small colonies is preferable to fewer larger colonies as it ensures a homogenous response to doxycycline.
2. On the day of the experiment, 6 h before cell dissociation: add fresh media (N2B27/2iLIF) to all ESCs, add Dox (1:1000 dilution) to the tetO Gata4 ESCs, and top up the medium (TSF4H) on TSCs.
3. Six hours following Dox addition, plating of the experiment is started.
4. Begin by adding 0.5 mL of anti-adherence rinsing solution to the desired number of AggreWells and centrifuge the plate at 2000 × g (RCF) for 5 min. Check to ensure no bubbles are present and set aside at room temperature until cells have been quantified. If bubbles are present, centrifuge once more for another 5 min.
5. Prepare a volume of FC that is 2× that of the number of AggreWells (for 1 well prepare 2 mL) in a Falcon tube and

equilibrate in the incubator with a loose cap. This media will be used for washes of the AggreWells and for cell resuspension following the combination of each of the three cell types.

6. Prepare 1–2 wells with 1 mL of gelatin and set aside at room temperature. This is required for MEF depletion from the TSC suspension.
7. Dissociate ESCs and TSCs as described above. Following dissociation, resuspend TSCs in 2 mL of TSF4H and plate in the gelatinized dish prepared in the previous step.
8. Allow MEFs to stick down for 20–30 min in the incubator. During this time, quantify the ESC lines. Then collect the TSC suspension.
9. Quantify the remaining cells with a hemocytometer (use 10 uL of the cell suspension). Place the Falcon tube with remaining cells in the incubator until needed.
10. Calculate volumes yielding 6,000 ESCs, 6,000 tetO-Gata4 ESCs, and 19,200 TSCs (per AggreWell) and transfer these to a 15 mL Falcon tube.
11. Centrifuge for 4 min at 0.2 × g (RCF).
12. In the meantime, aspirate the rinsing solution from the AggreWell and rinse the wells once with 1 mL of PBS, and once with 0.5 mL of equilibrated FC (**step 5**). Aspirate again and add 0.5 mL equilibrated FC.
13. Place the AggreWell plate back in the incubator until the cells are ready to be plated.
14. Add Rock inhibitor at a final concentration of 2 uM to the remaining medium in the incubator (1 mL per AggreWell) and use this to resuspend the cells at the end of the centrifugation (**step 11**).
15. Add cells to the AggreWell dropwise on the center of each well.
16. Centrifuge the plate for 3 min at 100 × g (RCF) and place the plate in the incubator (day 0).
17. The next day (day 1), cells should have come together and single cells are no longer observed under the microscope. On this day, perform two media changes. Begin by using a pipette to carefully remove 1 mL of media from each AggreWell. Gently replace this with 1 mL fresh FC equilibrated for ~20 min prior in the incubator (no Rock inhibitor added). The media changes need to be performed very gently exercising great care as not to disturb the structures forming within the wells. Media is removed and added slowly by placing the pipette tip against the side of the wells, avoiding formation of bubbles.

18. Place plate in the incubator for a minimum of 5 min before performing another media change as above.
19. On day 2, change media once with 1 mL of fresh FC medium (as in **step 17**).
20. On day 3, remove 1 mL media from AggreWell and add 1.5 mL of IVC-20% FBS equilibrated for ~20 min prior in the incubator.
21. On day 4, ETiX-embryoids in the AggreWell are transferred to CELLSTAR 6 well multiwell plate for suspension culture (Greiner Bio-One 657185). Prepare 5 mL of IVC-30% FBS per well.
22. Equilibrate IVC-30% FBS in the incubator for a minimum of 20 min.
23. All ETiX-embryoids are collected from AggreWell for analysis at 4 days of development and analyzed under a stereomicroscope. It is recommended to collect one individual AggreWell at a time. To collect cut the end of the tip off from a P1000 and gently pipette to dislodge structures from AggreWells. Collect in a Falcon tube. Wash each AggreWell with 1 mL PBS and combine this with the rest.
24. Add structures to a plastic dish and carefully examine the morphology. Select ETiX-embryoids with cylindrical morphology and two clearly defined cellular compartments (an ESC compartment and TSC compartment) surrounded by an outer cell layer, the VE-like layer. The ESC compartment should be epithelialized with a lumen. The TSC compartment is more variable in appearance and therefore, a wider range of appearances for the TSC compartment is acceptable (Fig. 1).
25. Transfer selected ETiX-embryoids with the ESC and TSC compartments surrounded by a VE-like layer to equilibrated IVC-30% FBS media to continue their culture.
26. When selecting at D5: (i) the ESC and TSC compartments should have a single merged lumen; (ii) the beginning of gastrulation on one side of the ETiX-embryoids should be apparent (a thickening of the proximal region of the ESCs on one side); (iii) the AVE should have migrated to the ESC-TSC boundary and be opposite to the forming streak. ETiX-embryoids with the AVE stuck at the tip of the structure or not at the boundary should be excluded. If the ESC line used contains a reporter line this can be used as an additional validation of asymmetry (Fig. 2).
27. On day 5, transfer each ETiX-embryoid to a single well of 48-well, non-adherent dish with 250 uL of DRH or EUCM medium.

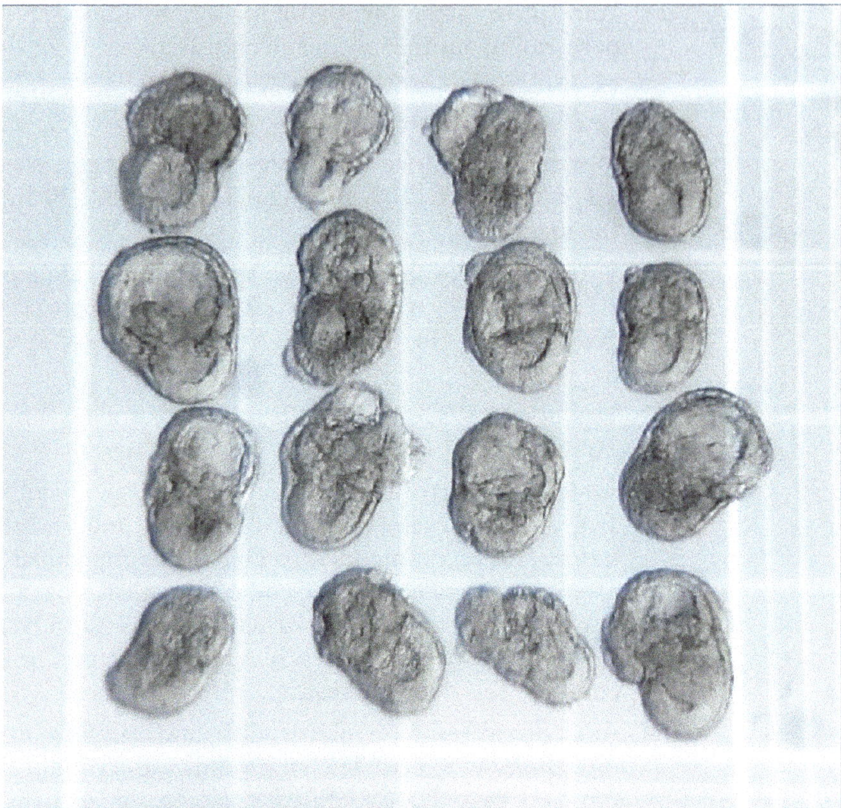

Fig. 1 A selection of ETiX-embryoids on day 4. ESC compartments are lumenized with a cup-shaped morphology while TSC compartments show greater morphological variability. A layer of VE-like cells enveloping the two compartments is clearly visible

28. On day 6, change media on each ETiX-embryoid using 250 uL of DRH/EUCM.
29. On day 7, ETiX-embryoids resemble E8.0 natural embryos with a clear anterior-posterior axis and bifurcating neural folds extending into a neural tube and culminating in a tail bud. Move the ETiX-embryoids to a rotating bottle culture chamber apparatus in DRH media supplemented with of 3.0 mg/mL of D-(+)-Glucose (Sigma G8644). Each rotating bottle contains 2 mL of medium and 3 ETiX-embryoids.
30. On day 8, change media to DRH medium supplemented with 3.5 mg/mL of D-(+)-Glucose. In each rotating bottle, culture 2 in 3 mL of medium.
31. At the end of day 8, dissect and fix the ETiX-embryoids in 4% PFA for 20 min at room temperature in preparation for downstream analyses (Fig. 3).

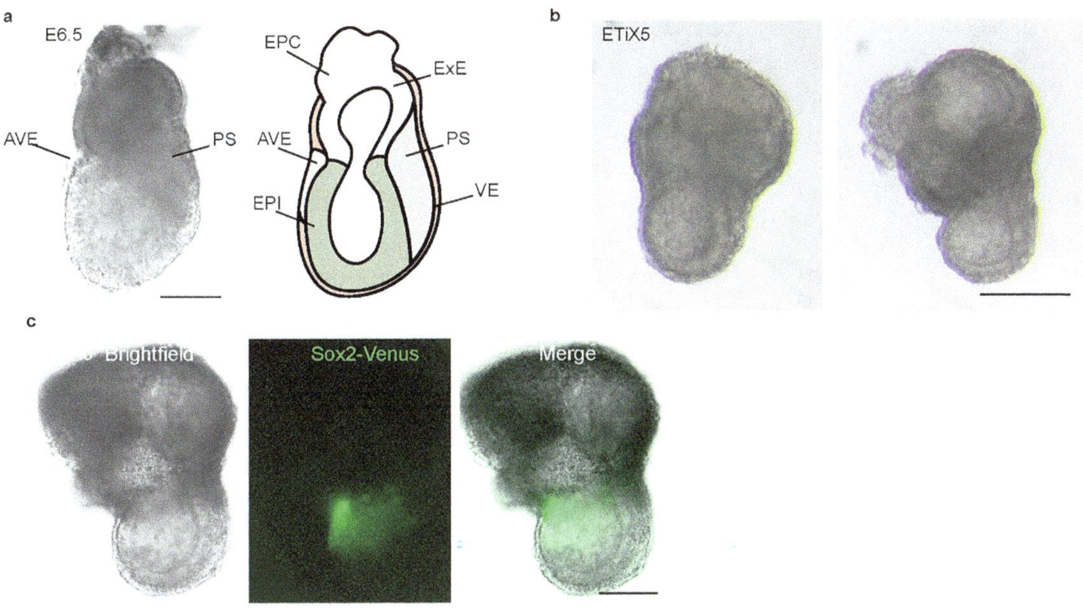

Fig. 2 (**a**) E6.5 natural embryo (left) and diagram highlighting key morphological features (right). (**b**) Representative examples of ETiX-embryoids selected at day 5. (**c**) ETiX-embryoid at day 5 made with ESCs containing a Sox2-Venus reporter. Sox2 signal displays a gradient of expression across the epiblast with the strongest signal apparent in the anterior

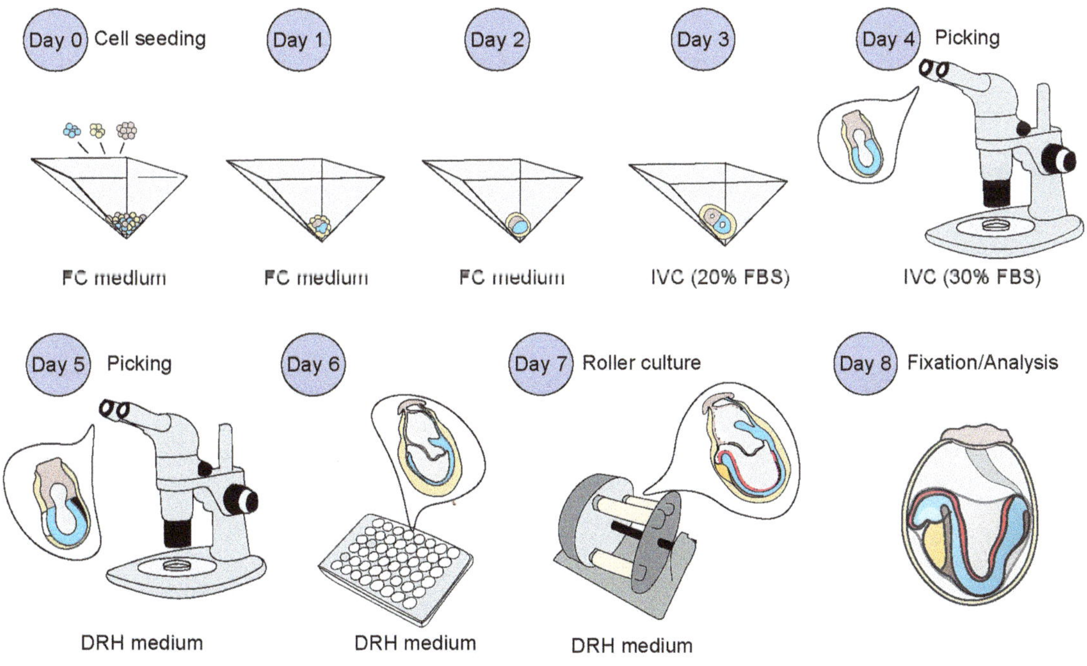

Fig. 3 Schematic depicting the preparation of ETiX-embryoids

4 Notes

1. Ensure cells are not displaying signs of stress (i.e., abnormal morphology, cell debris, excess differentiation or cell death) and are cycling normally prior to the beginning of an experiment. Passage at least once after thawing a fresh vial before using for experiment.
2. Where possible use cells with young passage numbers.
3. When passaging and prior to quenching the Trypsin activity ensure a single-cell suspension has been achieved. If unsure, check cells briefly under an optical microscope.
4. When returning the passaged cells to the incubator, close the door gently as vertical vibrational movements can transmit through the culture dish and cause unwanted cell crowding (particularly if incubator door contains magnetic strip). Avoid opening/closing the door of the incubator following passaging if possible.
5. For TSCs that are more than 50% confluent, perform media changes as follows: remove 50% of the media and top up with fresh TSF4H.

Conflict of Interest

A patent application has been submitted by MZG, GA, and CH based on these results.

References

1. ten Berge D et al (2008) Wnt signaling mediates self-organization and axis formation in embryoid bodies. Cell Stem Cell 3:508–518
2. van den Brink SC et al (2014) Symmetry breaking, germ layer specification and axial organisation in aggregates of mouse embryonic stem cells. Development 141:4231–4242
3. Xu P-F et al (2021) Construction of a mammalian embryo model from stem cells organized by a morphogen signalling centre. Nat Commun 12:3277
4. Beccari L et al (2018) Multi-axial self-organization properties of mouse embryonic stem cells into gastruloids. Nature 562:272–276
5. Veenvliet J et al (2020) Mouse embryonic stem cells self-organize into trunk-like structures with neural tube and somites. Science (1979) 370: eaba4937
6. Amadei G et al (2021) Inducible stem-cell-derived embryos capture mouse morphogenetic events in vitro. Dev Cell 56:366–382.e9
7. Bedzhov I, Leung CY, Bialecka M, Zernicka-Goetz M (2014) In vitro culture of mouse blastocysts beyond the implantation stages. Nat Protoc 9:2732–2739
8. Sturm K, Tam PP (1993) Isolation and culture of whole postimplantation embryos and germ layer derivatives. Methods Enzymol 225: 164–190

Methods in Molecular Biology (2024) 2767: 135–159
DOI 10.1007/7651_2023_491

Published online: 08 June 2023

Embryonic Spinal Cord Innervation in Human Trunk Organogenesis Gastruloids: Cardiac Versus Enteric Customization and Beyond

Zachary T. Olmsted, Maria Belen Paredes-Espinosa, and Janet L. Paluh

Abstract

Trunk-biased human gastruloids provide the ability to couple developmentally relevant spinal neurogenesis and organ morphogenesis via spatiotemporal self-organization events from derivatives of the three germ layers. The multi-lineage nature of gastruloids provides the full complexity of regulatory signaling cues that surpasses directed organoids and lays the foundation for an ex vivo self-evolving system. Here we detail two distinct protocols for trunk-biased gastruloids from an elongated, polarized structure with coordinated organ-specific neural patterning. Following an induction phase to caudalize iPSCs to trunk phenotype, divergent features of organogenesis and end-organ innervation yield separate models of enteric and cardiac nervous system formation. Both protocols are permissive to multi-lineage development and allow the study of neural integration events within a native, embryo-like context. We discuss the customizability of human gastruloids and the optimization of initial and extended conditions that maintain a permissive environment for multi-lineage differentiation and integration.

Key words Cardiac, Enteric, Gastruloid, Innervation, Multi-cellular, Multi-lineage, Organoid, Spinal cord

1 Introduction

Embryological morphogenesis with spatiotemporal self-organization of biological cells creates the foundational bottom-up process that underlies organogenesis and ultimately human development and can be mimicked ex vivo using stem cell technologies [1]. Human pluripotent stem cell-derived models of embryogenesis in a dish are revolutionizing the field of developmental biology and informing regenerative medicine therapies. While organoids remain the predominant stem cell model of the developing human brain [2], gastruloids that preserve the germ layers now dominate the study of the trunk [3]. This is due to the need to capture myriad interconnected organs and organ systems that are contained within, such as the spine, spinal cord, and paraxial musculature as well as the cardiovascular, cardiopulmonary, gastrointestinal, hepatobiliary, and urogenital systems, to name a few. Although it is currently difficult to capture the full gamut of these

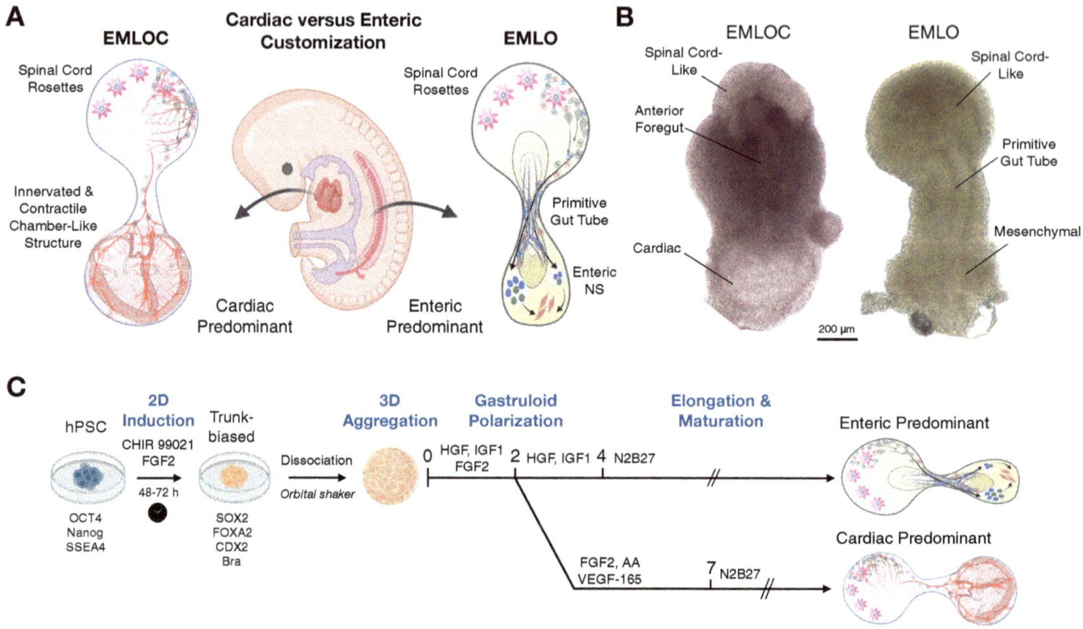

Fig. 1 Overview of cardiac versus enteric customization in trunk-biased EMLO and EMLOC gastruloids. (**a**) Schematic representation of EMLOC (left) and EMLO (right) gastruloid models with respect to the developing human embryo (center). (**b**) Phase contrast images of EMLOC (left) and EMLO (right) gastruloids matured to larger organoid-like multi-lineage systems. Labeled scale bar is provided. (**c**) Overview of protocols to generate EMLOs and EMLOCs from common initial steps

discrete yet interconnected systems within any single in vitro model from a human iPSC (hiPSC) starting point, key lineages and events with adjacent spatiotemporal developmental context share multi-lineage priming. In this chapter, we discuss two protocols to generate human trunk-biased gastruloids of enteric or cardiac nature that achieve customized morphogenic features and neural patterning phenotypes following establishment of the multi-cellular, multi-lineage context required for proper induction, patterning, and self-organization (Fig. 1a, b) [4, 5]. The two protocols begin from a common starting point for axial induction of trunk phenotype by FGF2 and the Wnt agonist CHIR 99021, but they are subsequently distinguished by signaling factor type and temporal exposure profile to determine the final multi-cellular composition and architecture (Figs. 1c and 2). The resulting cellular aggregates are referred to in the literature as elongating multi-lineage organized (EMLO) gastruloids that have neuro-enteric predominance [4] and EMLO cardiac-optimized (EMLOC) gastruloids that exhibit hallmark morphogenic features of cardiogenesis, chamber formation, and contractility along with patterned neural integration [5]. Both trunk-biased systems share developmental processes of polarization, elongation, and stereotyped multi-lineage specialization with neural patterning. However, EMLO

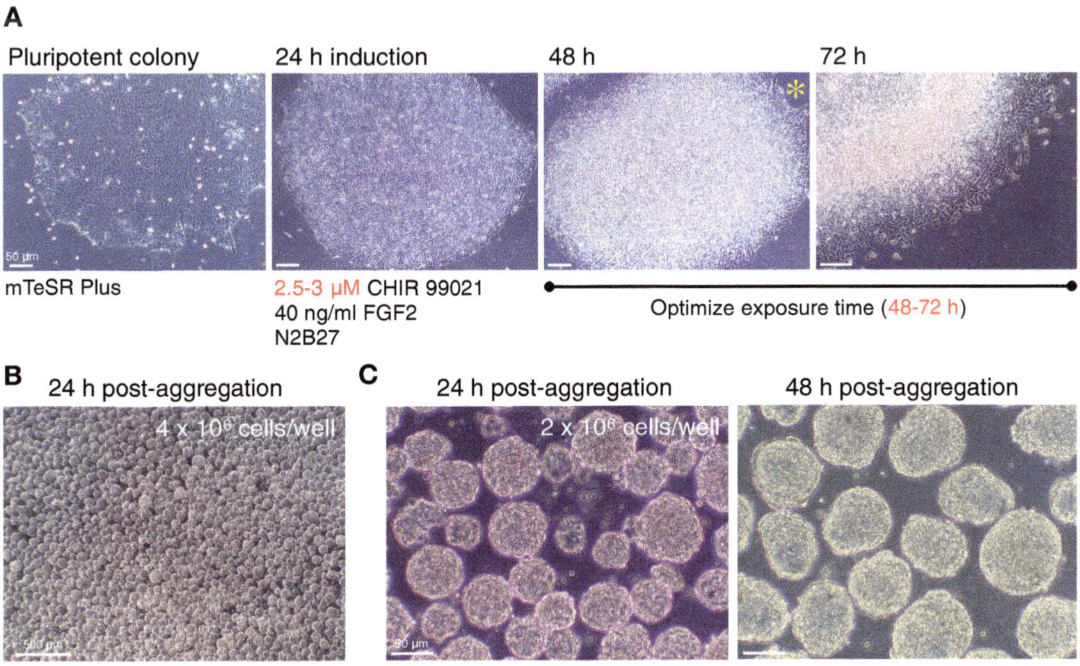

Fig. 2 EMLO and EMLOC gastruloid formation proceeds through common initial conditions. (**a**) Phase-contrast images of 2D induction from intact pluripotent colonies in mTeSR Plus (leftmost image) to trunk-biased axial progenitors at 24, 48, and 72 h (right images). Colonies are induced by CHIR 99021 and FGF2 in N2B27 basal medium. Optimal time of exposure prior to single-cell dissociation, and aggregation is cell-line dependent (48–72 h). Yellow asterisk indicates ideal colony morphology for dissociation and EMLO/EMLOC formation. CHIR 99021 should also be optimized between 2.5 and 3 μM. (**b**) Low magnification phase-contrast image of initial aggregates formed by single cell suspensions on orbital shaker. Aggregate density shown is from 4×10^6 cells/well in low-adhesion 6-well plates. Aggregate size range from 50 to 100 μm (50–100 cells/aggregate) is necessary for optimal EMLO/EMLOC formation in subsequent steps. (**c**) High-magnification phase-contrast image of aggregates formed by single-cell suspensions on orbital shaker at 24 h (left) and 48 h (right) post-aggregation. Aggregate density shown is from 2×10^6 cells/well in low-adhesion 6-well plates [5]. Initial polarized morphology can be observed at 48 h. Labeled scale bars are provided

and EMLOC gastruloids differ in final cellular composition, predominant organ precursor type, and ultimate application.

Neuro-enteric EMLO gastruloids are initiated from trunk-biased hiPSCs that aggregate spontaneously in orbital shaking culture without embedding in hydrogel (Fig. 2). Exposure to FGF2, HGF, and IGF-1 over four days beginning at the time of aggregation drives polarization into two general domains with neural and mesendodermal fate [6]. After removal of growth factors, EMLOs differentiate and mature in N2B27 basal medium and can be maintained at least until day 40 (Fig. 3). By day 16, EMLOs exhibit remarkable cellular diversity and organization as assessed by visual inspection, immunofluorescence (IF), and single-cell RNA sequencing (scRNAseq) for cell and tissue type-specific gene expression profiles (Figs. 4 and 6a). A primitive gut tube forms

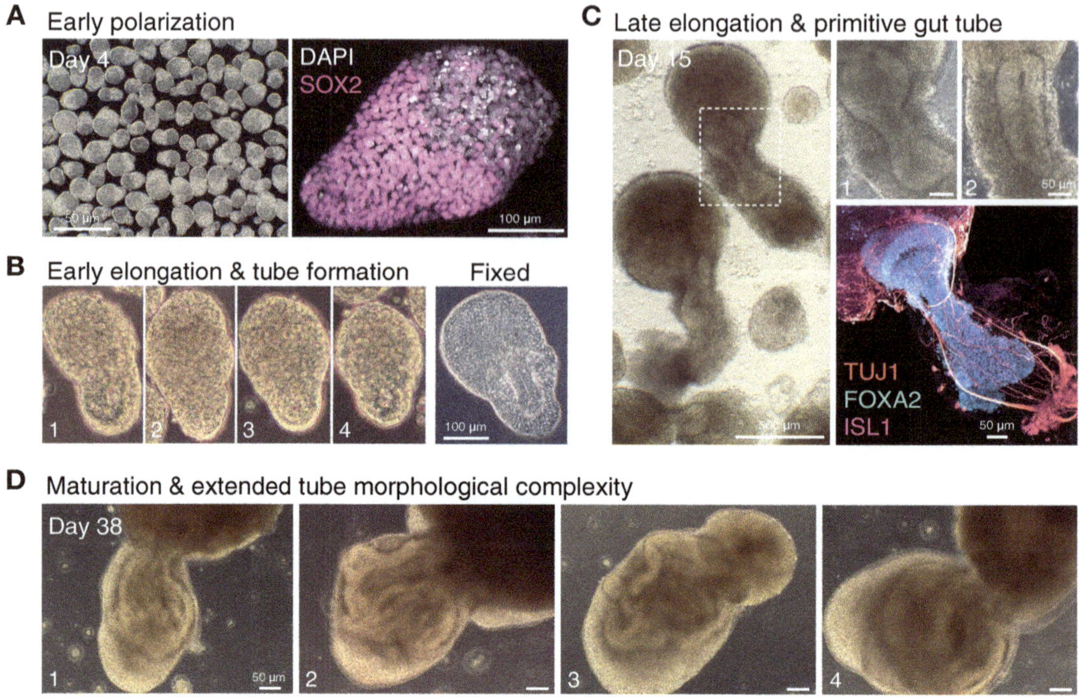

Fig. 3 EMLO initial polarization and formation hallmarks by visual inspection and immunofluorescence. (**a**) Low-magnification phase-contrast image of polarized EMLO aggregates at day 4 post-aggregation in orbital shaker culture (left). Morphological signs of polarization are mirrored by IF biomarker expression. SOX2 is shown (right). (**b**) High-magnification phase-contrast image of polarized EMLO gastruloid elongation and tube formation. Four live-cell examples are shown (left), along with a formalin-fixed sample (right). (**c**) Late elongation and primitive gut tube in day 15 EMLO gastruloids. Low-magnification (left) and high-magnification (top right) phase-contrast images of gut tube are shown. Total Z-stack projection of EMLO using TUJ1, FOXA2, and ISL1 biomarkers demonstrates neuronal organization around this structure (bottom right). (**d**) Maturation and extended gut tube morphological complexity in day 38 EMLO gastruloids. Four live-cell examples are shown. Labeled scale bars are provided

and elongates, bound by basement membrane that surrounds a central lumen (Fig. 3b–d). The gut tube associates with peripheral neurons derived from neural crest cells that migrate out of the neural compartment in which central spinal neurons also develop and reside (Fig. 4). Peripheral neurons organize into ganglionic architectures, similar to enteric nervous system organization, within a mesenchymal cellular milieu surrounding the gut tube (Fig. 4e). In this manner, the EMLO gastruloids capture diverse lineages including definitive endoderm, spinal cord neural progenitors, central spinal neurons, neural crest, splanchnic mesenchyme, and ectoderm (Fig. 6a). Within each lineage, additional subtype diversity is achieved such as the array of ventral/dorsal spinal interneurons, motor neurons, and sensory neurons along with both ectodermal and non-ectodermal neural crest cell derivatives [4].

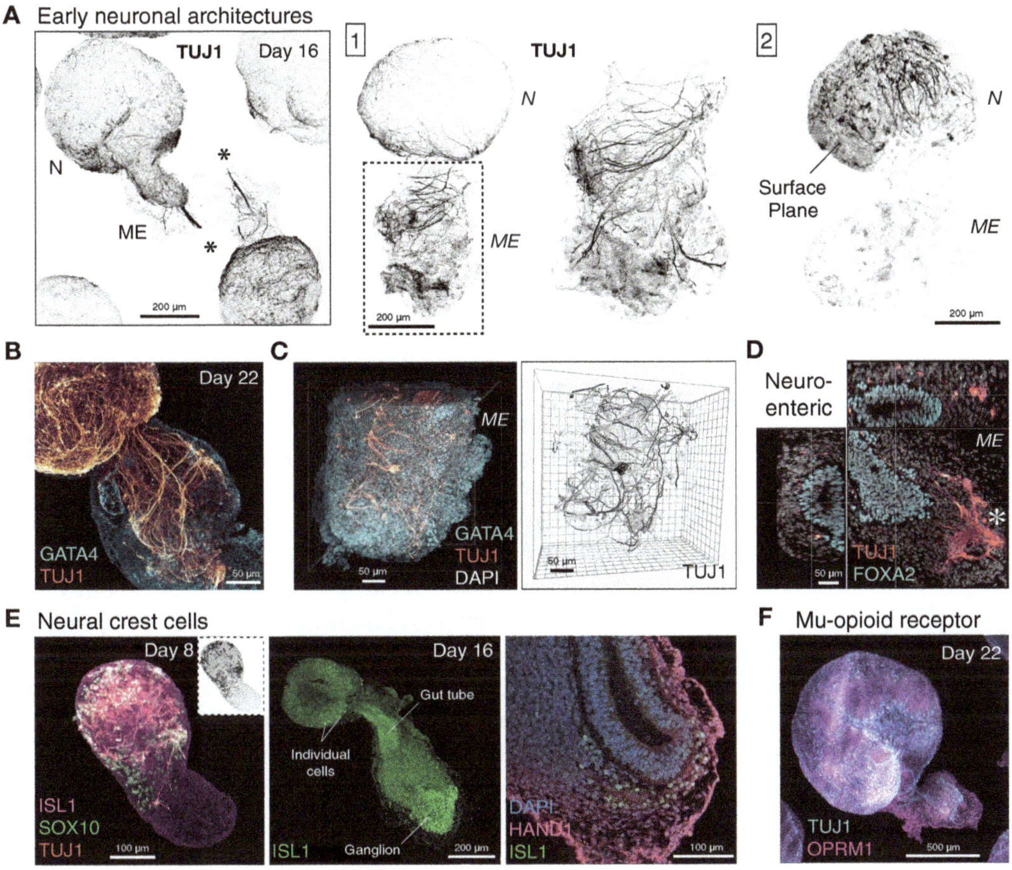

Fig. 4 EMLO-specific features of neuro-enteric integration through neural crest cells. (**a**) Early neuronal architectures (TUJ1) form around the primitive gut tube in EMLO gastruloids. A full field of day 16 EMLOs is shown along with two isolated samples demonstrating neuronal organization in the mesendodermal (*ME*) compartment (#1, left; shown with high magnification) and along the surface of the neural (*N*) compartment (#2, right). (**b**) Neurons integrate within the mesendodermal tissue shown with TUJ1 and GATA4 biomarkers, respectively. (**c**) 3D Imaris reconstruction of the ME compartment and neurons by GATA4 and TUJ1 (left). TUJ1 shown with inverted LUT (right). (**d**) Multi-dimensional visualization of primitive gut tube (FOXA2) and surrounding neurons (TUJ1) demonstrates close neuro-enteric interaction and ganglion formation (white asterisk). Total EMLO Z-stack projection is provided above in Fig. 3c. (**e**) SOX10+ NCCs originate in the neural compartment and migrate anteriorly from polarized neural to mesendodermal compartments. ISL1 and TUJ1 co-stain shown. Inset is SOX10 inverted LUT (day 8 EMLO, left). A subset of NCCs generate TUJ1+ neurons. ISL1+ cells localize to the neural compartment (individual cells), segments of the gut tube, and distal ganglion pool in day 16 EMLO (low-magnification IF image, center). High-magnification IF image depicts gut tube ISL1 signal and surrounding ISL1+ cells within HAND1+ mesenchyme (right image). (**f**) A neuronal subset in day 22 EMLO gastruloids shows mu-opioid receptor OPRM1 co-expression with TUJ1 similar to sensory and enteric organs of the trunk and spine. Labeled scale bars are shown

The EMLO gastruloid thereby achieves a combined human central-peripheral nervous system model of early organogenesis with enteric predominance and enables the study of neuro-enteric interacting events with multi-lineage contributions.

In vivo, the anterior foregut acts as a developmental organizer of cardiogenesis wherein the first and second heart fields along with other contributing cell types arise from splanchnic mesoderm [7]. Since these features are captured in EMLO gastruloids, tailored changes to signaling cues and temporal exposure profile can be made to optimize cardiogenesis in EMLOCs [4, 5]. In this protocol [8], hiPSCs are induced and aggregated under the same conditions (Figs. 1c and 2). However, from day 2 to day 7 after aggregate formation, polarization medium is discontinued and exchanged for cardiac induction medium containing FGF2, VEGF-165, and ascorbic acid that are known angiocrine and pro-cardiogenic factors [9], followed by maturation in N2B27 with no added factors. By this approach, a cardiac region originates anterior to gut tube primordium and recapitulates aspects of early human heart development such as heart tube formation and specialization, cardiomyocyte differentiation and morphogenetic remodeling phases, epicardial development, and ventricular wall morphogenesis in chamber-like structures (Fig. 5). The EMLOC gastruloid system enables the tracking of multiple events in cardiomyocyte differentiation from splanchnic mesoderm to spontaneous contractility and chamber maturation. EMLOC gastruloids also capture distinct neural, cardiac, and epithelial lineages including the following: neuroectoderm and spinal cord neural progenitors, spinal neurons, peripheral glia, cardiomyocytes, epicardial cells, cardiac fibroblasts, anterior foregut, intermediate mesoderm, metanephric mesoderm, genitourinary/renal epithelium, and vascular endothelial cells (Fig. 6b). Similar to EMLO gastruloids, the neuronal classes generated consist of autonomic, sensory, and motor neurons but with a lower proportion of motor neurons in EMLOCs. The cardiogenic region is progressively populated by neurons in a spatial architecture that mirrors the known distribution of neurons in the innervated human heart [5] (Fig. 5f), yielding a valuable model for interrogating autonomic neuronal-cardiac interactions that are only primitively defined with respect to the skeletal neuromuscular junction.

The two protocols detailed in this chapter are initiated through a common induction method for trunk-biasing but are then coaxed toward separate outcomes [8, 10]. Shared features such as polarization and elongation give way to divergent features of organogenesis type and end-organ innervation. Small spheroids evolve into highly specialized cellular assemblies with neuro-enteric versus neuro-cardiac predominance. A hallmark feature of both strategies is the creation of a permissive environment to allow neurogenesis and non-neural organogenesis that in other protocols are mutually exclusive. While we emphasize the neural integration patterns, this does not imply the exclusion of other cell or tissue types such as mesenchyme by our strategy. The trunk-biased gastruloid approach therefore offers the desirable customizability of phenotypes, given that the enabling conditions are discovered and optimized.

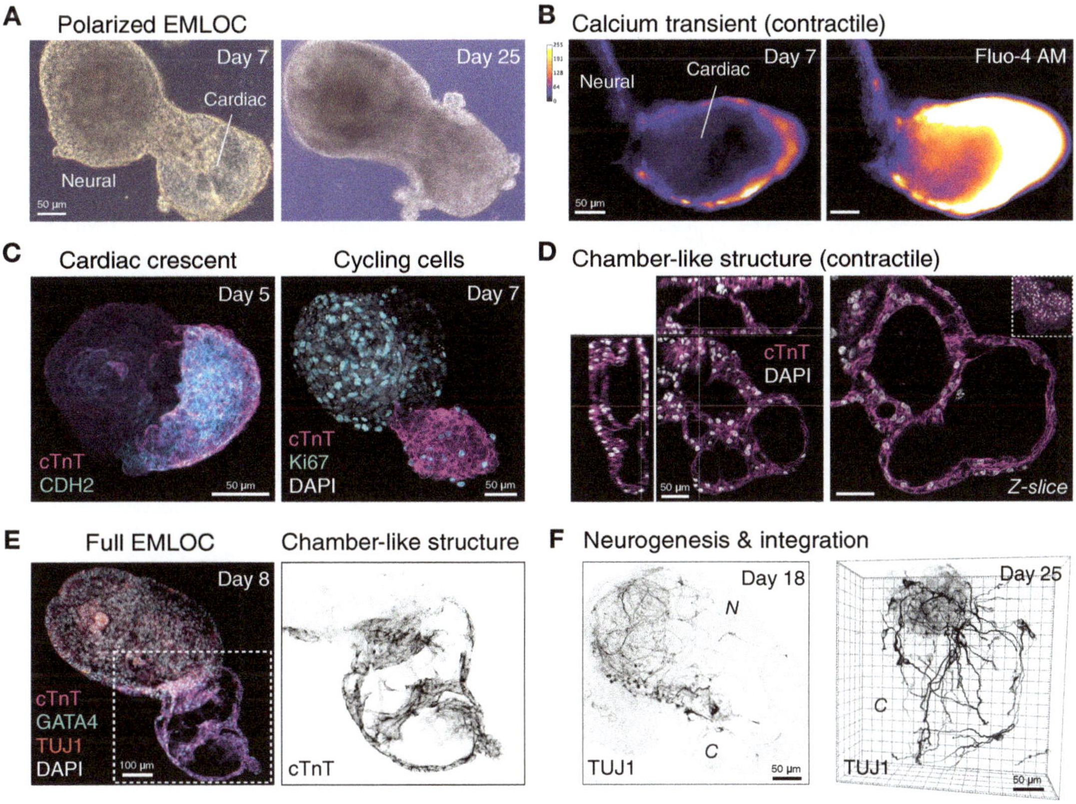

Fig. 5 EMLOC-specific features of neuro-cardiac integration through chamber-like structures. (**a**) Phase-contrast images of polarized EMLOC with contractile cardiogenic compartment at days 7 (left) and 25 (right). (**b**) Fluo-4 AM live-cell calcium transient in contractile cardiogenic region of day 7 EMLOC before (left) and during contraction (right) (Thermo Fisher Scientific, cat. #F14201). (**c**) Cardiogenesis in EMLOC gastruloids proceeds through a cTnT+, CDH2+ cardiac crescent-like structure (day 5, left). A subset of cells remains cycling by Ki-67 in neural and elongated cardiogenic compartments (day 7, right). (**d**) Septated chamber-like structures shown in two Z-planes with multi-dimensional visualization (left) and 3D surface reconstruction (right inset). (**e**) Low-magnification Z-slice of cTnT+, GATA4+ chamber-like structure at timepoint of early neurogenesis (day 8; left). High-magnification chamber-like structure is shown with inverted cTnT immunostain (right). (**f**) TUJ1+ neurons project from neural compartment (*N*) into cardiac compartment (*C*). Inverted TUJ1 signal shown in day 18 (left) and day 25 EMLOCs (right; Imaris reconstruction) [5]. Labeled scale bars are shown

2 Materials

2.1 Pluripotent Stem Cells, Culture, and Passaging

1. hiPSC cell lines: H3.3.1, H3.1.1 [11, 12].
2. hESC-qualified Matrigel (Corning, cat. #08-774-552); thaw on ice overnight in a refrigerator at 4 °C and maintain on ice during handling to prevent premature solidification.
3. DMEM/F-12 (Thermo Fisher Scientific, cat. #11320033); store at 4 °C.

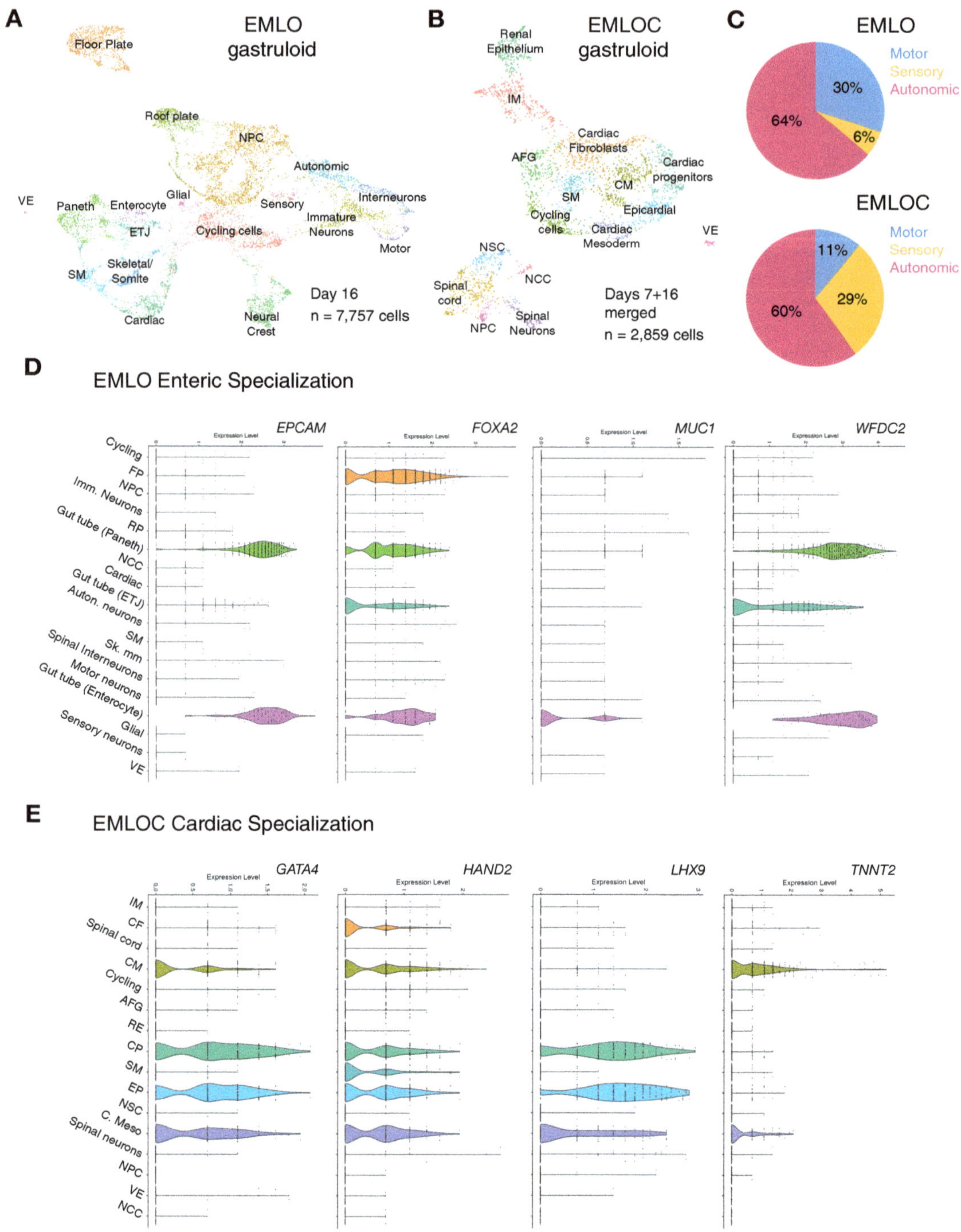

Fig. 6 Permissive differentiation conditions establish distinct specialized lineages in EMLO and EMLOC gastruloids assessed by scRNAseq. (**a**) UMAP plot of scRNAseq data from day 16 EMLO gastruloids ($n = 7757$ cells). (**b**) UMAP plot of merged scRNAseq data from day 7 and day 16 EMLOC gastruloids ($n = 2859$ cells total). (**c**) Relative proportion of neuronal classes generated in EMLO (top) versus EMLOC

4. mTeSR Plus (STEMCELL Technologies, cat. #05825); store at 4 °C and warm aliquot to room temperature without water bath prior to use.
5. Gentle Cell Dissociation Reagent (STEMCELL Technologies, cat. #07174); store at room temperature.
6. mFreSR (STEMCELL Technologies, cat. #05854).
7. 35 mm tissue culture-treated dish (CELLTREAT, cat. #229635).
8. 6-well plates (CELLTREAT, cat. #229105).

2.2 EMLO and EMLOC Gastruloid Formation (General Reagents)

1. HBSS CM-free (Thermo Fisher Scientific, cat. #14175079); store at 4 °C.
2. Anti-Adherence Rinsing Solution (STEMCELL Technologies, cat. #07010); store at room temperature.
3. Accutase (STEMCELL Technologies, cat. #07920); store at 4 °C.
4. 6-well plates (CELLTREAT, cat. #229105).
5. 100 mm petri dish (Fisher, cat. #80826).

2.3 EMLO and EMLOC Gastruloid Formation (Differentiation Media)

Refer to the original sources for additional images and discussion [4, 5]

1. *N2B27 basal medium*: 1:1 DMEM/F-12 (Thermo Fisher Scientific, cat. #11320033), Neurobasal Plus (Thermo Fisher Scientific, cat. #A3582901), with 2% B-27 Plus (Thermo Fisher Scientific, cat. #17504044), 1% N-2 (Thermo Fisher Scientific, cat. #17502048), 1× GlutaMAX (100×) (Thermo Fisher Scientific, cat. #35050061), 1× MEM Non-Essential Amino Acids (100×) (Thermo Fisher Scientific, cat.

Fig. 6 (continued) gastruloids (bottom) based on scRNAseq. *TUBB3+* neuronal cells were quantified along with co-expressed *MNX1+* (motor class), *POU4F1+* (sensory class), *ASCL1+* (autonomic class). (**d**) Violin plots of gene expression across all clusters in EMLOs depict enteric endodermal specialization. *EPCAM*, *FOXA2*, *MUC1*, and *WFDC2* gene expressions are shown. Clusters: cycling cells, floor plate (FP), neural progenitor cells (NPC), immature neurons, roof plate (RP), gut tube (paneth cell type), neural crest cells (NCC), cardiac-type, gut tube (esophageal-tracheal junction, ETJ), autonomic neurons, splanchnic mesoderm (SM), spinal interneurons, spinal motor neurons, gut tube (enterocyte type), glial cells, sensory neurons, and vascular endothelial cells (VE). (**e**) Violin plots of gene expression across all clusters in EMLOCs depict cardiac specialization. *GATA4*, *HAND2*, *LHX9*, and *TNNT2* gene expressions are shown. Clusters: intermediate mesoderm (IM), cardiac fibroblasts (CF), spinal cord/spinal radial glia, cardiomyocytes (CM), cycling cells, anterior foregut (AFG), renal epithelium (RE), cardiomyocyte progenitors (CP), splanchnic mesoderm (SM), epicardial cells (EP), trunk neural stem cells (NSC), cardiac mesoderm (C. Meso), spinal neurons, neuronal progenitor cells (NPC), vascular endothelial cells (VE), and neural crest cells/peripheral glia (NCC/Schwann cells)

#11140050), and 1× Penicillin-Streptomycin (100×) (Thermo Fisher Scientific, cat. #15140122) in a sterile container. Filter-sterilize the medium using a 0.22 μm filter. The prepared N2B27 basal medium can be stored at 4 °C for up to 2 weeks.

Reagent	Final concentration	Amount
DMEM/F-12	n/a	235 mL
Neurobasal Plus	n/a	235 mL
B-27 Plus	2%	10 mL
N-2	1%	5 ml
GlutaMAX (100×)	1×	5 mL
MEM Non-Essential Amino Acids (100×)	1×	5 mL
Penicillin-Streptomycin (100×)	1×	5 mL
Total	n/a	500 mL

2. *Axial Induction Medium*: N2B27 basal medium supplemented with 40 ng/mL FGF2 (100 μg/mL stock) (R&D Systems, cat. #233-FB) and 2.5–3 μM CHIR 99021 (10 mM stock; optimize concentration per cell line) (Tocris Bioscience, cat. #4423). Axial Induction Medium should be prepared fresh with supplements, usually 10 mL at a time. During the 48 h treatment period, replace the entire medium at the 24 h mark to ensure optimal conditions.

Reagent	Final concentration	Amount
N2B27	n/a	10 mL
FGF2	40 ng/mL (100 μg/mL stock)	4 μL
CHIR 99021	3 μM (10 mM stock)	3 μL
Total	n/a	10 mL

3. *Polarization Medium*: N2B27 basal medium supplemented with 10 ng/mL FGF2 (100 μg/mL stock), 2 ng/mL IGF1 (100 μg/mL stock) (R&D Systems, cat. #291-G1), 2 ng/mL of HGF (100 μg/mL stock) (R&D Systems, cat. #294-HG), and 50 μM Y-27632 (10 mM stock) (Tocris Bioscience, cat. #1254). Polarization Medium should be prepared fresh with supplements, usually 10–20 mL at a time. During the treatment period, replace one-half volume of the aggregation medium at the 24 h mark. The replacement medium should be modified to contain 2× the concentration of FGF2, IGF1, and HGF to maintain steady-state levels but excluding Y-27632. For EMLO gastruloids, Polarization Medium is

added without FGF2 from 48 h post-aggregation to the end of treatment at 96 h (2 ng/mL IGF1 and 2 ng/mL HGF only). For EMLOC gastruloids, Polarization Medium is exchanged for Cardiac Induction Medium at 48 h post-aggregation, as detailed below.

Reagent	Final concentration	Amount
N2B27	n/a	10 mL
FGF2	10 ng/ml (100 μg/mL stock)	1 μL
IGF1	2 ng/mL (100 μg/mL stock)	200 nL
HGF	2 ng/mL (100 μg/mL stock)	200 nL
Y-27632	50 μM (10 mM stock)	50 μL
Total	n/a	10 mL

4. *Cardiac Induction Medium (EMLOC-specific)*: N2B27 basal medium supplemented with 30 ng/mL FGF2 (100 μg/mL stock), 0.5 mM of ascorbic acid (50 mM stock) (Tocris Bioscience, cat. #4055), and 5 ng/mL VEGF-165 (100 μg/mL stock) (R&D Systems, cat. #293-VE). Cardiac Induction Medium should be prepared fresh with supplements, usually 10–20 mL at a time. At 48 h post-aggregation, perform a full medium change to Cardiac Induction Medium. Maintain the EMLOCs in this medium for a total of 5 d, with one medium change after 72 h or longer from the first addition (5 d post-aggregation).

Reagent	Final concentration	Amount
N2B27	n/a	9.9 mL
FGF2	30 ng/mL (100 μg/mL stock)	3 μL
Ascorbic acid	0.5 mM (50 mM stock)	100 μL
VEGF-165	5 ng/mL (100 μg/mL stock)	0.5 μL
Total	n/a	10 mL

2.4 Immunofluorescence Buffers

1. Phosphate-buffered saline (1 L of 1× PBS): dissolve 8 g of sodium chloride (NaCl) in 800 mL deionized water. Next dissolve 0.2 g of potassium chloride (KCl), 1.44 g of disodium hydrogen phosphate (Na_2HPO_4), and 0.24 g of monopotassium phosphate (KH_2PO_4). Adjust the pH to 7.4. Add distilled water to a final volume of 1 L. Filter-sterilize using 0.22 μm filter and store at room temperature.

2. Fixation solution: 10% neutral-buffered formalin; store at room temperature.
3. Permeabilization solution: 1× PBS containing 2% Triton-X-100 (PBST); store at 4 °C.
4. Blocking solution: 1% bovine serum albumin (BSA) in PBST; store at 4 °C.
5. Phosphate buffer (0.1 M PB): 0.025 M NaH_2PO_4, 0.075 M Na_2HPO_4, pH 7.4 containing 0.1% Triton-X-100; store at room temperature; dilute stock 0.1 M PB for 0.02 M PB used in some buffers; store at 4 °C.
6. Clearing solution: 88% Histodenz (w/v) (Sigma-Aldrich, cat. #D2158) in 0.02 M PB; store at room temperature.
7. Primary antibodies [4, 5].
8. Secondary antibodies [4, 5].
9. NucBlue fixed cell ReadyProbes reagent (Thermo Fisher Scientific, cat. #R37606).
10. Glass microscope slides.
11. 22 × 22 mm coverslip.
12. Clear nail polish.
13. 12-well plates.

2.5 Single-Cell RNA Sequencing

1. *Cold-Activated Protease Solution*: dissociation solution containing psychrophilic, cold-activated *Bacillus licheniformis* protease is used to quickly dissociate integrated samples to single cells for scRNAseq analysis with reduced artifacts [13]. Single cells are stored in CryoStor CS10 cryopreservation medium at −80 °C and then in liquid nitrogen prior to sequencing, or can be directly sequenced without freezing steps.

Reagent	Final concentration	Amount
1× PBS	n/a	9.4 mL
Psychrophilic *Bacillus licheniformis* protease	10 mg/mL	100 mg
DNase	125 U/mL	1250 U
$CaCl_2$	5 mM (100 mM stock)	0.5 mL
Total	n/a	10 mL

2. 1× PBS containing 10% fetal bovine serum (FBS) (Thermo Fisher Scientific, cat. #26140079); store at 4 °C.
3. CryoStor CS10 cryopreservation medium (BioLife Solutions, cat. #210373); store at 4 °C.

4. RPMI1640 (Thermo Fisher Scientific, cat. #11875093) containing 10% fetal bovine serum (FBS) (Thermo Fisher Scientific, cat. #26140079); store at 4 °C.
5. NextGem v3.1 single cell reagent kit.
6. Roche KAPA universal qPCR kit.
7. Nunc cryostorage tubes.
8. Nalgene Mr. Frosty freezing container.
9. 40 μm cell strainer.

2.6 Equipment

1. CHEMcell VERSA-ORB2 orbital shaker.
2. Zeiss Invertoskop 40 C mounted with an Olympus DP22 color camera.
3. Leica confocal TCS SP5 II system.
4. Logos Biosystems LUNA II.
5. 10× Genomics Chromium platform.
6. Agilent Fragment Analyzer.
7. NovaSeq 6000 (28 × 91).

2.7 Software

1. cellSens acquisition software.
2. Leica Application Suite Advanced Fluorescence software (v2.7.3.9723).
3. Imaris image analysis software (https://imaris.oxinst.com/).
4. 10× Genomics Cell Ranger analysis software (V4.0.0).
5. Seurat (Satija lab, https://satijalab.org/seurat/).
6. 10× Genomics Loupe browser freeware.

3 Methods

3.1 Coating Cultureware with hESC-Qualified Matrigel

Time: 1 h.
Duration: 24 h

1. Store and handle hESC-qualified Corning Matrigel according to manufacturer's instructions. We prefer to aliquot 100 μL of undiluted Matrigel on ice and store at −20 °C for later use.
2. UV-sterilize a 35-mm culture dish with the top off in a laminar flow tissue culture hood for 30 min.
3. Pre-chill the sterilized dish at −20 °C for ~10 min prior to use.
4. Thaw a single aliquot of Matrigel on ice. The remaining Matrigel can be stored short term at 4 °C.
5. Dilute Matrigel stock to a working concentration of 1:100 in ice-cold DMEM/F-12 medium (e.g., 20 μL Matrigel stock in 2 mL DMEM/F-12).

6. Add 1 mL of diluted Matrigel per chilled dish.
7. Incubate the dish at 37 °C in a humidified incubator with 5% CO_2 for at least 1 h prior to use in order to polymerize the surface coating (*see* **Note 1**).

3.2 hiPSC Culture and Pluripotency Maintenance

Time: 15 min.
Duration: ~3–7 days

1. Thaw hiPSCs cryopreserved in mFreSR cryopreservation medium by transferring a vial containing 1 mL of cell suspension from liquid nitrogen storage to a 37 °C water bath.
2. While the cell suspension is thawing, remove DMEM/F-12 from the 35 mm dish freshly coated with Matrigel. Rinse 1× with 1 mL of DMEM/F-12 and replace with 2.5 mL of mTeSR Plus hiPSC pluripotency medium supplemented with 1× penicillin-streptomycin (unless otherwise specified in this protocol) (*see* **Note 2**).
3. In a laminar flow tissue culture hood, carefully add 1 mL of thawed cells in mFreSR to the fresh mTeSR Plus (3.5 mL total) in a dropwise manner, ensuring an even distribution of cells across the dish's surface.
4. Incubate the seeded cells overnight at 37 °C in a humidified incubator with 5% CO_2.
5. The following day, visually inspect for stem cell colony adherence, and then rinse the cells twice with 1 mL of prewarmed DMEM/F-12 from a 37 °C water bath.
6. Add 2 mL of room-temperature mTeSR Plus to the cells and return them to the incubator for expansion.
7. Monitor the cells visually for proper colony. Replace mTeSR Plus every other day or as required until the cultures reach the desired confluency for routine passaging (~70% confluency). This is slightly more confluent than hiPSC cultures used for axial induction (~50–60% confluency). Aim for well-defined, undifferentiated colony edges and minimal colony-to-colony contact.

3.3 Passaging of hiPSCs Prior to Induction

Time: 30 min.
Duration: 2–4 days.

This description is for passaging from a 35 mm culture dish to 6-well plate.

1. Remove media from the well to be passaged and immediately add 1 mL of Gentle Cell Dissociation Reagent (GCDR) to

the empty well. Incubate at room temperature for ~3 min (*see* **Notes 3** and **4**).

2. Gently aspirate the GCDR without dislodging the cells and immediately add 3 mL mTeSR Plus to the well that is being passaged.
3. Dislodge colonies by using a 5 mL serological pipette oriented orthogonally to the plate's plane. Perform side-to-side scraping motions across the entire well area. Rotate the plate 90° and repeat the scraping to ensure that the majority of cells are dislodged.
4. Pipette the suspension up and down twice using a P-1000 blue tip to further reduce clump size.
5. Gently transfer 0.5 mL of dislodged cells to each well of a Matrigel-treated 6-well plate already containing 1 mL of mTeSR Plus. Add cells dropwise quickly and serially to achieve an even distribution in each well (1.5 mL total). Follow the detailed Matrigel coating protocol for a 35 mm culture dish, treating each well of a 6-well plate equivalently.
6. Incubate the plate overnight, undisturbed at 37 °C to allow colonies to settle and adhere.
7. The next day, visually inspect cultures for adherence of small colonies. If positive, remove the 1.5 mL mTeSR Plus.
8. Rinse 2× in 1 mL DMEM/F-12 to remove non-adherent cell debris.
9. Add 2 mL fresh mTeSR Plus to each well.
10. Return the plate to the incubator. Perform fresh media changes every 2 d until the colonies are expanded and cultures are ~50–60% confluent. This confluency percentage is optimal for cell induction to generate gastruloids. Non-induced wells can be maintained in mTeSR Plus medium and used for further passaging to a separate 6-well plate for future experiments.

3.4 2D Induction of hiPSC Colonies to Axial Trunk-Biased Phenotype

Time*:* 20 min.

Duration: 2–3 days.

hiPSC colonies at ~50–60% confluency in single wells of a 6-well plate are ready for 2D induction (Fig. 2a).

1. Prepare working volume (2 mL per well) of Axial Induction Medium (N2B27 with 3 μM CHIR 99021, 40 ng/mL FGF2) and prewarm to room temperature without a water bath (*see* **Note 5**).
2. Remove the mTeSR Plus medium and rinse cells 2× with DMEM/F-12.
3. Add 2 mL of Axial Induction Medium to the hiPSC wells designated for gastruloid generation and return plate to incubator.

4. After 24 h, exchange the medium with 2 mL fresh Axial Induction Medium and return plate to incubator for an additional 24 h.

3.5 Single-Cell Suspension, Transition to Shaking Culture, and Early Polarization

Time: 45 min.
Duration: 2 days.

Forty-eight hours after the initial induction as 2D colonies in Axial Induction Medium (Fig. 2a), cells can be used to generate 3D aggregates (Fig. 2b, c).

1. Prepare a fresh low-adhesion 6-well plate by treatment with 1 mL Anti-Adherence Rinsing Solution per well for 10 min.
2. Remove the Anti-Adherence Rinsing Solution from the wells and rinse 2× with 1 mL HBSS (CM-free). On the second rinse, do not remove the HBSS to prevent drying.
3. Remove the Axial Induction Medium from the wells containing primed colonies and rinse 2× with 1 mL HBSS (CM-free).
4. Dilute Accutase 1:1 with HBSS (CM-free) for enzymatic dissociation to single cells and add 1 mL per well. Incubate at 37 °C for 5 min (cell line-dependent incubation time).
5. Gently remove dissociation solution, add 1 mL of N2B27 (no supplements) to the empty well, and dislodge cells using a side-to-side scraping motion with a serological pipette. Triturate the suspension manually with a P-1000 pipette (~6 times) to generate a single-cell suspension (*see* **Note 6**).
6. Combine 2–3 Accutase-treated hiPSC wells and transfer the volume to a 15-mL conical tube and centrifuge at 350 × g for 5 min to pellet the single cells.
7. Aspirate the supernatant, resuspend the cell pellet in Polarization Medium (N2B27 with 10 ng/mL FGF2, 2 ng/mL IGF1, 2 ng/mL HGF, 50 μM ROCK inhibitor Y-27632) at the appropriate cell density, remove remaining HBSS (CM-free) from treated wells, and transfer the cell suspension to the pre-treated low-adherence well plate (*see* **Note 7**).
8. Place the plate on an orbital shaker at 80 rpm clockwise in a humidified incubator with 5% CO_2.
9. Visually inspect the orbital shaking cultures at 24 h post-aggregation. At this stage, round aggregates of similar size distribution (~50–100 μm) should be visible (Fig. 2b). To change the culture medium, pool the aggregates in a 15 mL conical tube and allow them to settle by gravity for ~10 min (*see* **Note 8**).
10. Aspirate one-half volume of the media and replace with fresh modified Polarization Medium (N2B27 supplemented with

higher 20 ng/mL FGF2, 4 ng/mL IGF1, 4 ng/mL HGF without ROCK inhibitor Y-27632) (*see* **Notes 9** and **10**).

11. Return the culture to the orbital shaker at 80 rpm clockwise in a humidified incubator with 5% CO_2 (*see* **Note 11**).

At this point, the EMLO and EMLOC protocols bifurcate to capture distinct features of organogenesis with either cardiac or enteric innervation predominance (Fig. 1c). Follow either "EMLO gastruloid induction" (Subheading 3.6) or "EMLOC gastruloid induction" (Subheading 3.7) in the sections to follow. The subsequent section "Multi-lineage permissive differentiation and maturation" (Subheading 3.8) then occurs similarly in both EMLO and EMLOC protocols.

3.6 EMLO Gastruloid Induction Continued (Neuro-enteric)

Time: 15 min.
Duration: 2 days

1. 48 h post-aggregation (Fig. 2c), pool aggregates in a 15 mL conical tube and allow them to settle by gravity for 10 min. Completely aspirate the Polarization Medium and replace the full volume with Polarization Medium, excluding FGF2 (2 ng/mL IGF1, 2 ng/mL HGF only; *see* **Note 12**).
2. Return 6-well plates to the orbital shaker at 80 rpm clockwise in a humidified incubator with 5% CO_2.
3. Visually monitor cultures for morphological signs of polarization and elongation, such as protrusions from the main body of the aggregate (Fig. 3a, b).

3.7 EMLOC Gastruloid Induction (Neuro-cardiac)

Time: 15 min.
Duration: 5 days

1. 48 h post-aggregation (Fig. 2c), initiate cardiac induction by pooling aggregates in a 15 mL conical tube and allow them to settle by gravity for 10 min as described above. Completely aspirate the Polarization Medium and rinse with 10 mL HBSS (CM-free). Let the aggregates resettle and aspirate the HBSS (CM-free) (*see* **Note 12**).
2. Resuspend in Cardiac Induction Medium (N2B27 supplemented with 5 ng/mL VEGF-165, 30 ng/mL FGF2, 0.5 mM ascorbic acid).
3. Return cells to the orbital shaker at 80 rpm clockwise in a humidified incubator with 5% CO_2 (*see* **Note 13**).
4. The Cardiac Induction Medium can be replaced with fresh media at 4–5 d post-aggregation (*see* **Note 14**).
5. Visually monitor cultures for early polarization, cardiac crescent formation, and progression to contractile chamber-like structures (Fig. 5).

3.8 Multi-lineage Permissive Differentiation and Maturation in EMLO and EMLOC Gastruloids

Time: 15 min.
Duration: 18+ days

1. After the respective timing of each induction protocol, collect gastruloids in a 15-mL tube and let settle at 37 °C for 10 min.
2. Remove medium and rinse with HBSS (CM-free). Let gastruloids settle and remove the HBSS (CM-free).
3. Exchange with N2B27 (no additives) and redistribute evenly to new cultureware, freshly treated with anti-adherence rinsing solution.
4. Place cultures on orbital shaker at 75 rpm clockwise in a humidified incubator with 5% CO_2.
5. Replenish N2B27 basal media every 3–5 d as needed to maintain the maturing gastruloids for the remainder of the protocol.
6. Monitor by visual inspection prior to biomarker and gene expression analysis (Figs. 3c, d and 5a).

3.9 Biomarker Characterization of Cell Types by Immunofluorescence

Time: 5–8 days

1. On the day of fixation, pool gastruloids in a 15-mL tube and rinse with 1× PBS. Let settle and aspirate supernatant.
2. Fix samples by adding 10% neutral-buffered formalin solution for 2 h, rocking at 4 °C.
3. After fixation, wash samples 3× for 5 min in 1× PBS.
4. To permeabilize, wash samples 3× for 20 min in PBST (1× PBS + 0.2% Triton-X-100), rocking at 4 °C.
5. Block samples overnight in 1% BSA in PBST (blocking solution, BSA/PBST), rocking at 4 °C.
6. Distribute samples at 1 mL/well in 1% BSA/PBST to individual wells of a 12-well plate as needed.
7. Primary antibodies: incubate samples in 1 mL 1% BSA/PBST blocking solution containing diluted primary antibodies. Rock for 48–72 h at 4 °C.
8. Transfer samples to 2 mL tubes and rinse 3× in 1% BSA/PBST.
9. Secondary antibodies: wash 3× for 5 min in PBST and incubate secondary antibodies in 1% BSA/PBST. Add nuclear counterstain (e.g., NucBlue fixed cell stain) and incubate overnight, rocking at 4 °C in the dark.
10. Wash 2× for 5 min with 1% BSA/PBST and then 2× for 5 min in PBST.
11. Equilibrate in 0.1 M phosphate buffer (PB: 0.025 M NaH_2PO_4, 0.075 M Na_2HPO_4, pH 7.4) containing 0.1%

Triton-X-100, which is added to prevent sample adhesion to the plastic side wall. Repeat washes 3× for 5 min in this buffer.

12. Post-fix with 10% neutral-buffered formalin for 20 min at 4 °C.
13. Dilute fixation solution by washing 3× in 0.02 M PB containing 0.1% Triton-X-100.
14. To clear samples, replace phosphate buffer with 200 μL of 88% Histodenz solution (w/v). Incubate for 24–48 h at 4 °C in the dark.
15. Using a wide bore pipette tip, transfer the samples in 50–80 μL Histodenz solution directly to glass slide and apply 22 × 22 mm coverslip. Let set overnight in the dark at room temperature.
16. When dry, seal with clear nail polish and image using a confocal microscope (Figs. 4 and 5).

3.10 Gene Expression Analysis and Characterization by scRNAseq

Viable gastruloids are prepared for scRNAseq by dissociation to single cells using a cold activated protease method, similarly to that previously described [4, 13]. scRNAseq data is provided in Fig. 6.

1. Collect ~35 EMLOs or EMLOCs in a 2 mL centrifuge tube and let them settle.
2. Aspirate culture medium and add 1 mL ice-cold dissociation solution.
3. Incubate on ice for 15 min, triturating with a P-1000 pipette every 30–60 s. Confirm single-cell dissociation by microscopic examination.
4. Add 1 mL of ice-cold 1× PBS/10% FBS, centrifuge at 1200 × *g* for 5 min, and aspirate supernatant.
5. Resuspend cells in 1 mL of 1× PBS/10% FBS. Count the cells, centrifuge and aspirate the supernatant, and then resuspend at ~1 × 10^6 cells/mL in CryoStor CS10 cryopreservation medium.
6. Filter cells through a 40 μm cell strainer and transfer to a 1.8 mL Nunc cryostorage tube. Store at −80 °C until sequencing.
7. On the day of cell capture, thaw the dissociated sample in a 37 °C water bath for 2 min with gentle shaking.
8. Transfer cells to a 15 mL tube and add prewarmed media (RPMI1640/10% FBS) dropwise to a final volume of 10 mL.
9. Centrifuge cells at 300 × *g* for 5 min. Repeat wash steps 3×.
10. After the final spin, resuspend the cell pellet in 500 μL media and count cells using an automated cell counter (e.g., Logos Biosystems LUNA II with 0.4% Trypan blue exclusion).

11. Load single cells onto a 10× Genomics Chromium platform using the NextGem v3.1 single-cell reagent kit following the manufacturer's instructions.
12. Generate cDNAs from captured transcripts, prepare double-stranded DNA libraries, and quantify using the Agilent Fragment Analyzer and Roche KAPA universal qPCR kit.
13. Dilute samples to 250 pM, denature, and sequence on a NovaSeq 6000 (28 × 91).
14. Demultiplex samples using 10× Genomics Cell Ranger analysis software (V4.0.0) to produce quality reports, fastq files, and per-cell-per-transcript count matrices.
15. Review data quality (e.g., mean-reads per cell, fraction of reads in cell, and mapping rates to the reference genome).
16. Analyze data using Seurat and/or 10× Genomics Loupe Browser freeware.

3.11 Future Directions

Gastruloids represent one customizable tool in the armamentarium of developmental embryology models that range from traditional organoids and assembloids to bioengineered manufacturing platforms to state-of-the-art embryoids and stem cell-derived embryos. While not covered in this chapter, all of these contribute unique features and specializations that may be optimal for a particular problem or objective. Achieving neural innervation of non-neural tissue structures has remained a challenge and priority of the stem cell field. The gastruloid represents a straightforward, cost-effective, and reproducible yet perturbable model that retains a fundamental blueprint of cell lineage epigenetics in multi-system integration. Directed cardiac organoids and intestinal organoids have inherently lacked concomitant neurodevelopment due to lineage restriction [16–18]. While EMLO and EMLOC gastruloids begin from small cellular aggregates, they soon evolve from this early stage to produce expansive, interconnected organogenesis systems. Perhaps the single most impressive feature is the spatiotemporal organization and structural positioning during formation and maturation that is required to accurately capture development, and not just the generation of diverse cell types. However, equally important to recapitulating embryology is the ability to capture new architectures and interactions that may exist outside of the typical in vivo developmental paradigm as the cells within in vitro systems solve new problems. This in turn may inform new biomanufacturing approaches relying on cells and cell collectives with direct relevance to regenerative medicine and human disease.

EMLO and EMLOC gastruloid protocols begin with the same induction phase to caudalize iPSCs to trunk phenotype, but then diverge based on exposure to distinct developmental signaling cues. As two distinct models, they capture unique cell type diversity, multi-cellular events, and interactions that are distinguishing. While both protocols generate classes of motor, sensory, and autonomic neurons, the proportion of neuronal class varies somewhat. That is, the enteric-predominant EMLOs on average contain a higher proportion of *MNX1+/TUBB3+* expressing cells (motor type) compared to EMLOCs (Fig. 6c). By contrast, EMLOCs contain a higher proportion of *POU4F1+/TUBB3+* cells (sensory type). Initial neurogenesis begins at a similar timepoint in both EMLOs and EMLOCs (~days 7 to 8 in formation), but neuronal interaction in EMLOCs occurs subsequent to, and not preceding, chamber formation. In EMLOs, neural interaction with the gut tube occurs sooner. EMLOC-derived neurons are rebuffed from entering the cardiac domain until the later stage. Such spatiotemporal control in EMLOCs may reflect the autonomic innervation of the heart in vivo, which occurs only after complex cardiac morphogenesis has been established [19]. EMLOs also generate a vast diversity of central spinal neurons including all interneuron subtypes and exquisitely capture aspects of neural crest cell biology from delamination to epithelial-to-mesenchymal (EMT) transition, migration of fate-biased progenitors, and ultimate contribution to mesenchyme, peripheral neurons, and peripheral glia in organization with a multi-layered gut precursor [4]. Each model exemplifies the importance of short-range input from surrounding cells and tissue structures in addition to long-range input by neurons and the dynamics of these interactions, but with varying application. For example, EMLOCs may be useful to investigate how autonomic innervation of the heart through peripheral neural circuitry fine-tunes cardiac pacemaking, conduction, and contractility through chronotropic and inotropic effects of neurons at the autonomic neuro-cardiac junction [20, 21]. On the other hand, EMLOs may be useful to investigate how sympathetic, parasympathetic, and visceral neurons organize with developing multi-layered gastrointestinal tissue to become largely autonomous, and how non-neural lineages contribute to enteric nervous system formation and function. Notably, in addition to the unique cardiogenic and neural profile, the EMLOC protocol also produces co-developed renal epithelial and mesenchymal derivatives that are absent in EMLOs by scRNAseq. A small cluster of vascular endothelial cells is observed in both EMLOs and EMLOCs (Fig. 6a, b).

Despite great advancements made in the gastruloid field in recent years, the technology has yet to be applied to its fullest potential. In the future, EMLOs and EMLOCs may be coupled with bioengineering technologies for development of 3D organogenesis and organ platforms to bring additional features of vascularization, organ growth, and maturation. Recently, traditional gastruloids formed from mouse embryonic stem cells (mESCs) have succeeded in capturing vascular endothelial networks, embryonic blood development [22, 23], somitogenesis [24], and even anterior forebrain tissue [25], in addition to cardiogenesis [9] and endoderm emergence, organization, and elongation [26, 27]. By optimizing such complexity in combination as interconnected systems, it is likely that current human EMLO and EMLOC models will advance to an even higher degree of complexity and organization. It would be fascinating, for example, to optimize EMLOCs to generate peripheral networks of vascular endothelial cells integrated with outflow tracts that have endothelial, smooth muscle, endocardial and neural crest cell contributions. How this is coordinated with embryonic blood cell production and with neuronal tract formation such as to form neuro-vascular bundles could be further pursued. The EMLO and EMLOC protocols were developed using hiPSCs, and so it would also be intriguing to investigate contributions from naïve versus primed human embryonic stem cells with respect to the resulting cell lineage and organization phenotypes including extraembryonic tissues. Application of spatial scRNAseq to more fully elucidate cellular diversity and organization is anticipated to provide complementary insights into IF biomarker and scRNAseq analysis alone.

4 Notes

1. Carry out **steps 1–6** on ice to avoid premature gelling or non-uniform matrix coating. Adjust the working concentration and coating time according to cell type and application, following the manufacturer's guidelines for Matrigel preparation and handling.
2. Prior to adding mTeSR Plus, allow the working volume to reach room temperature for ~15 min outside of the water bath.
3. Optimize GCDR incubation time based on cell line according to the manufacturer's instructions.
4. The stated GCDR incubation time allows pluripotent cells to be released in small "colony patches" rather than as single cells.
5. Optimize CHIR 99021 and FGF2 induction concentration and exposure time depending on cell line (Fig. 2a). We have used an optimal concentration of 3 μM CHIR 99021 for 2–3 days of induction. However, a concentration window of 2.5–3.5 μM CHIR 99021 can be empirically tested. By visual inspection, optimal colony induction is characterized by slight raised-edge character just before cells begin to migrate away from the colony border. Exposure time to signaling factors for optimal induction may vary between cell line and require optimization. Extended exposure to FGF and Wnt signaling agonists (e.g., 4–5 d induction period) typically generates more uniform cultures of restricted neuromesodermal progenitor cells that is not optimal for gastruloids [14, 15]. IF can be

performed on adherent colonies to evaluate co-expression of neuroectodermal biomarker SOX2, mesendodermal biomarker FOXA2/HNF-3β, and other biomarkers (Fig. 1c).

6. If cells detach and float after a 10 min incubation with 1:1 Accutase:HBSS (CM-free) dissociation solution, optimize the exposure time for the cell line. For the current experiment, salvage the cells: add 1 mL N2B27 (no supplements), detach remaining cells by orienting a serological pipette orthogonally to the plane of the plate, and perform a side-to-side scraping motion over the entire surface area of the well. Triturate the cell suspension manually and centrifuge at 350 × *g* for 5 min to remove dissociation solution. Resuspend the cell pellet in supplemented media. The combined suspensions used to generate aggregates should have ~2 × 10^6 cells total in 2 mL medium (1 × 10^6 cells/mL). Too many cells will interfere with correct ratio of factors. Too few cells will prevent any aggregate formation. Total cell number per well may range from 2–4 × 10^6 cells depending on cell line in 2 mL total volume. If single cells do not aggregate, it is likely that the cell number per well is too low prior to aggregation. Combine 2 or 3 wells to achieve at least 2 × 10^6 cells in 2 mL medium. Typically, the combination of primed cultures from 2 wells is sufficient but this too is cell line dependent. If 4 × 10^6 cells are used and the resulting aggregates form at very high density (Fig. 2b), the cultures can be split 1:2 at this time in a low adhesion 6-well plate or to 100 mm petri dishes.
7. To initiate gastruloid aggregation, ensure a combined suspension of at least 2 × 10^6 cells in 2 mL medium per well (1 × 10^6 cells/mL). Adjust cell numbers if necessary to facilitate aggregate formation. Typically, the combination of primed cultures from 2 wells is sufficient but this too is cell line dependent.
8. Avoid over-settling of cultures to prevent aggregate fusion. Manually remove larger aggregates if needed.
9. If aggregates do not polarize, use IF to evaluate co-expression of the pluripotency marker SOX2 (posterior; region will mature to have neural identity), FOXA2/HNF-3β, and GATA4 (anterior; region will mature through cardiac crescent to have primarily cTnT+ identity). Cell number in the initial aggregate is also key to establish the appropriate length scale (50–100 cells/aggregate). Cell number can be assessed by direct and/or indirect methods: (1) determine cell number directly by fixation of a subset of aggregates 24 h after transitioning to orbital shaker, staining with DAPI and counting cells from imaging data using confocal microscope-generated Z-stacks. Determine the average cell number; (2) indirect bulk method to determine the concentration of aggregates

from a subset of the population by hemocytometer, dilute in buffer to achieve a known absolute number of aggregates, dissociate to single cells, and count again with hemocytometer or an automated cell counting system. Divide the total number of cells by the number of aggregates used to determine an average value at this stage. A final measure to take is to reduce the aggregate density by 2× to increase accessibility to polarization factors. This is done by doubling the total culture volume by splitting aggregates 1:2 to new wells in 2 mL per well.

10. Double the concentration of growth factors to maintain the same overall concentration in wells (assuming original recombinant proteins are still present).
11. Optimize the rotational speed of the orbital shaker depending on the cell line.
12. Use fresh plates at this stage to minimize non-viable cell debris.
13. If EMLOCs do not contract, verify early polarization biomarkers and cardiogenic biomarker expression cTnT at the cardiac crescent formation stage (day 4–7 window post-aggregation). Performing less media changes in Cardiac Induction Medium may be beneficial. Purchase fresh stocks of induction factors and/or optimize ascorbic acid concentration. Attempt L-ascorbic acid phosphate Mg salt from different supplier (Wako, cat # 013-12061) [9]. Prepare fresh reagents.
14. To promote intra-aggregate cell-cell and paracrine signaling, it may be beneficial to replace procardiogenic media only once during the 5-day cardiac induction period (days 2–7 post-aggregation).

Acknowledgments

This work was supported by multiple New York state awards. EMLO work was funded by the NY State Department of Health (NYS DOH) Spinal Cord Injury Research Board (NYSCIRB), Projects to Accelerate Research Translation (PART) award C33278GG and SUNY Polytechnic SEED award 917035-21 and used published lines developed through previous New York State stem cell research (NYSTEM) funding. The EMLOC research was supported by SUNY Polytechnic SEED award 917035 and CATN2 award 1180838-1-92476. The scRNAseq was performed at the SUNY Buffalo Genomics and Bioinformatics Core.

Conflicts of Interest The authors declare the following potential conflict of interest on patent pending EMLOC technology: 63/311,498, "A combined human gastruloid model for cardiogenesis and neurogenesis."

References

1. Shao Y, Fu J (2022) Engineering multiscale structural orders for high-fidelity embryoids and organoids. Cell Stem Cell 29:722–743
2. Shi Y, Wu Q, Wang X (2021) Modeling brain development and diseases with human cerebral organoids. Curr Opin Neurobiol 66:103–115
3. Olmsted ZT, Paluh JL (2021a) Stem cell neurodevelopmental solutions for restorative treatments of the human trunk and spine. Front Cell Neurosci 15:667590
4. Olmsted ZT, Paluh JL (2021b) Co-development of central and peripheral neurons with trunk mesendoderm in human elongating multi-lineage organized gastruloids. Nat Commun 12:3020
5. Olmsted ZT, Paluh JL (2022a) A combined human gastruloid model of cardiogenesis and neurogenesis. iScience 25:104486
6. Faustino Martins JM, Fischer C, Urzi A et al (2020) Self-organizing 3D human trunk neuromuscular organoids. Cell Stem Cell 26:172–186
7. Anderson C, Khan MAF, Wong F et al (2016) A strategy to discover new organizers identifies a putative heart organizer. Nat Commun 7:12656
8. Olmsted ZT, Paredes-Espinosa MB, Paluh JL (2022b) Generation of human elongating multi-lineage organized cardiac gastruloids. STAR Protoc 3:101898
9. Rossi G, Broguiere N, Miyamoto M et al (2021) Capturing cardiogenesis in gastruloids. Cell Stem Cell 28:230–240
10. Olmsted ZT, Paluh JL (2021c) Generation of human elongating multi-lineage organized (EMLO) gastruloids. Research Square. https://doi.org/10.21203/rs.3.pex-1441/v1
11. Chang EA, Tomov ML, Suhr ST et al (2015) Derivation of ethnically diverse human induced pluripotent stem cell lines. Sci Rep 5:15234
12. Tomov ML, Olmsted ZT, Dogan H et al (2016) Distinct and shared determinants of cardiomyocyte contractility in multi-lineage competent ethnically diverse human iPSCs. Sci Rep 6:37637
13. Adam M, Potter AS, Potter SS (2017) Psychrophilic proteases dramatically reduce single-cell RNA-seq artifacts: a molecular atlas of kidney development. Development 144:3625–3632
14. Gouti M, Tsakiridis A, Wymeersch FJ et al (2014) In vitro generation of neuromesodermal progenitors reveals distinct roles for wnt signaling in the specification of spinal cord and paraxial mesoderm identity. PLoS Biol 12:e1001937
15. Olmsted ZT, Stigliano C, Badri A et al (2020) Fabrication of homotypic neural ribbons as a multiplex platform optimized for spinal cord delivery. Sci Rep 10:12939
16. Silva AC, Matthys OB, Joy DA et al (2021) Co-emergence of cardiac and gut tissues promotes cardiomyocyte maturation within human iPSC-derived organoids. Cell Stem Cell 28:2137–2152
17. Drakhlis L, Biswanath S, Farr C-M et al (2021) Human heart-forming organoids recapitulate early heart and foregut development. Nat Biotechnol 39:737–746
18. Hofbauer P, Jahnel SM, Papai N et al (2021) Cardioids reveal self-organizing principles of human cardiogenesis. Cell 184:3299–3317
19. George RM, Maldonado-Velez G, Firulli AB (2020) The heart of the neural crest: cardiac neural crest cells in development and regeneration. Development 147:dev188706
20. Rajendran PS, Challis RC, Fowlkes CC et al (2019) Identification of peripheral neural circuits that regulate heart rate using optogenetic and viral vector strategies. Nat Commun 10:1944
21. Fedele L, Brand T (2020) The intrinsic cardiac nervous system and its role in cardiac pacemaking and conduction. J Cardiovasc Dev Dis 7:54
22. Rossi G, Giger S, Hubscher T, Lutolf MP (2022) Gastruloids as in vitro models of embryonic blood development with spatial and temporal resolution. Sci Rep 12:13380
23. Ragusa D, Suen C-W, Cortes GT et al (2022) Dissecting infant leukemia developmental origins with a hemogenic gastruloid model. bioRxiv. https://doi.org/10.1101/2022.10.07.511362
24. Veenvliet JV, Bolondi A, Kretzmer H et al (2020) Mouse embryonic stem cells self-organize intro trunk-like structures with neural tube and somites. Science 370:eaba4937
25. Girgin MU, Broguiere N, Mattolini L, Lutolf MP (2021) Gastruloids generated without exogenous Wnt activation develop anterior neural tissues. Stem Cell Rep 16:1–13
26. Vianello S, Lutolf MP (2021) In vitro endoderm emergence and self-organisation in the absence of extraembryonic tissues and embryonic architecture. bioRxiv. https://doi.org/10.1101/2020.06.07.138883
27. Farag N, Schiff C, Nachman I (2023) Coordination between endoderm progression and gastruloid elongation controls endodermal morphotype choice. bioRxiv. https://doi.org/10.1101/2023.02.07.527329

Methods in Molecular Biology (2024) 2767: 161–173
DOI 10.1007/7651_2023_486

Published online: 19 May 2023

Use of Epigenetic Cues and Mechanical Stimuli to Generate Blastocyst-Like Structures from Mammalian Skin Dermal Fibroblasts

Sharon Arcuri, Georgia Pennarossa, Sergio Ledda, Fulvio Gandolfi, and Tiziana A. L. Brevini

Abstract

Mammalian embryogenesis is characterized by complex interactions between embryonic and extra-embryonic tissues that coordinate morphogenesis, coupling bio-mechanical and bio-chemical cues, to regulate gene expression and influence cell fate. Deciphering such mechanisms is essential to understand early embryogenesis, as well as to harness differentiation disorders. Currently, several early developmental events remain unclear, mainly due to ethical and technical limitations related to the use of natural embryos.

Here, we describe a three-step approach to generate 3D spherical structures, arbitrarily defined "epi-Blastoids," whose phenotype is remarkably similar to natural embryos. In the first step, adult dermal fibroblasts are converted into trophoblast-like cells, combining the use of 5-azacytidine, to erase the original cell phenotype, with an ad hoc induction protocol, to drive erased cells into the trophoblast lineage. In the second step, once again epigenetic erasing is applied, in combination with mechanosensing-related cues, to generate inner cell mass (ICM)-like spheroids. More specifically, erased cells are encapsulated in micro-bioreactors to promote 3D cell rearrangement and boost pluripotency. In the third step, chemically induced trophoblast-like cells and ICM-like spheroids are co-cultured in the same micro-bioreactors. The newly generated embryoids are then transferred to microwells, to encourage further differentiation and favor epiBlastoid formation. The procedure here described is a novel strategy for in vitro generation of 3D spherical structures, phenotypically similar to natural embryos. The use of easily accessible dermal fibroblasts and the lack of retroviral gene transfection make this protocol a promising strategy to study early embryogenesis as well as embryo disorders.

Key words 3D culture system, 5-aza-CR, epiBlastoids, Epigenetics, Fibroblasts, Human, ICM-like spheroids, Mechanical cues, Porcine, Trophoblast-like cells

1 Introduction

Blastocyst formation is considered the milestone of the developmental and stem cell biology [1, 2]. It represents the first morphogenetic process in which cell lineage specification occurs for the first time, through the delamination of epithelial trophectoderm cells, the expansion of a fluid-filled cavity, and the formation of the inner cell mass (ICM). These events are driven by complex interactions between embryonic and extra-embryonic tissues that coordinate early embryogenesis [3], coupling bio-mechanical and bio-chemical cues, to regulate gene expression and cell fate [4]. Dissecting the mechanisms involved in these processes is essential to understand the early stages of embryonic development and better elucidate these still unclear early events.

Recently, blastocyst in vitro models were generated and defined "blastoids," by assembling embryonic (ESCs) and extra-embryonic stem cells [5, 6] or ESC-derived trophoblast, hypoblast, and epiblast cells [7–9] in three-dimensional (3D) systems. However, given the ethical limitations in using embryos, coupled with the technical issues related to the isolation and maintenance of ESCs in several species [10–12], alternative approaches for blastoid generation would be advantageous.

Besides the widely accepted concept of unidirectional progression toward cell commitment and differentiation, several studies demonstrated that cell specification can be reversed and that a terminally differentiated adult cell can be pushed back toward a less committed, higher permissive state [13, 14]. In this context, induced pluripotent stem cell (iPSC) technology allows to directly generate pluripotent stem cells from somatic cells and, more recently, to produce blastocyst-like structures, defined iBlastoids [6, 7, 15–17]. It must be noted, however, that iPSC technology involves the use of retroviruses/lentiviruses to force the stable long-term expression of pluripotency-related genes and suffers from either low reprogramming efficiency or too many genomic modifications [18].

Alternative approaches have been recently proposed. In particular, the use of epigenetic compounds has been shown to interact with and modify the epigenetic signature of adult mature cells, converting them into primitive progenitors, multipotent, or high plasticity/pluripotent cells [19–29]. Accurate regulation of cell plasticity and/or differentiation can also be achieved using mechanosensing-related cues and mechanical forces [30–33], mimicking the key role exerted by extracellular matrix (ECM) in the control of cell fate definition in vivo [34–38].

Here, a three-step approach is reported that allows the generation of 3D spherical structures, arbitrarily defined "epiBlastoids", whose phenotype is remarkably similar to natural embryos (Fig. 1). In the first step, porcine and human adult dermal fibroblasts are converted into trophoblast-like cells, using 5-azacytidine to erase the original cell phenotype, followed by an ad hoc induction protocol to drive cells into the trophoblast lineage [39, 40]. In the second step, mechanosensing-related cues are added to chemically reprogrammed porcine and human fibroblasts for the generation of ICM-like organoids [41]. In the third step, chemically induced trophoblast-like cells are co-cultured with ICM-like organoids in the same micro-bioreactors; the newly generated embryoids are then transferred to microwells to encourage further differentiation and allow epiBlastoid formation.

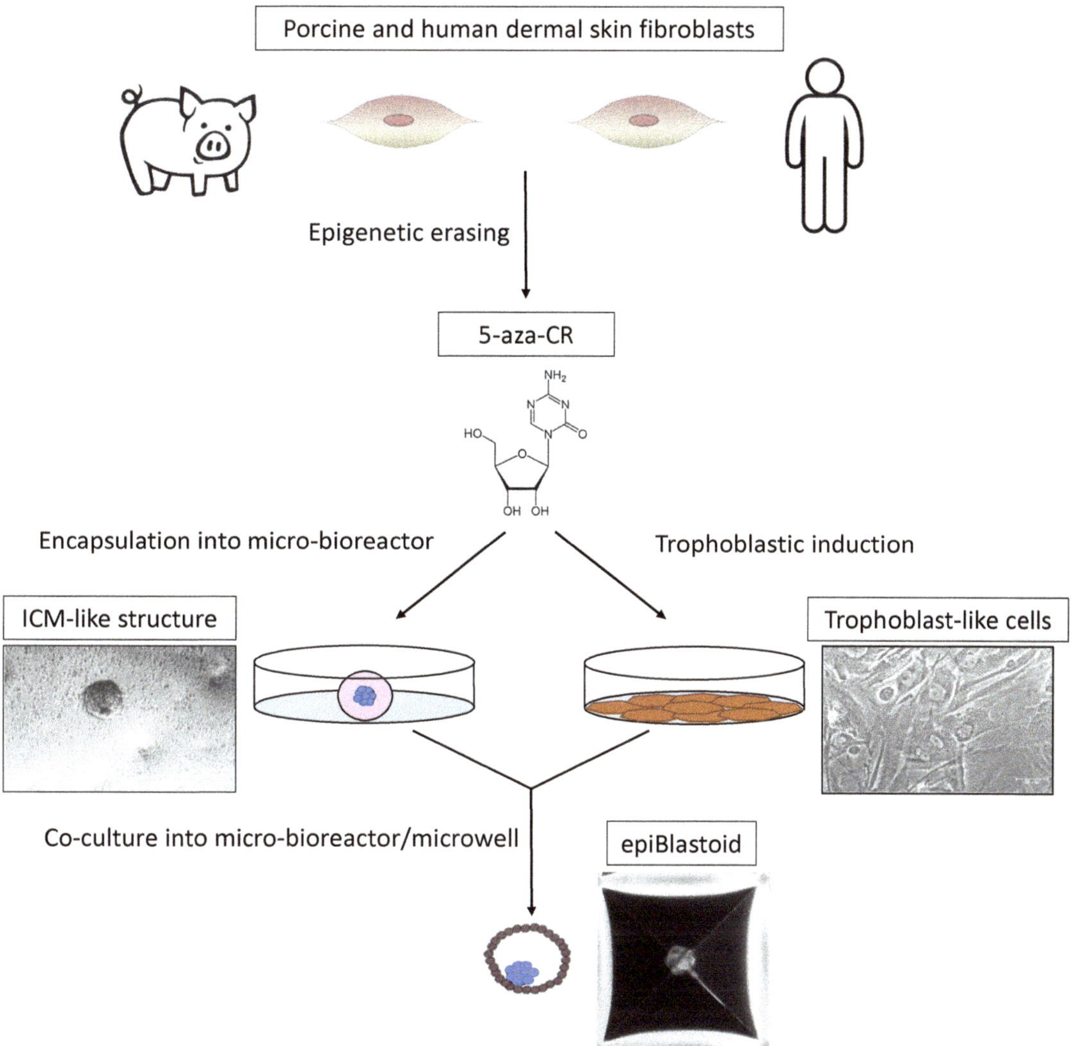

Fig. 1 Scheme illustrating epiBlastoid generation from porcine and human adult dermal fibroblasts

2 Materials

2.1 Mouse Embryonic Fibroblast (MEF) Thawing and Inactivation

1. MEF cells.
2. Water bath.
3. 15 mL sterile tubes.
4. Centrifuge.
5. 35 mm Petri dishes.
6. T25 flasks.
7. 0.22 μm filter.
8. Inverted microscope.

9. CO_2 incubator.
10. Trypsin-EDTA solution.
11. Dulbecco's phosphate-buffered saline (PBS) containing 1% antibiotics.
12. MEF culture medium: 88% (v/v) Dulbecco's modified Eagle medium, 10% (v/v) fetal bovine serum (FBS), 2 mM (v/v) L-glutamine solution, 1% antibiotics.
13. MEF inactivating medium: 6 mL of MEF culture medium supplemented with 10 μg/mL mytomicin C.
14. ESC culture medium without bFGF: 40% Ham's F-10 Nutrient mix, 40% DMEM low glucose, 10% knockout serum replacement, 5% FCS, 1% antibiotics, 2 mM L-glutamine, 1% nucleoside mix, 1% MEM non-essential amino acids solution, 0.1 mM 2-mercaptoethanol, 1 unit/mL LIF.

2.2 Isolation of Porcine Skin Dermal Fibroblasts

1. Skin biopsies collected from adult pig.
2. 50 mL sterile tubes.
3. 15 mL sterile tubes.
4. Sterile surgical instruments.
5. 100 mm Petri dishes.
6. 35 mm Petri dishes.
7. T25 flasks.
8. CO_2 incubator.
9. Inverted microscope.
10. Dulbecco's phosphate-buffered saline (PBS) containing 2% antibiotics.
11. 0.1% porcine gelatin: dissolve 0.1 g of porcine gelatin in 100 mL of sterile water. Autoclave and stock at +4 °C.
12. Fibroblast culture medium: 77% (v/v) Dulbecco's modified Eagle medium, 20% (v/v) fetal bovine serum (FBS), 2 mM (v/v) L-glutamine solution, 1% antibiotics.

2.3 Isolation of Human Skin Dermal Fibroblasts

1. Skin biopsy collected from human adult donors.
2. 50 mL sterile tubes.
3. 15 mL sterile tubes.
4. Sterile surgical instruments.
5. 35 mm Petri dishes.
6. 100 mm Petri dishes.
7. T25 flasks.
8. CO_2 incubator.

9. Inverted microscope.
10. Dulbecco's phosphate-buffered saline (PBS) containing 2% antibiotics.
11. 0.1% porcine gelatin: dissolve 0.1 g of porcine gelatin in 100 mL of sterile water. Autoclave and stock at +4 °C.
12. Fibroblast culture medium: 77% (v/v) Dulbecco's modified Eagle medium, 20% (v/v) fetal bovine serum (FBS), 2 mM (v/v) L-glutamine solution, 1% antibiotics.

2.4 Seeding of Porcine and Human Dermal Fibroblasts

1. Porcine or human skin dermal fibroblasts.
2. Water bath.
3. 15 mL sterile tubes.
4. Cell counter.
5. Centrifuge.
6. Inverted microscope.
7. Four-well multi dishes.
8. 0.1% porcine gelatin: dissolve 0.1 g of porcine gelatin in 100 mL of sterile water. Autoclave and stock at +4 °C.
9. Fibroblast culture medium: fresh culture medium: 88% (v/v) Dulbecco's modified Eagle medium, 10% (v/v) fetal bovine serum (FBS), 2 mM (v/v) L-glutamine solution, 1% antibiotics.
10. Dulbecco's phosphate-buffered saline (PBS) containing 2% antibiotics.
11. Trypsin-EDTA solution.
12. CO_2 incubator.

2.5 Porcine and Human Fibroblast Exposure to 5-Azacytidine (5-aza-CR) and Trophoblast Induction

1. Water bath.
2. Inverted microscope.
3. 1 mM 5-aza-CR: dissolve 0.0024 g of 5-aza-CR in 10 mL of warm Dulbecco's modified Eagle medium; vortex the solution and sterilize using 0.22 μm filter.
4. Fibroblast culture medium: fresh culture medium: 88% (v/v) Dulbecco's modified Eagle medium, 10% (v/v) fetal bovine serum (FBS), 2 mM (v/v) L-glutamine solution, 1% antibiotics.
5. ESC culture medium: 40% Ham's F-10 nutrient mix, 40% DMEM low glucose, 10% knockout serum replacement, 5% FCS, 1% antibiotics, 2 mM L-glutamine, 1% nucleoside mix, 1% MEM non-essential amino acids solution, 0.1 mM 2-mercaptoethanol, 1 unit/mL LIF, 5 ng/mL human basic fibroblast growth factor (bFGF).

6. Dulbecco's phosphate-buffered saline (PBS) containing 2% antibiotics.
7. Trophoblast induction medium: 10 mL of conditioned MEF-inactivated medium, 10 μL of BMP4 (50 ng/mL), 10 μL A83-01 (1 μM), 10 μL PD173074 (0.1 μM) (*see* **Notes 1** and **2**).
8. Tri-gas incubator.

2.6 Exposure of Porcine and Human Fibroblasts Encapsulated in PTFE Micro-bioreactor to 5-aza-CR

1. Porcine or human skin dermal fibroblasts.
2. Water bath.
3. 35 mm bacteriology Petri dishes.
4. 100 mm Petri dishes.
5. 15 mL sterile tubes.
6. Cell counter.
7. Centrifuge.
8. Inverted microscope.
9. PTFE powder.
10. 1000 μL pipette tip.
11. 200 μL pipette tip.
12. Trypsin-EDTA solution.
13. Dulbecco's phosphate-buffered saline (PBS) containing 2% antibiotics.
14. Sterile water.
15. 5-aza-CR stock solution: dissolve 0.0024 g of 5-aza-CR in 10 mL of warm Dulbecco's modified Eagle medium; vortex the solution and sterilize using 0.22 μm filter.
16. Fibroblast culture medium: fresh culture medium: 88% (v/v) Dulbecco's modified Eagle medium, 10% (v/v) fetal bovine serum (FBS), 2 mM (v/v) L-glutamine solution, 1% antibiotics.
17. CO_2 incubator.

2.7 Generation of Porcine and Human epiBlastoids

1. Water bath.
2. 35 mm bacteriology petri dishes.
3. 100 mm petri dishes.
4. 15 mL sterile tubes.
5. Cell counter.
6. Sterile needle.
7. Centrifuge.
8. Inverted microscope.

9. PTFE powder.
10. 1000 μL pipette tip.
11. 200 μL pipette tip.
12. Stereomicroscope.
13. Aggrewells.
14. Trypsin-EDTA solution.
15. Dulbecco's phosphate-buffered saline (PBS) containing 2% antibiotics.
16. Sterile water.
17. G1-PLUS medium.
18. G2-PLUS medium.
19. PODS Activin A.
20. CO_2 incubator.

3 Methods

3.1 MEF Thawing and Inactivation

1. Thaw a cryo-vial containing MEFs using a 37 °C water bath.
2. Add 2 mL of fibroblast culture medium in a 15 mL sterile tube.
3. Transfer MEFs to a 15 mL sterile tube containing 2 mL of fibroblast culture medium.
4. Centrifuge at 300 g for 5 min.
5. Resuspend the pellet in 2 mL of fibroblast culture medium.
6. Seed into 35 mm Petri dish and incubate at 37 °C in CO_2 incubator.
7. Change fibroblast culture medium every day.
8. When MEFs reach confluency, remove culture medium from the dish.
9. Wash 3 times in sterile PBS supplemented with 1% antibiotic antimycotic solution.
10. Add 0.5 mL of trypsin-EDTA solution and incubate at 37 °C until cell monolayer begins to detach from the bottom of the tissue culture dish and cells dissociate.
11. Neutralize trypsin-EDTA using 1.5 mL of MEF culture medium.
12. Dislodge cells by repeatedly and gently pipetting.
13. Plate cells in a new culture dish using a 1:2 passage ratio and culture at 37 °C in 5% CO_2 incubator.
14. Change fibroblast culture medium every day and passage every 2 days.

15. Seed MEFs at a density of 2×10^4 in T25 flask and incubate for 24 h in CO_2 incubator.
16. Rinse cells 3 times in sterile PBS and expose sub-confluent monolayers to the MEF inactivating medium for 3 h.
17. Eliminate MEF inactivating medium and incubate cells with ESC medium without b-FGF for 24 h.
18. Collect conditioned medium, filter using 0.22 μm filter, and store at −20 °C until use.

3.2 Isolation of Porcine Skin Dermal Fibroblasts

1. Collect porcine skin biopsy and transfer to a 50 mL sterile tube containing sterile PBS with 2% antibiotics.
2. Prepare 2 mL of 0.1% of porcine gelatin in a 35 mm Petri dish. Wait for 45 min to coat the surface at room temperature.
3. Wash the biopsy in sterile PBS and cut it into small fragments of approximately 2 mm^3 (*see* **Note 1**).
4. Eliminate gelatin immediately prior to plate skin fragments.
5. Transfer five skin fragments to a 35 mm Petri dish.
6. Add 20 μL of fibroblast culture medium onto each fragment and incubate at 37 °C in CO_2 incubator.
7. Add around 20–50 μL of fibroblast culture medium daily for the first 5 days.
8. Add 1 mL in each petri dish and culture until cells reach confluency.
9. When cells reach confluency, trypsinize (see Subheading 3.1, **steps 9–12**) and transfer cells to a new T25 flask.
10. Incubate at 37 °C.
11. Passage every 3 days with a ratio of 1:3.

3.3 Isolation of Human Skin Dermal Fibroblasts

1. Collect human skin biopsy and transfer to a 50 mL sterile tube containing sterile PBS with 2% antibiotics.
2. Prepare 2 mL of 0.1% of porcine gelatin in a 35 mm petri dish. Wait for 20 min to coat the surface at room temperature.
3. Wash the biopsy in sterile PBS and cut it into small fragments of approximately 2 mm^3 (*see* **Note 3**).
4. Eliminate gelatin immediately prior to plate skin fragments.
5. Transfer five skin fragments to a 35 mm Petri dish.
6. Add 20 μL of fibroblast culture medium onto each fragment and incubate at 37 °C in CO_2 incubator.
7. Add around 20–50 μL of fibroblast culture medium daily for the first 5 days.

8. Add 1 mL in each petri dish and culture until cells reach confluency.
9. When cells reach confluency, trypsinize (see Subheading 3.1, **steps 9–12**) and transfer cells to a new T25 flask.
10. Incubate at 37 °C.
11. Passage every 3 days with a ratio of 1:2.

3.4 Seeding of Porcine and Human Dermal Fibroblasts

1. Add 0.5 mL of 0.1% of porcine gelatin in each well of four-well multidish and wait for 20 min to coat the surface at room temperature.
2. Eliminate gelatin and let it dry for 45 min.
3. Remove the medium from T25 flask, and then trypsinize using 0.6 mL of trypsin-EDTA (*see* **Note 3**).
4. Neutralize trypsin-EDTA using 5.4 mL of fibroblast culture medium.
5. Collect cell suspension in a 15 mL sterile tube.
6. Count cells using a counting chamber under an optical inverted microscope at room temperature. Calculate the volume of medium needed to resuspend cells to obtain 7.8×10^4 cells/cm^2.
7. Centrifuge cells at 300 g for 5 min.
8. Resuspend the pellet in a calculated volume by pipetting carefully.
9. Plate cells in four-well multidish.
10. Incubate at 37 °C in 5% CO_2 incubator for 24 h.

3.5 Porcine and Human Fibroblast Exposure to 5-aza-CR and Trophoblast Induction

1. Dilute 1 μL of 5-aza-CR stock solution in 1 mL of fibroblast culture medium to obtain 5-aza-CR working solution with 1 μM final concentration (*see* **Note 3**).
2. Remove the medium from the dish.
3. Add 0.5 mL of fibroblast culture medium containing 5-aza-CR in three wells. Use one as a negative control (*see* **Note 4**).
4. Incubate at 37 °C in 5% CO_2 incubator for 18 h.
5. Remove 5-aza-CR.
6. Rinse cells 3 times in sterile PBS and incubate with 0.5 mL of ESC culture medium at 37 °C in CO_2 incubator for 3 h.
7. Rinse cells 3 times in sterile PBS.
8. Expose cells to the trophoblast induction medium and incubate in a hypoxic environment (5% O_2) in the tri-gas incubator for 11 days (*see* **Note 5**).
9. Refresh trophoblast induction medium every other day.

3.6 Exposure of Porcine and Human Fibroblasts Encapsulated in PTFE Micro-bioreactor to 5-aza-CR

1. Start from porcine or human fibroblast cultures described in Subheadings 3.2 and 3.3.
2. Remove the medium from T25 flask and trypsinize using 0.6 mL of trypsin-EDTA (*see* Subheading 3.4, **steps 3–4**).
3. Collect cell suspension in a 15 mL sterile tube.
4. Count cells using a counting chamber under an optical inverted microscope at room temperature. Calculate the volume of medium needed to resuspend cells in order to obtain 1×10^4 cells/30 μL.
5. Centrifuge cells at 300 g for 5 min.
6. Dilute 1 μL of 5-aza-CR stock solution in 1 mL of fibroblast culture medium to obtain 5-aza-CR working solution with 1 μM of final concentration (*see* **Note 4**).
7. Resuspend the pellet in a calculated volume of 5-aza-CR working solution by pipetting carefully.
8. Fill a 35 mm bacteriology petri dish with PTFE powder and dispense 30 μL of single droplet containing 1×10^4 cells onto the powder bed.
9. Gently rotate the 35 mm bacteriology petri dish in a circular motion to completely cover the surface of the liquid drop with the PTFE powder particles and form a liquid marble micro-bioreactor (*see* **Note 6**).
10. Pick up the PTFE micro-bioreactor using a 1000 μL pipette tip cut at the edge and lay it onto a clean 35 mm bacteriology petri dish. Place 35 mm bacteriology petri dish containing the micro-bioreactor within a 100 mm petri dish containing sterile water to prevent evaporation.
11. Incubate at 37 °C in 5% CO_2 incubator for 24 h.

3.7 Generation of Porcine and Human epiBlastoids

1. Detach day 11 trophoblast-like cells from the four-well multidish.
2. Remove trophoblast induction medium.
3. Wash 3 times in sterile PBS supplemented with 1% antibiotic antimycotic solution.
4. Add 0.1 mL of trypsin-EDTA solution and incubate at 37 °C until cell monolayer begins to detach from the bottom of the tissue culture dish and cells dissociate.
5. Neutralize trypsin-EDTA using 0.5 mL of MEF inactivated medium.
6. Dislodge cells by repeatedly and gently pipetting.
7. Collect detached cells in conical tubes.

8. Count cells and calculate the volume of medium required to obtain 3×10^4 cells/30 μL.
9. Centrifuge at 300 g for 5 min and resuspend pelleted trophoblast-like cells in the calculated volume of G1-PLUS medium.
10. Break PTFE micro-bioreactors by puncturing it with a needle, collect the formed ICM-like spheroids with a 200 μL pipette tip under a stereomicroscope, and transfer them to G1-PLUS culture medium.
11. Fill a 35 mm bacteriology petri dish with PTFE powder.
12. Produce a droplet of 40 μL containing 3×10^4 cells (30 μL) and one ICM-like spheroid (10 μL).
13. Dispense the 40 μL droplet onto PTFE powder bed to generate new micro-bioreactors (*see* Subheading 3.6, **step 9**).
14. Pick up the new PTFE micro-bioreactors using a 1000 μL pipette tip cut at the edge and lay it onto a clean 35 mm bacteriology petri dish. Place 35 mm bacteriology petri dish containing the micro-bioreactor within a 100 mm petri dish containing sterile water to prevent evaporation.
15. Incubate at 37 °C in 5% CO_2 incubator for 48 h.
16. Break the micro-bioreactors, collect the formed epiBlastoid, and place it in microwells for culture using G2-PLUS medium.
17. Incubate at 37 °C in 5% CO_2 for 24 h.
18. After 24 h, add 10^5 PODS Activin A/150 μL of G2-PLUS medium.
19. Culture at 37 °C in 5% CO_2 incubator for 4 days.

4 Notes

1. Make sure that cells are in a sub-confluency state before 5-aza-CR treatment.
2. Aliquot BMP4, A83-01, PD173074 just prior use.
3. It is crucial to cut very small pieces of skin tissues to facilitate fibroblasts to grow out of the tissue fragments.
4. Prepare 5-aza-CR solution just prior use.
5. It is very important to add induction factors just before use.
6. Carefully rotate the 35 mm bacteriology petri dish in a circular motion to completely cover the surface of the liquid drop with the PTFE powder particles and form a liquid marble micro-bioreactor.

Acknowledgments

This work was supported by Carraresi Foundation, PSR2021, and HORIZON-WIDERA-2021 project n#101079349 (OH-Boost). Authors are members of the Trans-COST Actions Taskforce on Covid-19.

References

1. Evans MJ, Kaufman MH (1981) Establishment in culture of pluripotential cells from mouse embryos. Nature 292:154–156
2. Martin GR (1981) Isolation of a pluripotent cell line from early mouse embryos cultured in medium conditioned by teratocarcinoma stem cells. Proc Natl Acad Sci 78:7634–7638
3. Liu L, Warmflash A (2021) Self-organized signaling in stem cell models of embryos. Stem Cell Reports 16:1065–1077
4. Pennarossa G, Gandolfi F, Brevini TAL (2021) Biomechanical signaling in oocytes and parthenogenetic cells. Front Cell Dev Biol 9:1–7
5. Harrison SE, Sozen B, Christodoulou N, et al (2017) Assembly of embryonic and extraembryonic stem cells to mimic embryogenesis in vitro. Science (80-) 356
6. Weatherbee BAT, Cui T, Zernicka-Goetz M (2021) Modeling human embryo development with embryonic and extra-embryonic stem cells. Dev Biol 474:91–99
7. Sozen B, Jorgensen V, Weatherbee BAT et al (2021) Reconstructing aspects of human embryogenesis with pluripotent stem cells. Nat Commun 12:1–13
8. Yanagida A, Spindlow D, Nichols J, et al (2021) Naive stem cell blastocyst model captures human embryo lineage segregation. Cell Stem Cell 28:1016–1022.e4
9. Yu L, Wei Y, Duan J et al (2021) Blastocyst-like structures generated from human pluripotent stem cells. Nature 591(7851):620–626
10. Nicolas P, Etoc F, Brivanlou AH (2021) The ethics of human-embryoids model: a call for consistency. J Mol Med 99:569–579
11. Brevini TAL, Pennarossa G, Maffei S et al (2012) Pluripotency network in porcine embryos and derived cell lines. Reprod Domest Anim 47:86–91
12. Gandolfi F, Arcuri S, Pennarossa G et al (2019) New tools for cell reprogramming and conversion: possible applications to livestock. Anim Reprod 16(3):475–484
13. Pennarossa G, Gandolfi F, Brevini TAL (2020) All roads lead to Rome: the many ways to pluripotency. J Assist Reprod Genet 37:1029–1036
14. Brevini TAL, Pennarossa G, Manzoni EFM et al (2016) The quest for an effective and safe personalized cell therapy using epigenetic tools. Clin Epigenetics 8:119
15. Amadei G, Lau KYC, Jonghe J De, et al (2021) Inducible stem-cell-derived embryos capture mouse morphogenetic events in vitro. Dev Cell 56:366–382.e9
16. Bao M, Cornwall-Scoones J, Zernicka-Goetz M (2022) Stem-cell-based human and mouse embryo models. Curr Opin Genet Dev 76: 101970
17. Liu X, Tan JP, Schröder J et al (2021) Modelling human blastocysts by reprogramming fibroblasts into iBlastoids. Nature 591:627–632
18. Shao L, Wu WS (2010) Gene-delivery systems for iPS cell generation. Expert Opin Biol Ther 10:231–242
19. Pennarossa G, Maffei S, Campagnol M et al (2013) Brief demethylation step allows the conversion of adult human skin fibroblasts into insulin-secreting cells. Proc Natl Acad Sci U S A 110:8948–8953
20. Pennarossa G, Maffei S, Campagnol M et al (2014) Reprogramming of pig dermal fibroblast into insulin secreting cells by a brief exposure to 5-aza-cytidine. Stem Cell Rev Reports 10:31–43
21. Brevini TAL, Pennarossa G, Rahman MM et al (2014) Morphological and molecular changes of human granulosa cells exposed to 5-Azacytidine and addressed toward muscular differentiation. Stem Cell Rev Reports 10:633–642
22. Mirakhori F, Zeynali B, Kiani S et al (2015) Brief azacytidine step allows the conversion of suspension human fibroblasts into neural progenitor-like cells. Cell J 17:153–158
23. Tan SJ, Fang JY, Wu Y et al (2015) Muscle tissue engineering and regeneration through epigenetic reprogramming and scaffold manipulation. Sci Rep 5:16333

24. Brevini TAL, Pennarossa G, Acocella F et al (2016) Epigenetic conversion of adult dog skin fibroblasts into insulin-secreting cells. Vet J 211:52–56
25. Chandrakanthan V, Yeola A, Kwan JC et al (2016) Pdgf-ab and 5-Azacytidine induce conversion of somatic cells into tissue-regenerative multipotent stem cells. Proc Natl Acad Sci U S A 113:E2306–E2315
26. Manzoni EFM, Pennarossa G, DeEguileor M et al (2016) 5-azacytidine affects TET2 and histone transcription and reshapes morphology of human skin fibroblasts. Sci Rep 6:37017
27. Pennarossa G, Santoro R, Manzoni EFM et al (2018) Epigenetic erasing and pancreatic differentiation of dermal fibroblasts into insulin-producing cells are boosted by the use of low-stiffness substrate. Stem Cell Rev Reports 14:398–411
28. Pennarossa G, Manzoni EFM, Ledda S et al (2019) Use of a PTFE micro-bioreactor to promote 3D cell rearrangement and maintain high plasticity in epigenetically erased fibroblasts. Stem Cell Rev Reports 15:82–92
29. Pennarossa G, Arcuri S, De Iorio T et al (2023) Combination of epigenetic erasing and mechanical cues to generate human epiBlastoids from adult dermal fibroblasts. J Assist Reprod Genet. https://doi.org/10.1007/s10815-023-02773-4
30. Vining KH, Mooney DJ (2017) Mechanical forces direct stem cell behaviour in development and regeneration. Nat Rev Mol Cell Biol 18:728–742
31. Matamoro-Vidal A, Levayer R (2019) Multiple influences of mechanical forces on cell competition. Curr Biol 29:R762–R774
32. Yim EKF, Sheetz MP (2012) Force-dependent cell signaling in stem cell differentiation. Stem Cell Res Ther 3(5):41
33. Kumar A, Placone JK, Engler AJ (2017) Understanding the extracellular forces that determine cell fate and maintenance. Development 144:4261–4270
34. Bissell MJ, Rizki A, Mian IS (2003) Tissue architecture: the ultimate regulator of breast epithelial function. Curr Opin Cell Biol 15: 753–762
35. Christman JK (2002) 5-Azacytidine and 5-aza-2′-deoxycytidine as inhibitors of DNA methylation: mechanistic studies and their implications for cancer therapy. Oncogene 21: 5483–5495
36. Pennarossa G, Iorio T De, Gandolfi F, et al (2022) Impact of aging on the ovarian extracellular matrix and derived 3D scaffolds. Nanomaterials (Basel) 12:345
37. Pennarossa G, Iorio T De, Gandolfi F, et al (2021) Ovarian decellularized bioscaffolds provide an optimal microenvironment for cell growth and differentiation in vitro. Cells 10:1–19
38. Pennarossa G, Ghiringhelli M, Gandolfi F et al (2021) Creation of a bioengineered ovary: isolation of female germline stem cells for the repopulation of a decellularized ovarian bioscaffold. Methods Mol Biol 2273:139–149
39. Arcuri S, Gandolfi F, Somigliana E et al (2021) A two-step protocol to erase human skin fibroblasts and convert them into trophoblast-like cells. Methods Mol Biol 2273:151–158
40. Arcuri S, Pennarossa G, Gandolfi F et al (2021) Generation of trophoblast-like cells from hypomethylated porcine adult dermal fibroblasts. Front Vet Sci 8:1–10
41. Pennarossa G, Ledda S, Arcuri S et al (2020) A two-step strategy that combines epigenetic modification and biomechanical cues to generate mammalian pluripotent cells. J Vls Exp:1–22

Methods in Molecular Biology (2024) 2767: 175–188
DOI 10.1007/7651_2022_469

Published online: 12 Febuary 2023

Directed Differentiation of Human Pluripotent Stem Cells to Cytotrophoblast and Syncytiotrophoblast

Asha Shaji Antony, A. S. Devika, and Smita Sudheer

Abstract

Human pluripotent stem cells (hPSCs) form an ideal system to study the formation of placental cells, from an undifferentiated human embryonic stem cell state. The conventional human in vitro model systems to study the human placenta cannot be employed for understanding placental dysfunctions or the development of specialized placental cell types. Hence, human PSCs make an ideal model system to study human placental development and disorders. Here, we describe an efficient and validated protocol to reproducibly study the formation of human cytotrophoblasts (CTBs) and syncytiotrophoblast (STBs) from undifferentiated hPSCs. CTBs are the trophoblast stem cells that can differentiate into specialized placental cell types such as STBs. The multinucleated STB plays vital role in the exchange of nutrients and gases across the placenta and secretes several hormones during pregnancy, such as human chorionic gonadotropin β (hCGβ). Here we describe two methods of seeding the hPSCs: chemical (clumps method) and enzymatic methods (single cells) to differentiate them to CTB and STB, activating BMP (B) signaling and inhibiting ACTIVIN/NODAL and FGF signaling pathways (2i), thus naming our protocol as "B2i" (Sudheer et al., Stem Cells Dev 21:2987–3000, 2012). This protocol forms the perfect model system for understanding in vitro placentation, modeling diseases arising from abnormal placentation that cause complications such as miscarriage, preeclampsia or intrauterine growth restriction (IUGR), and drug discovery for placental disorders.

Key words B2i, Cytotrophoblast, Differentiation, Human pluripotent stem cells, Syncytiotrophoblast

1 Introduction

Trophoblast (TB) is the first differentiated lineage that is specified during blastocyst formation of a developing mammalian embryo. The morula undergoes compaction and cavitation to form a blastocyst, in which the cells segregate into the inner cell mass (ICM) and an outer lining of blastomeres called the trophectoderm (TE) [2–4]. ICM develops into the embryo proper, and the TE develops into trophoblast (TB), which gives rise to the placenta that provides nutrition and protection to the developing embryo [5, 6]. TB lineage commitment is the earliest cell fate decision of a developing mammalian embryo;

Supplementary Information The online version contains supplementary material available at https://doi.org/10.1007/7651_2022_469.

however, the cues that initiate this decision vary across species [7]. TB differentiates and gives rise to cytotrophoblast (CTB), the multipotent trophoblast stem cells, which can acquire two mutually exclusive cell fates—one that retains the epithelial nature, formed by the fusion of CTB, called the syncytiotrophoblast (STB) [8], and the second one that follows the epithelial-mesenchymal transition of the CTB, called the extravillous trophoblast (EVT) [9, 10].

Regulatory and ethical concerns limit our ability to understand the development of trophoblast to a fully functional placenta in human, which would be helpful in studying human placental development and disorders. Most of our knowledge on placental development comes from mouse [11, 12], but studies have shown major gene expression differences between mouse and human pre-implantation embryos [13, 14]. Although the best human in vitro models to study the development of placental cells from TB have been choriocarcinoma cells, immortalized extravillous cytotrophoblast cell lines, and primary TB cultures from placenta [15, 16], they have major drawbacks, as none of them can be used for studying the further development of specialized placental cells such as STB and EVT cell types [16, 17].

Human pluripotent stem cells (hPSCs) that include human embryonic stem cells (hESCs) [3] and human-induced pluripotent stem cells (hiPSCs) [18] have unlimited capacity to self-renew and generate derivatives of all three germ layers, primordial germ cells, and cells of the extra embryonic lineage, trophectoderm and primitive endoderm [1, 19–22].

BMP4 plays an instructive role in TB specification of hPSCs [19] and henceforth has been employed in all TB differentiation protocols. Initial attempts to differentiate hPSCs to trophoblast included differentiation of hPSCs in mouse embryonic fibroblasts (MEF) feeder-conditioned media with an addition of high or low concentration of BMP4 [19]. The absence of FGF in the BMP4 containing MEF-conditioned medium led to rapid progression toward trophoblast [23]. The MEF-conditioned medium contains serum, which has several unknown factors that could influence the differentiation. Further, luciferase reporter assays show that both TGFβ and FGF signals synergize to inhibit BMP signaling in human ESCs and sustain pluripotency [24]. These findings propelled the discovery that BMP activation, along with combined inhibition of FGF (SU5402) and TGFβ signaling (SB431542), leads hPSCs towards cytotrophoblast and then syncytiotrophoblast (BSBSU) [1]. BMP signaling can support differentiation toward embryonic lineages and FGF inhibition is the key to this directed differentiation that avoids all the embryonic lineages [1]. Later, Amita et al. [22] used alternative inhibitors for TGFβ and FGF signaling for trophoblast differentiation (BAP protocol). Irrespective of the difference in the inhibitors used, BMP activation, with 2 inhibitors (TGFβ and FGF inhibitors), directs hPSCs toward syncytiotrophoblast (B2i). Here, we describe this efficient protocol to differentiate hPSCs into cytotrophoblast and further to syncytiotrophoblast.

2 Materials

2.1 Cell Banking and Storage

The human embryonic stem cell (hESCs) (BJNhem19 (JNCASRe001-A)) line was procured from Dr. Maneesha Inamdar, Jawaharlal Nehru Centre for Advanced Scientific Research (JNCASR), India. The human-induced pluripotent stem cell (hiPSCs) (D14C2) line was a kind gift from Dr. R. V. Shaji, Centre for Stem Cell Research, (CSCR), InStem, India. hESCs and hiPSCs are stored in cryotubes in liquid nitrogen.

2.2 Reconstitution of Recombinant Growth Factors and Inhibitors

2.2.1 hbFGF

Centrifuge the vial containing human bFGF (Peprotech, 100-18B) before opening to pull down the contents. In order to prepare a stock concentration of 10 μg/mL, reconstitute the 50 μg lyophilized powder in 5 mL of 1× DPBS (Thermo Scientific, 14190144) containing 0.1% carrier protein BSA (Thermo Scientific, 15260037) (0.1% BSA is prepared in 1× DPBS and filtered through syringe filter of pore size 0.2 μm (Corning, CLS431222-50A)). For a working concentration of 10 ng/mL, thaw the vial on ice and add 1 μL stock solution to 1 mL medium.

2.2.2 hBMP-4

To reconstitute hBMP-4 (Gibco, PHC9534), first centrifuge the vial prior to opening to bring down the lyophilized powder. Reconstitute in 4 mM HCl containing 0.1% BSA (Thermo Scientific, 15260037) to make a stock solution of 10 μg/mL. Aliquot the working stocks in appropriate volumes into polypropylene tubes and store at −20 °C to −80 °C for 6 months. For preparing a working concentration of 10 ng/mL, thaw the vial on ice and add 1 μL stock to 1 mL of medium. Store this vial at 2–4 °C for a week and avoid freeze-thaw cycles.

2.2.3 ALK 4/5/7 Inhibitor—SB431542

To prepare 10 mM stock solution of SB431542 (Peprotech, SM-3014193-B), calculate the volume of DMSO (Himedia, TC 185) to be added to 10 mg of the lyophilized powder of SB431542 (Formula weight 384.4 g/mol) according to the formula:

$$\text{Voulme (litre)} = \text{Mass (gm)} \big/ \text{Concentration (M)} \times \text{Formula weight (g/mol)}$$

2.6 mL DMSO in 10 mg SB431542 gives a stock solution of 10 mM. Aliquot the working stocks at appropriate volumes into polypropylene tubes and store at −20 °C for 6 months. To prepare a working concentration of 20 μM, thaw the vial on ice and add 2 μL of the stock to 1 mL of media. Store this vial at 2–4 °C for a week and avoid freeze thaw cycles.

2.2.4 ALK 4/5/7 Inhibitor—A 83-01

To prepare 10 mM stock solution of A 83-01 (Peprtotech, SM-9094360-A), calculate the volume of DMSO to be added to 10 mg of the lyophilized powder of A83-01 (molecular weight 384.4 g/mol) using the formula mentioned in Subheading 2.2.3.

1.6 mL DMSO in 10 mg A 83-01 gives a stock solution of 10 mM. Aliquot into appropriate working stocks in polypropylene tubes and store at −20 °C for a maximum of 6 months. For preparing a working concentration of 20 μM, thaw the vial on ice and add 2 μL of the stock solution to 1 mL medium. Maintain the vial at 2–4 °C for a week and avoid freeze-thaw cycles.

2.2.5 MEK/ERK Pathway Inhibitor (FGFR-1/3/2/4 Inhibitor)—PD173074

For preparing 5 mM stock solution of PD173074 (Peprotech, SM-2191178-A), determine the volume of DMSO to be added to 10 mg of the lyophilized powder of PD173074 (formula weight 523.7 g/mol) according to the formula given in Subheading 2.2.3.

Use DMSO to make a stock solution of 0.1 mM. Aliquot the working stocks at appropriate volumes into polypropylene tubes and store at −20 °C for a maximum of 6 months. For making a working concentration of 0.1 μM, thaw the vial on ice and add 1 μL stock solution to 1 mL of medium. Store the vial at 2–4 °C for a week and avoid freeze-thaw cycles.

2.2.6 MEK/ERK Pathway Inhibitor (FGFR-1 Inhibitor)—SU5402

To prepare 10 mM stock solution of SU5402 (Peprotech, SM-2159233-A), calculate the volume of DMSO to be added to 5 mg of the lyophilized powder of SU5402 (formula weight 296.3 g/mol) according to the formula mentioned in Subheading 2.2.3.

1.69 mL DMSO in 5 mg SU5402 gives a stock solution of 10 mM. Aliquot the working stocks at appropriate volumes into polypropylene tubes and store at −20 °C for a maximum of 6 months. To make a working concentration of 20 μM, thaw the vial on ice and add 2 μL stock solution to 1 mL medium. Store this vial at 2–4 °C for a maximum of 1 week and avoid freeze-thaw cycles.

2.2.7 ROCK Inhibitor—Y27632

To prepare 30 mM stock solution of Y27632 (Peprotech, SM-1293823-B), calculate the volume of DMSO to be added to 10 mg of the lyophilized powder of Y27632 (formula weight 320.3 g/mol) according to the formula given in Subheading 2.2.3.

1.041 mL DMSO in 10 mg Y27632 gives a stock solution of 30 mM. Aliquot the working stock at appropriate volumes into polypropylene tubes and store at −20 °C for 1 year, protected from light. For preparing a working concentration of 10 μM, thaw the vial on ice and add 0.33 μL of the stock to 1 mL of medium. Store this vial at 2–4 °C for a maximum of one week and avoid freeze-thaw cycles.

2.3 Preparation of Extracellular Matrices

2.3.1 Vitronectin

Thaw the vial containing Vitronectin (VTN-N) (Gibco, A14700) at room temperature, and aseptically prepare working aliquots and store at −80 °C in polypropylene tubes for later use.

2.3.2 Matrigel

Thaw the tube containing Matrigel (Corning, 356234) in ice at 4 °C overnight, and aseptically prepare working aliquots and store at −20 °C in polypropylene tubes for later use.

2.4 Media Preparation, Aliquoting, and Storage

2.4.1 Maintenance Medium (Essential 8 (E8) Media)

hPSCs are maintained in Essential 8 medium (ThermoFisher Scientific, A1517001). Complete E8 medium (E8C) (basal medium + supplement) is prepared according to the manufacturer's instructions and aliquoted into appropriate volumes and stored at 2–8 °C for a maximum of one week.

2.4.2 Differentiation Medium (N2B27)

For differentiation, cells are grown in the chemically defined N2B27 medium. All the following components of the medium are added and mixed, followed by filter-sterilization using syringe filters of pore size 0.2 μm (Corning, CLS431222-50A).

N2B27 medium (50 mL)

DMEM/F12	46.6 mL (Thermo, Life Technologies, 11330032)
N2 Supplement (100×)	500 μL (Thermo, Life Technologies, 17502048)
B27 Supplement (50×)	1 mL (Thermo, Life Technologies, 12587010)
Bovine Serum Albumin Fraction V (7.5% soln.)	340 μL (Thermo Scientific, 15260037)
200 mM L-Glutamine	500 μL (Thermo Scientific, 25030149)
Penicillin/Streptomycin	500 μL (Thermo Scientific, 15140122)
MEM Non-Essential Aminoacids	500 μL (Thermo Scientific, 11140050)
55 mM β-Mercaptoethanol	91 μL (Thermo, Life Technologies, 21985023)

2.5 Reagents for Cell Dissociation

2.5.1 EDTA (0.5 mM)

Prepared by 1:1000 dilution of 0.5 M EDTA (Thermo, Life Technologies, 15575- 020) in 1× DPBS without calcium and magnesium (Thermo, Life Technologies, 14190136)

2.5.2 Accutase

Thaw the StemPro Accutase (Thermo Fisher, A1110501) stored at −80 °C at 2–8 °C overnight. Aliquot the Accutase into appropriate volumes and store at 2–8 °C for a maximum of 24 months. For long-term storage, the aliquots can be stored at −20 °C.

2.6 Reagents and Materials for qRT-PCR

QIAzol Lysis Reagent (QIAGEN, 79306) was used for total RNA isolation and iScript cDNA Synthesis Kit (Bio-Rad, 1708891) for cDNA preparation. qRT-PCR was done using the 2× PowerUp™ SYBR™ Green Master Mix (Applied Biosystems, A25776). For quantifying RNA, NanoDrop One Spectrophotometer (Thermo Fisher Scientific) was used and qRT-PCR was executed in the thermal cycler, Roche Light Cycler II 480.

3 Methods

3.1 Extracellular Matrix Coating of Cell Culture Plates

3.1.1 Vitronectin (VTN-N) Coating

1. Thaw the VTN-N vial stored at −80 °C at room temperature.
2. The final coating concentration of VTN-N is 0.5 μg/cm^2. Calculate the working concentration and dilution factor needed to coat the wells in accordance to the following formulae:

$$\text{Working Concentration} = \text{Coating Concentration} \times \frac{\text{Culture Surface area}}{\text{Volume required for surface area}}$$

$$\text{Dilution Factor} = \frac{\text{Stock concentration (0.5 mg/mL)}}{\text{Working Concentration}}$$

3. Calculation of the working concentration for one well in a 6-well plate: the surface area ~10 cm^2 and the volume required to cover the surface area is 1 mL. Hence the working concentration is 5 μg/mL and the dilution factor is 100×.
4. To coat 1 well of a 6-well plate, aseptically add 10 μL VTN-N to 1 mL DPBS without calcium and magnesium, so that the working concentration is 5 μg/mL.
5. Gently resuspend by pipetting up and down. Coat the diluted VTN-N at 1 mL/well by pipetting out to the center of the wells and moving the plate to ensure an even distribution across the well surface.
6. Incubate the plates for 1 h at 37 °C.
7. For later use, after incubation, seal the VTN-N-coated plates tightly with parafilm to prevent evaporation and store at 2–8 °C for a maximum of 1 week. Allow the coated plates to come to room temperature for about 1 h, prior to use.

8. For immediate use, tilt the plate and aspirate off the excess VTN-N solution without scratching the coated surface. Add 1.5 mL E8C medium to each well and store at 37 °C and 5% CO_2 until the cells are seeded into them.

3.1.2 Matrigel Coating

1. Thaw Matrigel overnight by submerging the entire bottle in ice in a cold room or at 4 °C. Use pre-chilled micropipette tips, serological pipettes, and tubes for diluting and aliquoting Matrigel (*see* **Note 1**).
2. The protein concentration of Matrigel varies across lots. Calculate the concentration of Matrigel required to coat and the appropriate volume of basal medium (dilution factor) accordingly. The final coating-concentration to be used is 8.7 μg/cm^2. Once thawed, avoid freeze-thaw cycles. Aliquot working stocks and store at −20 °C.
3. For coating the plates, add appropriate volume of ice-cold, serum-free basal medium (DMEM/F12) using a pre-chilled pipette tip to the fresh or frozen Matrigel aliquot. Gently mix by pipetting up and down (*see* **Note 2**), while the tube is on ice. Then transfer the diluted Matrigel to the center of the well (1 mL/well of a 6 well plate) and swirl gently to ensure a uniform coating.
4. For later use, wrap the Matrigel-coated plates tightly with parafilm to prevent drying up and store at 2–8 °C for a maximum of 1 week. Prior to use, allow the coated plates to come to room temperature for about 1 h.
5. For immediate use, incubate the plates at 37 °C for an hour for gelation. Tilt the plate and aspirate off the excess Matrigel solution. Add 1.5 mL E8C medium to each well and store at 37 °C and 5% CO_2 until cells are seeded into them.

3.2 Passaging and Seeding of hPSCs

3.2.1 Chemical Method (Dissociation into Clumps Using 0.5 mM EDTA)

1. Prepare the ECM-coated plates (Vitronectin or Matrigel) and incubate for 1 h at room temperature.
2. Cells are passaged when they reach 60–70% confluence. Aspirate off the media and wash with pre-warmed 0.5 mM EDTA. Incubate the cells with 0.5 mM EDTA for 3–4 min at room temperature (*see* **Note 3**).
3. After 2.5 min, observe under a phase-contrast microscope and as soon as the cells begin to separate uniformly (small gaps start appearing inside the colonies and the colony edges begin to fold up), aspirate away the EDTA solution and flush the wells with 1 mL E8C medium with 10 μM ROCK inhibitor (ROCKi) very gently to dislodge the colonies into an optimal size (*see* **Note 4**).
4. Remove the excess ECM from the pre-prepared plates.

5. To seed the cells, pipette up the cell suspension and dispense it dropwise into the new ECM-coated wells. For maintenance, the cells are seeded at a ratio of 1:6 to 8 in Vitronectin-coated plates in E8C media. For differentiation, the cells are seeded at 1:12 ratio on Matrigel-coated plates.
6. Place the new plate in the incubator at 37 °C and 5% CO_2 and move the plate in a '+' pattern in the case of a dish or back and forth twice or thrice and then once sideways to ensure even distribution of the colonies throughout the wells.
7. The next day, medium is replaced with fresh medium without ROCKi. Medium is replaced every day and when the cells are 80–90% confluent, they are passaged (maintenance).

3.2.2 Enzymatic Method (Single Cell Suspension by Accutase Treatment)

1. Prepare the ECM-coated plates (Vitronectin or Matrigel) and incubate for 1 h at room temperature.
2. Initial ROCK inhibition: Aspirate off the media and wash with 1 mL of 1× DPBS. Add 2 mL of fresh E8C medium supplemented with 10 μM of ROCKi and incubate it for 1 h in a CO_2 incubator at 37 °C and 5% CO_2.
3. Cell dissociation:
   (a) After the incubation, aspirate off the ROCKi-containing E8C medium and wash with 1× DPBS. Add 1 mL Accutase and incubate at 37 °C for 15 min.
   (b) Neutralize Accutase by diluting it with equal volume (1 mL) of E8C medium supplemented with 10 μM ROCKi to the cell suspension and gently pipette up and down until the cells are in cell suspension.
   (c) Transfer the cell suspension to a 15 mL conical tube containing 3 ml E8 media supplemented with 10 μM ROCKi and centrifuge at 200 g for 5 min.
   (d) Remove supernatant and resuspend the cells in fresh E8C medium supplemented with 10 μM ROCKi.
   (e) Cell counting: Take 20 μL of the cell suspension in a 0.5 mL microfuge tube and add 30 μL of PBS and 50 μL of 0.4% trypan blue solution to the cell suspension to create a dilution factor of 5. Mix thoroughly and incubate for 5 min. Transfer a small amount of the trypan blue-cell suspension to a chamber on the hemocytometer and count the viable cells (non-viable cells stain blue, viable cells will remain opaque) in four corner squares. Calculate the total number of cells/ml using the following formula:

$$\text{Cells per mL} = \frac{\text{the average count per square} \times \text{the dilution factor}}{10^{-4}}$$

4. Remove the ECM from the pre-prepared plates.
5. After calculating the cell concentration, seed the desired number of cells on the ECM-coated dish(es) with E8C medium and ROCKi and incubate at 37 °C in a humidified atmosphere of 5% CO_2 in air. For differentiation, the cells are seeded at 4×10^4 cells/cm^2 density on Matrigel-coated plates. For maintenance, the seeding ratio can be varied and the cells are seeded on vitronectin-coated plates.
6. The next day, replace the medium with fresh medium without ROCKi. Medium is replaced every day and when the cells are 80–90% confluent, they are passaged (maintenance).

3.3 Differentiation of hPSCs to Trophoblast Lineage

This differentiation protocol (Fig. 1) has been reproduced in both human embryonic stem cells (Figs. 2 and S1) and human-induced pluripotent stem cells (hiPSCs) (Figs. 3 and S2). Two different methods of initial seeding of the hPSCs were adopted—the conventional chemical (clumps) method, using EDTA (Figs. 2 and S1) and the enzymatic (single) cells method, using Accutase (Figs. 2 and S2). The latter method may be used, when the beginning of the experiment requires a finite number of cells. Depending on the cell lines and the media conditions, there could be slight difference in the time taken for the cytotrophoblast (CTB) (2–3 days) and syncytiotrophoblast (STB) induction (4–6 days) (Fig. 1) [1, 22]. Here, the trophoblast lineage is characterized by the sustained expression of the pan-trophoblast marker, *KRT7* from the CTB stage onward; transient upregulation of the markers, *CDX2*, *TFAP2A*, *VGLL1*, *P63* at the CTB stage; and the expression of the markers, *CGB*, *ERVWE1*, *GCM1*, and *HERV-FRD1* at the STB stage (Fig. 1). The hPSC-induced syncytiotrophoblast expresses hCGβ and secretes hCG hormone [1].

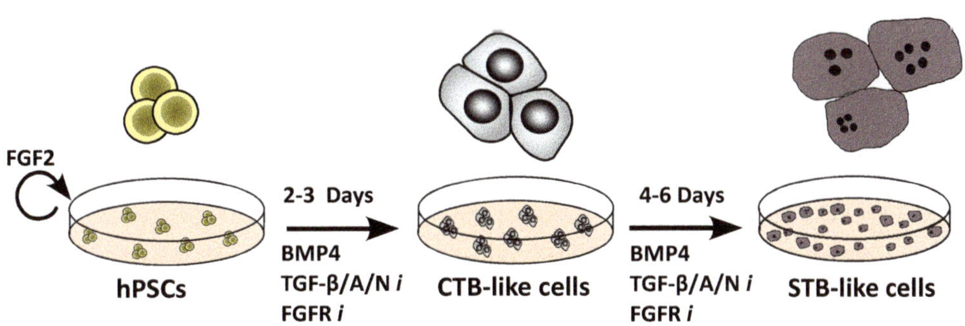

Fig. 1 Graphical representation of the protocol for the directed differentiation of hPSCs to cytotrophoblast (CTB) and syncytiotrophoblast (STB). BMP4, 10 ng/mL; TGF β/ACTIVIN/NODAL inhibitor (TGFβ/A/N *i*), SB431542, 20 μM; FGF Receptor inhibitor (FGFR *i*), 20 μM; cytotrophoblast (CTB); syncytiotrophoblast (STB)

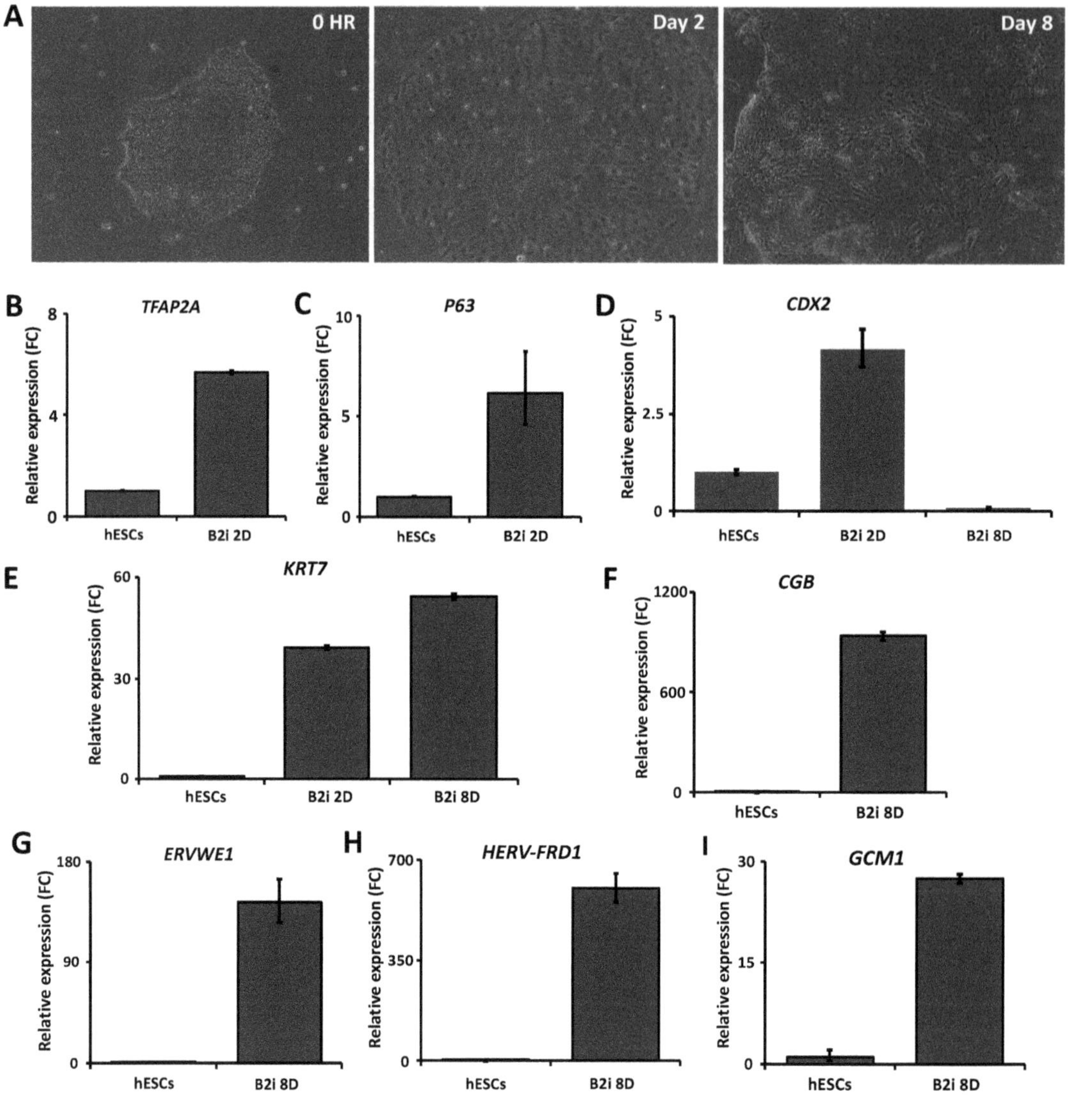

Fig. 2 B2i treatment of hESCs seeded using the chemical (clumps) method differentiates them to trophoblast. (**a**) Temporal changes observed in the cell morphology of hESCs before (0 h) and during the B2i treatment (after 2 and 8 days). Quantitative RT-PCR (qRT-PCR) analysis of the expression of the CTB markers (**b**–**d**), Pan-trophoblast marker (**e**), and STB markers (**f**–**i**) in the undifferentiated hESC (BJNhem 19) and the hESC-derived CTB and STB-like cells. The data is normalized against the housekeeping gene, *GAPDH*. Error bars: mean ± SD ($n = 2$)

3.3.1 Differentiation Using Chemical (Clumps) Method

1. To induce differentiation of hPSCs to trophoblast lineage, the cells are seeded in Matrigel-coated plates at a passage ratio of 1: 12 in E8C medium and cultured until the wells are approximately 50–60% confluent. Medium is replaced daily with fresh E8C medium. When the cells are confluent, the medium is aspirated and the cells are rinsed once with 1× DPBS prior to the treatment with the differentiation medium.

2. The differentiation medium consists of N2B27, supplemented with 10 ng/mL hBMP-4, 20 μM SB431542, and 20 μM SU5402 (B2i) (Fig. 1).
3. The differentiation medium is replaced daily, with freshly added B2i. During the treatment, the cell morphological changes initiate in the periphery of the colonies. These changes start becoming visible after 1 day of treatment itself, and by 2 days, the difference between hPSCs and the B2i-treated cells becomes more prominent (Figs. 2a and S1a). Cells become considerably bigger in size and gain a cobble stone-like appearance (Figs. 2a and S1a). The expression of the pan-trophoblast marker, *KRT7* (Figs. 2e and S1c), and the cytotrophoblast (CTB) markers, *TFAP2A*, *P63*, *CDX2*, and *VGLL1* after 2 days of the treatment (Figs. 2b–d and S1b) reflect the cells to be committed to the trophoblast lineage and are in the trophoblast stem cell (CTB) state.
4. On day 8 of the treatment, the cell morphology becomes completely distinct from that of the initial undifferentiated state. By now, they appear flattened, with multinucleated cells (syncytium) (Figs. 1, 2a, b and S1a). The multinucleated cells can be observed, when they are stained with CDH1 or ACTB [1]. In some parts of the colonies, the cells tend to lose their opaqueness and exhibit a slightly transparent morphology. The induction of the marker gene, *CGB* (Figs. 2f and S1d), and the secretion of the hCG hormone [1] are indicative of the cells to be syncytiotrophoblast (STB). At this stage, they express other STB markers such as *GCM1*, *EVRWE1*, and *HERV-FRD1* (Figs. 2g–i and S1d).

3.3.2 Differentiation Using Enzymatic (Single Cell) Method

1. hPSCs are seeded on Matrigel-coated dish(es) at a cell density of 4×10^4 cells/cm^2, in E8C medium supplemented with 10 μM ROCKi.
2. After 24 h, E8C medium is removed and washed with 1× DPBS. Cells are then exposed to the differentiation medium (N2B27 medium supplemented with B2i).
3. The further treatment methods and gene expression analysis (Figs. 3a–g and S2a–d) are the same as that followed for the clumps method (*see* Subheading 3.3.1, **step 2–4**). Unlike the clumps method, in the single cells method, the differentiation and morphological changes in the colonies are more uniform.

This BMP4-mediated differentiation method can also be carried out in the presence of another ALK4/5/7 inhibitor, A83-01 with a concentration of 1 μM (instead of SB431542), and PD173074 (at a concentration of 0.1 μM), an FGFR1/3/2/4 inhibitor (instead of the FGFR1 inhibitor SU5402) [22, 23]. The duration of the treatment and the time points of the expression of

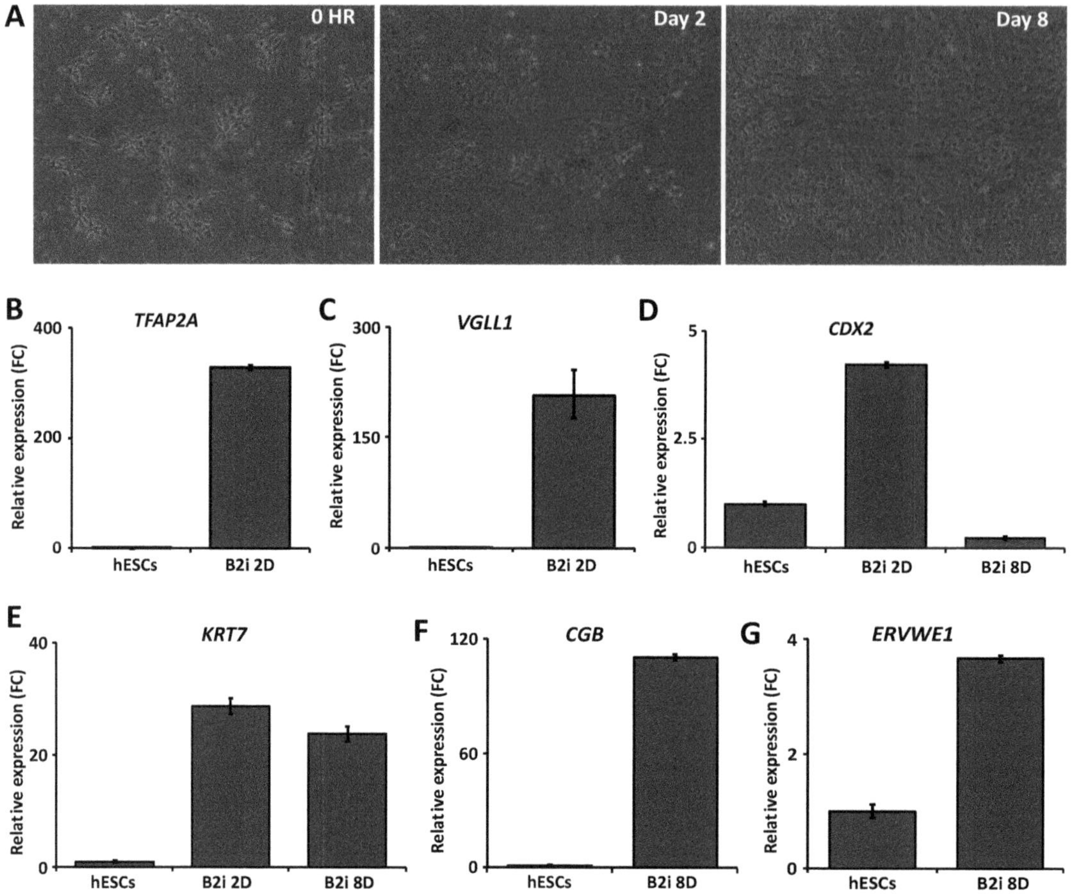

Fig. 3 B2i treatment of hESCs seeded using the enzymatic (single cells) method differentiates them to trophoblast. (**a**) Temporal changes observed in the cell morphology of hESCs before (0 h) and during the B2i treatment (after 2 and 8 days) of human-induced pluripotent cells (hiPSCs: D14C2). (**b**–**g**): qRT-PCR analysis of the expression of the CTB markers (**b**–**d**), Pan-trophoblast marker (**e**), and STB markers (**f**–**g**) in the undifferentiated hiPSC and the hiPSCs-derived CTB and STB-like cells. The data is normalized against the housekeeping gene, *GAPDH*. Error bars: mean $\pm$ SD ($n = 2$)

the CTB and STB markers could vary, depending on the medium for the differentiation, the small molecules, the hPSC line, and the ECM used for the differentiation [1, 22, 23].

3.4 Quantitative Real-Time PCR (qRT-PCR)

Cells are harvested using QIAzol Lysis reagent and the total RNA is isolated according to the manufacturer's protocol. The isolated RNA is quantified using NanoDrop One Spectrophotometer and converted to cDNA. Quantitative real-time PCR is done with gene-specific primers with Tm 58 °C (Table S1) in a thermal cycler. Data analysis is done using the *dd*Ct method, with the house-keeping genes, *GAPDH* or *ACTB*.

4 Notes

1. Gelation of Matrigel matrix begins above 10 °C; therefore, use of pre-chilled micropipette tips and serological pipettes is highly recommended.
2. Take care to hold the tube above the Matrigel level in the tube. The warmth from fingers could cause gelation of Matrigel.
3. Take immense care to limit the exposure of cells to EDTA to a maximum of 4 min at RT, as overexposure to EDTA can result in overdissociation to single cells, affecting cell viability adversely.
4. Calcium and magnesium ions in E8 medium neutralize the remaining EDTA.

Acknowledgments

We are very much grateful to Dr. R. V Shaji, CSCR, Vellore, India, for his support and the generous gift of the hiPSCs (D14C2). We thank Dr. Raghu Bhushan, YRC, Yenepoya (Deemed to be) University, Mangalore, India, for reading the chapter and helpful comments. SM.S. acknowledges the financial support from the Department of Biotechnology (DBT) (BT/PR29717/PFN/20/1361/2018), Indian Council of Medical Research (ICMR) (2019-3008/SCR/ADHOC/BMS) and Council of Scientific and Industrial Research (CSIR) (09/1108(13739)/2022-EMR-1), New Delhi, India.

Authors' Contributions A.S.A. and D.A.S. conducted the experiments and analysis. A.S.A., D.A.S., and SM.S. contributed to the analysis, validation, and writing of the original draft. SM.S. reviewed and edited the chapter and contributed to the conceptualization, methodology, supervision, project administration, and funding acquisition.

References

1. Sudheer S, Bhushan R, Fauler B, Lehrach H, Adjaye J (2012) FGF inhibition directs BMP4-mediated differentiation of human embryonic stem cells to syncytiotrophoblast. Stem Cells Dev 21:2987–3000
2. Evans MJ, Kaufman MH (1981) Establishment in culture of pluripotential cells from mouse embryos. Nature 292:154–156
3. Thomson JA et al (1998) Embryonic stem cell lines derived from human blastocysts published by: American Association for the Advancement of Science Stable URL: http://www.jstor.org/stable/2897366 Linked references are available on JSTOR for this article: Embryonic Stem Cell Li. Science (80-) 282:1145–1147.
4. Senner CE, Hemberger M (2010) Regulation of early trophoblast differentiation - lessons from the mouse. Placenta 31:944–950

5. Cross JC (2000) Genetic insights into trophoblast differentiation and placental morphogenesis. Semin Cell Dev Biol 11:105–113
6. Fisher SJ (2000) The placenta dilemma. Semin Reprod Med 18:321–326
7. Dyce J, George M, Goodall H, Fleming TP (1987) Do trophectoderm and inner cell mass cells in the mouse blastocyst maintain discrete lineages. Development 100:685–698
8. Gude NM, Roberts CT, Kalionis B, King RG (2004) Growth and function of the normal human placenta. Thromb Res. https://doi.org/10.1016/j.thromres.2004.06.038
9. Dasilva-arnold S, James JL, Al-khan A, Zamudio S, Illsley NP (2015) Differentiation of fi rst trimester cytotrophoblast to extravillous trophoblast involves an epithelial e mesenchymal transition. Placenta 36:1412–1418. https://doi.org/10.1016/j.placenta.2015.10.013
10. Davies JE et al (2016) Epithelial-mesenchymal transition during extravillous trophoblast differentiation. Cell Adhes Migr 10:310–321
11. Pfeffer PL, Pearton DJ (2012) Trophoblast development. Reproduction 143:231–246
12. Niakan KK, Eggan K (2013) Analysis of human embryos from zygote to blastocyst reveals distinct gene expression patterns relative to the mouse. Dev Biol 375:54–64
13. Blakeley P et al (2015) Defining the three cell lineages of the human blastocyst by single-cell RNA-seq. Development 142:3151–3165
14. Ringler GE, Strauss JF (1990) In vitro systems for the study of human placental endocrine function. Endocr Rev 11:105–123
15. Biancotti JC, Narwani K, Buehler N, Mandefro B, Golan-Lev T, Yanuka O, Clark A, Hill D, Benvenisty N, Lavon N (2009) Human embryonic stem cells as models for aneuploid chromosomal syndromes. Stem Cells 29:1–12
16. Genbacev O et al (2013) Human trophoblast progenitors: where do they reside? Semin Reprod Med 31:56–61
17. Leitch HG, Smith A (2013) The mammalian germline as a pluripotency cycle. Development 140:2495–2501
18. Takahashi K et al (2007) Induction of pluripotent stem cells from adult human fibroblasts by defined factors. Cell 131:861–872
19. Xu RH et al (2002) BMP4 initiates human embryonic stem cell differentiation to trophoblast. Nat Biotechnol 20:1261–1264
20. Harun R et al (2006) Cytotrophoblast stem cell lines derived from human embryonic stem cells and their capacity to mimic invasive implantation events. Hum Reprod 21:1349–1358
21. Udayashankar R et al (2011) Characterization of invasive trophoblasts generated from human embryonic stem cells. Hum Reprod 26:398–406
22. Amita M et al (2013) Complete and unidirectional conversion of human embryonic stem cells to trophoblast by BMP4. Proc Natl Acad Sci U S A 110:E1212–E1221
23. Das P et al (2007) Effects of FGF2 and oxygen in the BMP4-driven differentiation of trophoblast from human embryonic stem cells. Stem Cell Res 1:61–74
24. Xu RH et al (2008) NANOG is a direct target of TGFβ/Activin-mediated SMAD signaling in human ESCs. Cell Stem Cell 3:196–206

Methods in Molecular Biology (2024) 2767: 189–212
DOI 10.1007/7651_2023_487

Published online: 07 June 2023

Single-Cell mRNA-sncRNA Co-sequencing of Preimplantation Embryos

Savana Biondic, Cheng Zhao, Michael Hagemann-Jensen, Stewart J. Russell, Katherine Vandal, Jesica Canizo, Clifford L. Librach, and Sophie Petropoulos

Abstract

The development of single-cell multiomics has provided the ability to systematically investigate cellular diversity and heterogeneity in different biological systems via comprehensive delineations of individual cellular states. Single-cell RNA sequencing in particular has served as a powerful tool to the study of the molecular circuitries underlying preimplantation embryonic development in both the mouse and human. Here we describe a method to elucidate the cellular dynamics of the embryo further by performing both single-cell RNA sequencing (Smart-Seq2) and single-cell small non-coding RNA sequencing (Small-Seq) on the same individual embryonic cell.

Key words Blastocysts, Co-Seq, Embryos, Preimplantation development, Single-cell multiomics, Single-cell RNA sequencing, Small non-coding RNAs, Small-Seq, Smart-Seq2

1 Introduction

Assisted reproductive technologies (ARTs) such as in vitro fertilization (IVF) have become increasingly utilized to treat human infertility. As such, there is a need for the advancement of knowledge regarding preimplantation development. While many developmental processes such as embryonic genome activation (EGA), compaction, polarization, lineage specification, etc., have been thoroughly studied in both the mouse and human embryo [1], the role that small non-coding RNAs (sncRNAs) may play in these processes is yet to be examined in detail. sncRNAs are RNA species less than 300 nucleotides in length, of which there are multiple classes including microRNAs (miRNAs), small interfering RNAs (siRNAs), piwi-interacting RNAs (piRNAs), small nuclear RNAs (snRNAs), small nucleolar RNAs (snoRNAs), and transfer RNA

Supplementary Information The online version contains supplementary material available at https://doi.org/10.1007/7651_2023_487.

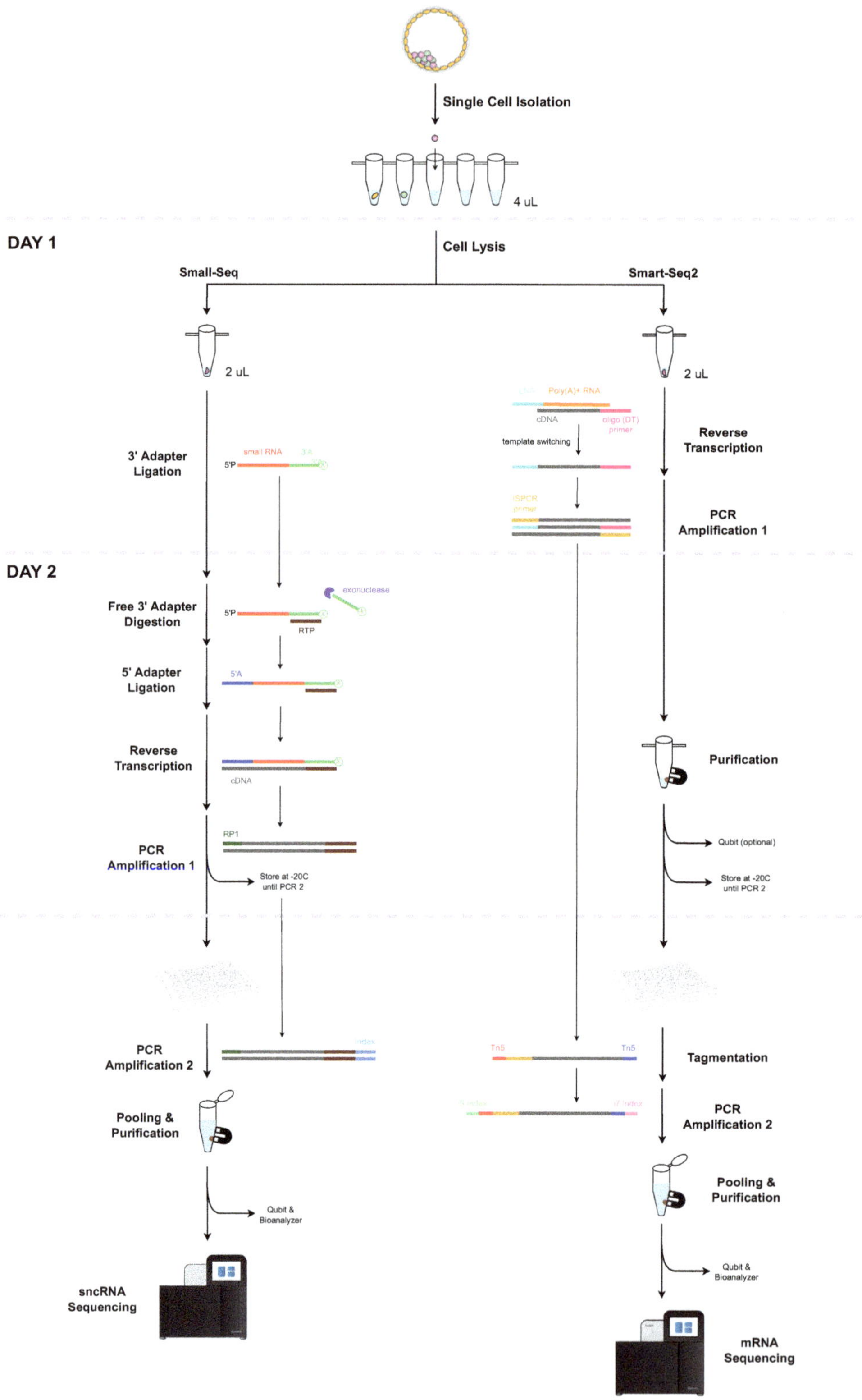

Fig. 1 Co-Seq Experimental Workflow. Overall strategy for profiling both the sncRNA and mRNA from individual blastomeres. Single cells are lysed, the lysates are split into two equal halves, and this is followed by sncRNA and mRNA library preparation. For sncRNA library preparation, a 3′ adaptor containing a 5′-adenylation and a 3′ dideoxycytidine is used to ligate to the 3′ end of sncRNA molecules. This is followed by digestion of

(tRNAs) [2]. In recent years, post-transcriptional gene expression regulation by sncRNAs has been shown to be involved in many essential mechanisms of mammalian cells [2], including pluripotency [3, 4], cellular differentiation [5, 6], and cell cycle regulation [7]. Given that many of these processes occur in the early embryo, it is plausible that sncRNAs may be critical for proper development during the preimplantation period.

Advancements in biotechnology, such as the emergence of robust methods enabling single-cell sncRNA sequencing [8], have made it possible to obtain a great deal of insight into the expression of sncRNAs during preimplantation development. However, during this period, cells of the embryo begin differentiating into the first embryonic lineages: the inner cell mass (ICM; which further differentiates into the epiblast (EPI) and primitive endoderm (PE)) and the trophectoderm (TE). Given that embryonic sncRNAs have not been studied on a single cell level, there are not yet sncRNA markers which are known to be highly expressed in each of these embryonic lineages, which would assist in distinguishing which profiles belong to which cell type. Here, we describe a detailed method for single-cell mRNA-sncRNA co-sequencing (Co-Seq), which leverages two previously published and validated protocols, Small-Seq [8] and Smart-Seq2 [9] (Fig. 1), to determine the parallel expression of both mRNA and sncRNA from the same individual cell.

2 Materials

For the original full length protocols and their material lists, see Hagemann-Jensen et al. [8] and Picelli et al. [9].

2.1 General

1. DNA-OFF™—*Takara Bio cat# 9036.*
2. RNase AWAY®—*VWR cat# 72830 022.*
3. Triton™ X-100 laboratory grade—*Sigma Life Science cat# X100-500ML.*
   CAUTION: Triton X-100 is harmful if swallowed and causes serious eye damage. Handle using appropriate safety equipment.
4. UltraPure™ Distilled Water (DNAse, RNAse, Free)—*Invitrogen cat# 10977-015.*
5. RNAse Inhibitor—*Takara Bio cat# 2313A.*

Fig. 1 (continued) unreacted 3′ adapter, ligation with the 5′ adaptor containing a 5′ amino-linker and a unique molecular identifier (UMI), reverse transcription, and PCR amplification. A total of 384 indexes, which are modified from standard Illumina TrySeq small RNA index primers, are used as barcodes. For Smart-Seq2 (mRNA) library preparation, reverse transcription is accompanied by template switching for the first-strand reaction. This is followed by PCR amplification and tagmentation. An Illumina dual-index strategy is used to barcode samples. Adapted from Hagemann-Jensen et al., 2018 (Small-seq) and Picelli et al., 2013 (Smart-Seq2)

6. Superscript II (includes DTT)—*Invitrogen cat# 18064-014. CAUTION: DTT is toxic when ingested. Avoid inhaling fumes or contact with the skin. Handle using appropriate safety equipment.*
7. AMPure XP Beads—*Beckman Coulter cat# A63881.*
8. 95% Denatured Ethyl Alcohol—*Chaptec cat# ABX5480. CAUTION: Flammable; handle using appropriate safety equipment.*
9. Elution Buffer (EB)—*Qiagen cat# 19086.*
10. Qubit™ 1× dsDNA Assay Kit—*Invitrogen cat# Q33231.*
11. Agilent high-sensitivity DNA kit—*Agilent Technologies cat# 5067-4626.*

2.2 Small-Seq

1. T4 RNA Ligase 2 truncated (includes PEG8000 and T4 RNA Ligase Buffer)—*New England Biolabs cat# M0373L.*
2. T4 RNA Ligase 1 (includes PEG8000 and T4 RNA Ligase Buffer)—*New England Biolabs cat# M0437M.*
3. Lambda Exonuclease—*New England Biolabs cat# M0262S.*
4. 5′ Deadenylase—*New England Biolabs cat# M0331S.*
5. ATP—*Thermo Scientific cat# R1441.*
6. Taq DNA Pol Buffer—*Thermo Scientific cat# 18067017.*
7. dNTP set (100 mM)—*Thermo Scientific cat# 10297018.*
8. Phusion Hot Start II DNA Pol (Includes Phusion Buffer)—*Thermo Scientific cat# F-549-L.*

2.3 Smart-Seq2

1. ERCC Spike-In Mix—*Invitrogen cat# 4456740.*
2. Deoxynucleotide (dNTP) Mix (10 mM)—*New England BioLabs cat# N0447S.*
3. SSRT II (FS) Buffer—*Invitrogen cat# 18064-014.*
4. Betaine (5 M)—*Sigma-Aldrich cat# B0300-5VL.*
5. $MgCl_2$ (1 M)—*Sigma-Aldrich cat# M1028-10X1ML.*
6. KAPA HiFi HotStart Ready Mix (2×)—*Kapa Biosystems cat# KK2602.*
7. Nextera XT DNA Library Preparation Kit—*Illumina cat# FC-131-1096.*
8. Nextera XT Index Kit V2 Set A-D—*Illumina cat# FC-131-2001, FC-131-2002, FC-131-2003, and FC-131-2004.*
9. UltraPure 10% SDS (Sodium Dodecyl Sulfate Solution)—*Invitrogen cat #15553-035.*

2.4 Consumables

1. Filtered pipette tips: 10, 20, 100, and 200 μL.
2. 200 μL non-filter pipette tip stacks (ex. Sarstedt cat# 70.3030.100).
3. 10 μL non-filter pipette tip stacks (ex. Sarstedt cat# 70.3010.100).
4. Eppendorf 0.2 mL PCR 8-Tube Strips—*Fisher Scientific cat# 95101002.*
5. Axygen® MaxyClear Microcentrifuge Tubes—*VWR cat# 10011-722.*
6. Qubit Assay Tubes—*Invitrogen cat# Q32856.*
7. Eppendorf 2 mL DNA/RNA LoBind Tubes, PCR Clean—*VWR cat# 22431048.*
8. Costar 96-Well "U" Bottom Plate—*Corning cat# 2797.*
9. Eppendorf LoBind twin.tec® PCR Plates—VWR cat# CA10049-106.
10. SecureSeal Thermal Adhesive Sealing Film for PCR Plates—*Fisher Scientific cat# T329-2.*
11. Index Replacement Caps—*Illumina cat# 15026762.*
12. TruSeq Index Plate Fixture Kit (2 fixtures)—*Illumina, catalog # FC-130-1005.*

2.5 Equipment

1. *Two Eppendorf Mastercycler X50s (*or with Eppendorf MasterCycler X50i).
2. Vornado Miniature Vortex Mixer—*Greiner bio-one cat# BV101B.*
3. Mini Centrifuge—*VWR cat# C0803.*
4. Magnetic Stand 96—*Ambion cat# AM10027.*
5. Mini-Plate Centrifuge—*VWR cat# C2001.*
6. Qubit 2.0 fluorometer—*Invitrogen cat# Q32866* (or alternative method for DNA concentration measurement).
7. Agilent 2100 Bioanalyzer—*Agilent Technologies, cat# 5067-4626.*
8. A compatible Illumina DNA sequencer (e.g., NovaSeq, NextSeq, and MiSeq).

2.6 Oligonucleotides

See Table 1.

2.7 Software

Various software including STAR [10], rsem [11], umi_tools [12], cutadapt [13], bowtie [14], samtools [15], picard (http://broadinstitute.github.io/picard), seqkit [15], BEDTools [16], edgeR [17], and MAST [18] may be used to analyze Smart-Seq2 and Small-Seq data. Here we developed a custom pipeline (https://github.com/zhaocheng3326/Pup_Savana_split_seq_protocol) that performs the data analysis task.

Table 1
Protocol DNA and RNA oligonucleotides from both Small-Seq (Small) and Smart-Seq2 (Smart), which are custom ordered from IDT

Oligo	Seq Protocol	Sequence (5′ → 3′)	HPLC purification	Working dilution (in H_2O)	Storage
5.8S rRNA masking	Small	ATC GGC AAG CGA CGC TCA GAC AGG CGT AGC CCC GGG AGG AAC CCG GGG CCG CAA GTG CGT TCG AAG TGT CGA TGA T/3Bio/	Y	100 μM	−20 °C for 6 months
3′ Adaptor (RA3)	Small	/5rApp/TG GAA TTC TCG GGT GCC AAG G/3ddC/	Y	100 μM	−20 °C for 6 months
5′ Adaptor (RA5)	Small	/5AmMC6/rGrU rUrCrA rGrArG rUrUrC rUrArC rArGrU rCrCrG rArCrG rArUrC rHrHrH rHrHrH rHrHrC rA	Y (RNAse-free)	100 μM	−80 °C for 6 months
RT primer (RTP)	Small	/5Biosg/CC TTG GCA CCC GAG AAT TCC rA	Y (RNAse-free)	100 μM	−80 °C for 6 months
PCR primer (RP1)	Small	AAT GAT ACG GCG ACC ACC GAG ATC TAC ACG TTC AGA GTT CTA CAG TCC GA	Optional	100 μM	−20 °C for 6 months
Locked nucleic acid (LNA)-containing TSO	Smart	AAGCAGTGGTATCAACGCAGAGTACATrGrG+G	Y	10 μM	−80 °C for 6 months
Oligo-dT_{30}VN	Smart	AAGCAGTGGTATCAACGCAGAGTACT30VN	Y	10 μM	−20 °C for 6 months
ISPCR primers	Smart	AAGCAGTGGTATCAACGCAGAGT	Y	10 μM	−20 °C for 6 months

3 Methods

3.1 Single-Cell Collection

Use filtered tips. Prepare a box of wet ice and a box of dry ice prior to beginning. Work surfaces should be free of RNase and DNA (cleaned with DNA-OFF™, RNAse-AWAY™) to minimize the risk of RNA degradation and cross-contamination.

1. Make enough Co-Seq Cell Lysis Buffer for the number of individual cells you expect to collect plus a few additional reactions (~10% of total, for example, if picking 96 cells, prepare enough for 110). Vortex the mix.

Co-Seq Cell Lysis Buffer

Reagent	Volume (μL)
0.4% Triton X-100	3.80
RNAse Inhibitor (40 U/μL)	0.20
Total volume	*4.00*

2. Distribute 4 μL of Co-Seq Cell Lysis Buffer into each of the wells of labeled PCR strips and place them on wet ice (*see* **Note 1**).
3. Commence the preferred method of zona pellucida removal, embryo dissociation, and single-cell collection from your specimen, placing the individual cells into the 4 μL of Co-Seq Lysis Buffer within the PCR strips. For a full protocol detailing picking single cells from preimplantation embryos, see Vandal et al., 2023, Methods in Molecular Biology, chapter entitled "Manual dissociation of mammalian preimplantation embryos for single-cell genomics".
4. Once a strip is filled, cap it and place it immediately on dry ice. Store all strips at −80 °C until prepared to proceed to the next steps.

3.2 Day 1: *Cell Lysis, 3′ Adapter Ligation (Small-Seq), Reverse Transcription, and PCR Amplification (Smart-Seq2)*

Use filtered tips. Prepare a box of wet ice prior to beginning. Work surfaces should be free of RNase and DNA (cleaned with DNA--OFF™, RNAse-AWAY™, and 70% ethanol) to minimize the risk of RNA degradation and cross-contamination (*see* **Note 2**).

1. Take out all non-enzymatic components of the Master Mixes below found in the −20 °C freezer and place on wet ice to thaw.
2. Prepare and label a new set of PCR strips, which will serve as the vessel in which to place the other half of each cell's lysate once split.

3. Take the cells to be prepped from the −80 °C freezer, and spin down the strips in a mini-centrifuge to thaw for approximately 30 s.
4. Place the strips containing the cells in the thermocycler at *72 °C for 20 min.*
5. During this time, prepare the Small-Seq 3′ Adapter (3′A) Ligation Master Mix (MM) and the Smart-Seq2 Reverse Transcription (RT) Master Mix by combining the reagents listed in the tables below. Make enough of each for the number of individual cells being prepped plus a few additional reactions (~10%). Vortex the mixes.

3′A Ligation MM (Small-Seq)

Reagent	Volume (μL)
rRNA masking oligo (100 μM)	0.05
RA3 (100 μM)	0.10
PEG 8000 (50%)	0.80
T4 RNA Ligase Buffer (10×)	0.40
T4 RNA Ligase 2 truncated (200 U/μL)	0.25
RNAse Inhibitor (40 U/μL)	0.10
H_2O	1.30
Total volume	*3.00*

RT MM (Smart-Seq2)

Reagent	Volume (μL)
dNTPs mix (10 mM)	1.00
Oligo dT (10 μM)	1.00
SSRT II (FS) Buffer (5×)	2.00
DTT (100 mM)	0.50
Betaine (5 M)	2.00
$MgCl_2$ (1 M)	0.10
RNAse Inhibitor (40 U/μL)	0.25
SSRT (Superscript) II	0.25
LNA (10 μM)	0.10
ERCC spike-ins (2.5×10^5 dilution)	0.10
H_2O	1.20
Total volume	*8.50*

6. Add 8.5 μL of the RT Master Mix (Smart-Seq2) into each well of the newly prepared strips.
7. Once the thermocycler program is complete, spin down the strips containing the cell lysate and place them immediately on ice.
8. Using a multichannel pipette set to 2 μL, mix the cell lysate thoroughly by pipetting up and down a minimum of 10×.
9. Transfer 2 μL of the 4 μL lysate to the newly prepared strips containing the RT Master Mix. From this point onward, this will be the half allocated to the Smart-Seq2 protocol. Pipette up and down 6× to mix.
10. Spin down and place the Smart-Seq2 strips into the thermocycler for reverse transcription with the following parameters: *90 min @42 °C, 10× of 2 min @50 °C, 2 min @42 °C, and then 15 min @70 °C* (~2.5 h).
11. To the strips containing the remaining half of the cell lysate, add 3 μL of the 3′A Ligation Master Mix. From this point onward, this will be the half allocated to the Small-Seq protocol. Pipette up and down 6× to mix.
12. Spin down and place the Small-Seq strips into the second thermocycler for 3′ Adapter ligation with the following parameters: *6 h @ 30 °C and then 10 h at 4 °C* (~16 h). For convenience, this step is usually carried out, so the samples are incubated overnight.
13. 5–10 min before the thermocycler RT program is complete, prepare the Smart-Seq2 PCR Amplification Master Mix by combining the reagents listed in the table below. Make enough for the number of individual cells being prepped plus a few additional reactions (~10%). Vortex the mix.

PCR Amplification MM (Smart-Seq2)

Reagent	Volume (μL)
KAPA HiFi HotStart ReadyMix (2×)	12.50
ISPCR Primers (10 μM)	0.20
H_2O	2.30
Total volume	*15.00*

14. Once the thermocycler program is complete, spin down the Smart-Seq2 strips and place them on ice.
15. Add 15 μL of the Amplification Master Mix to each well of the strip. Pipette up and down 6× to mix.
16. Spin down and place the Smart2-Seq strips into the thermocycler for PCR amplification with the following parameters:

3 min @98 °C, 20× of 20 s @98 °C, 15 s @67 °C, 6 min @72 °C, and then 5 min @72 °C (~2.5 h) (*see* **Note 3**). Follow this with an indefinite 4 °C hold. This step is usually carried out so the samples are incubated overnight. When the thermocycler program is complete, the Smart-Seq2 strips can be left in the indefinite 4 °C hold or can be stored in the fridge until ready to proceed with purification (Subheading 3.3 Day 2, **Steps 16–28**).

3.3 Day 2: 3′ Adapter Digestion, 5′ Adapter Ligation, Reverse Transcription and PCR Amplification (Small-Seq), and Purification (Smart-Seq2)

Use filtered tips. Prepare a box of wet ice prior to beginning. Work surfaces should be free of RNase and DNA (cleaned with DNA-OFF™, RNAse-AWAY™, and 70% ethanol) to minimize the risk of RNA degradation and cross-contamination.

1. 5–10 min before the 3′A Ligation thermocycler program is complete, prepare the Small-Seq 3′ Adapter Digestion Master Mix by combining the reagents listed in the table below. Make enough for the number of individual cells being prepped plus a few additional reactions (~10%). Vortex the mix.

3′A Digestion MM (Small-Seq)

Reagent	Volume (μL)
RTP (100 μM)	0.40
Lambda exonuclease (5 U/μL)	0.50
5′ Deadenylase (50 U/μL)	0.50
H_2O	1.60
Total volume	*3.00*

2. When the thermocycler program is complete, remove the Small-Seq strips immediately, spin them down, and place them on ice.
3. Add 3 μL of the 3′A Digestion MM to each well of the Small-Seq strips.
4. Spin down and place the Small-Seq strips into the thermocycler for 3′ Adapter Digestion with the following parameters: *15 min @ 30 °C, followed by 15 min @ 37 °C* (~30 min).
5. 5–10 min before the 3′A Digestion thermocycler program is complete, prepare the 5′A Ligation MM by combining the reagents listed in the table below. Make enough for the number of individual cells being prepped plus a few additional reactions (~10%). Vortex the mix.

5′A Ligation MM (Small-Seq)

Reagent	Volume (μL)
RA5 (100 μM)	0.10
ATP (100 mM)	0.07
T4 RNA ligase buffer (10×)	0.25
T4 RNA ligase 1 (30 units/μL)	0.15
H_2O	1.43
Total volume	*2.00*

6. Add 2 μL of the 5′A Ligation MM to each well of the Small-Seq strips.
7. Spin down and place the Small-Seq strips into the thermocycler for 5′ Adapter Ligation with the following parameters: *1 h @ 37 °C*.
8. 5–10 min before the 5′ Adapter Ligation program is complete, prepare the Small-Seq Reverse Transcription (RT) Master Mix by combining the reagents listed in the table below. Make enough for the number of individual cells being prepped plus a few additional reactions (~10%). Vortex the mix.

RT MM (Small-Seq)

Reagent	Volume (μL)
Taq DNA Pol Buffer (10×)	1.95
DTT (100 mM)	1.20
dNTP mix (25 mM/each)	0.30
RNAse Inhibitor (40 U/μL)	0.10
SuperScript II (200 U/μL)	0.50
H_2O	0.95
Total volume	*5.00*

9. Once the 5′A Ligation thermocycler program is complete, spin down the Small-Seq strips and place them on ice.
10. Add 5 μL of the RT MM to each well of the Small-Seq strips.
11. Spin down and place the Small-Seq strips into the thermocycler for Reverse Transcription with the following parameters: *1 h @ 42 °C, followed by 15 min @ 70 °C* (~1.5 h).
12. 5–10 min before the RT thermocycler program is complete, prepare the Small-Seq PCR Amplification Master Mix #1 by combining the reagents listed in the table below. Make enough for the number of individual cells being prepped plus a few additional reactions (~10%). Vortex the mix.

PCR Amp MM 1 (Small-Seq)

Reagent	Volume (μL)
Phusion buffer (5×)	5.00
Phusion Hot Start II	0.50
dNTP mix (25 mM/each)	0.15
RP1 (100 μM)	0.25
H_2O	4.10
Total volume	*10.0*

13. Once the RT thermocycler program is complete, spin down the Small-Seq strips and place them on ice.
14. Add 10 μL of the PCR Amp MM to each well of the Small-Seq strips.
15. Spin down and place the Small-Seq strips into the thermocycler for the first PCR amplification with the following parameters: 30 s @98 °C, 13× of 10 s @98 °C, 30 s @60 °C, 30 s @72 °C, and then 5 min @72 °C (~35 min). Follow this with an indefinite 4 °C hold.

    **Small-Seq Pause Point: PCR product can be stored at −20 °C for long-term usage, or you can proceed to Subheading 3.4. Small-Seq PCR 2 and Final Library Prep*

16. Before starting purification (*see* **Note 4**) of the Smart-Seq2 samples, let the Ampure XP beads equilibrate to room temperature (20 °C) for 30 min (*see* **Note 5**). In addition, prepare enough fresh 80% ethanol (EtOH) for the wash steps of the protocol.
17. Work at room temperature. In each of the wells of the Costar 96-Well "U" Bottom Plate, add 25 μL of the beads (such that they will be in a 1:1 ratio with the amount of PCR product).
18. When ready, take the Smart-Seq2 strips from the thermocycler or the fridge and spin them down.
19. Add the PCR products to the beads using a multichannel pipette, mixing up and down 10× to mix. Do this gently to avoid the production of bubbles in the wells.
20. Let this mixture sit at room temperature for 8 min. During this time, new strips can be labeled to which the cDNA will be moved once purified.
21. Next, place the Costar plate on the magnetic rack for 5 min.
22. Keeping the plate on the magnetic stand, remove the supernatant. Be careful not to make contact with or disturb the pellet of beads making contact with the magnet.

23. With the plate still on the magnetic stand, add 180 μL of 80% EtOH and incubate for 1 min. Afterward, remove the supernatant and repeat this wash. Once finished, be sure to remove as much EtOH from the well as possible (*see* **Note 6**).
24. Let the beads dry for approximately 5 min, or until microdroplets of EtOH are no longer visible. Try not to exceed approximately 8 min, as over-drying can make it difficult to resuspend beads and lead to sample loss.
25. Add 16.5 μL of elution buffer (EB) to each well.
26. Once all the wells have EB, take the plate off the magnet and use a multichannel pipette to mix each row 10× or until beads are fully resuspended. Let the mixture sit at room temperature for 2 min. We recommend covering the plate so as to avoid evaporation.
27. Place the plate back onto the magnetic stand to collect the beads for 1 min.
28. Collect 16 μL of the supernatant into the new strips labeled in **step 17**, ensuring no beads are taken up in the process (*see* **Note 7**). Qubit or a Bioanalyzer can be used at this point to determine individual sample concentrations (*see* **Note 8**).

**Smart-Seq2 Pause Point: Purified cDNA can be stored at −20 °C for long-term usage*
**At this point, downstream library preparations for Small-Seq and Smart-Seq2 can be carried out separately, at your convenience*

3.4 Small-Seq PCR 2 and Final Library Prep

Use filtered tips. Prepare a box of wet ice prior to beginning. Work surfaces should be free of RNase and DNA (cleaned with DNA-OFF™, RNAse-AWAY™, and 70% ethanol) to minimize the risk of RNA degradation and cross-contamination.

1. Prepare the Small-Seq PCR Amplification Master Mix #2 by combining the reagents listed in the table below. Make enough for the number of individual cells being prepped plus a few additional reactions (~10%). Vortex the mix.

PCR Amp MM 2 (Small-Seq)

Reagent	Volume (μL)
Phusion buffer (5×)	5.00
Phusion Hot Start II	0.25
dNTP mix (25 mM/each)	0.20
RP1 (100 μM)	0.20
H_2O	17.55
Total volume	*23.2*

2. Dispense 23 μL of the PCR Amp MM2 into the wells of a 96-well LoBind PCR plate.
3. Add 1 μL of each index primer (RPIX, 50 μM) to each well of the new plate, making sure to keep track of which specific indexes are in which locations.
4. When ready, take the Small-Seq strips containing the first PCR product from the −20 °C freezer and spin them down.
5. Transfer 1 μL of sample from the strips into the corresponding wells of the new plate containing the second PCR mix and the index primers. Pipette up and down 6× to mix.
6. Seal the plate with adhesive sealing film. Spin down and place the plate into the thermocycler for the second PCR amplification with the following parameters: *30 s @98 °C, 13× of 10 s @98 °C, 30 s @67 °C, 30 s @72 °C, and then 5 min @72 °C* (~35 min). Follow this with an indefinite 4 °C hold.

   **Small-Seq Pause Point: cDNA can be stored at −20 °C for long-term usage*

7. Once the thermocycler program is complete, spin down the Small-Seq plate containing the now-barcoded libraries.
8. Add (pool) a 2 μL aliquot of each PCR sample into a 2 mL LoBind tube.
9. Before starting purification of the Small-Seq pool, let the Ampure XP beads equilibrate to room temperature (20 °C) for 30 min. In addition, prepare enough fresh 80% ethanol (EtOH) for the wash steps of the protocol.
10. Divide the Small-Seq pool into a few equal sub-pools of about 25–50 μL by pipetting the determined amount into separate wells of a Costar 96-Well "U" Bottom Plate. Make sure the original pool is mixed very well.
11. Purify each sub-pool with a 1:1 volume of magnetic bead using the same procedure as above (Day 2: 3′ Adapter Digestion, 5′ Adapter Ligation, Reverse Transcription and PCR Amplification (Small-Seq), Purification (Smart-Seq2), **steps 13–25**). Resuspend the final purified product in water (DNAse- and RNAse-free) instead of EB.
12. Once purified, recombine each of the sub-pools in a new 2 mL DNA Lo-Bind tube. This is the final purified library pool, which along with the plate containing any remaining PCR product can be stored at −20 °C indefinitely.
13. Determine the concentration of the final purified library pool using the Qubit.
14. Run the final library pool on a bioanalyzer to see the profile and library size (bp).

15. Use the *Concentration* obtained with the Qubit and the *Average size* obtained on the Bioanalyzer to calculate the *Molarity* of the final library as follows (*see* **Note 9**):

$$\text{Molarity(nM)} = \frac{\text{Concentration(ng/}\mu\text{L)} \times 10}{\text{Average size(bp)} \times 660\,\text{(g/mol)}}$$

3.5 Smart-Seq2 Tagmentation, PCR 2, and Final Library Prep

Use unfiltered tips (stacks). Prepare a box of wet ice prior to beginning. Work surfaces should be free of RNase and DNA (cleaned with DNA-OFF™, RNAse-AWAY™, and 70% ethanol) to minimize the risk of RNA degradation and cross-contamination.

1. Take the ATM, TD, NPM (all from the Nextera XT DNA Library Preparation Kit) and the Nextera indexes out from the −20 °C freezer and place on wet ice to thaw approximately 30 min prior to beginning.
2. Into each well of a new 96-well LoBind PCR plate, dispense 2 μL of TD using a multichannel pipette.
3. Take the Smart-Seq2 strips containing the first PCR product from the −20 °C freezer and spin them down to thaw.
4. Distribute a predetermined amount of each sample into the TD-containing wells of the plate such that 1 ng of cDNA is added to the reaction (being careful not to mix up sample locations). This could be a universal volume if concentrations obtained after the first PCR amplification are generally consistent (as determined by the Qubit and/or bioanalyzer). If individual sample concentrations are quite variable, however, it is recommended that the volume of each sample be calculated in order to have approximately 1 ng of cDNA in every reaction. Volumes added should range between 1 μL and 5 μL. For example, dilute samples determined to have concentrations much higher than 1 ng/μL by the Qubit in EB and add 1 μL. If a concentration is lower, for example, 0.25 ng/μL, one should add 4 μL of sample.
5. Next, using a multichannel pipette, add 1 μL of ATM to the TD and cDNA in each well of the plate. Pipette up and down 10× to mix.
6. Seal the plate with adhesive sealing film. Spin down and place the plate into the thermocycler for the tagmentation with the following parameters: *5 min @ 55 °C.*
7. Once the thermocycler program is complete, spin down the Smart-Seq2 plate and immediately add 1 μL of 0.2% SDS to all the wells (*see* **Note 11**).
8. Seal and spin down the plate again, and incubate the mixture with the SDS at RT for exactly 5 minutes. Once 5 minutes has passed, move the plate onto ice.

9. Using a multichannel pipette, add 3 μL of NPM to each well of the Smart-Seq2 plate, keeping the plate on the ice.
10. Next, add water (DNAse- and RNAse-free) to each well such that the total volume of the sample + the water is 9 μL. This could be the same or a different volume for each sample depending on the inputted volume in **step 4** (for example, if you inputted 4 μL of sample, add 5 μL of water).
11. Using a multichannel pipette, add 1 μL of each of the pre-assigned Illumina adapters to each sample. Be careful to place index caps upside down on a Kimtech wipe and generally avoid any cross-contamination of indexes. If there is any chance of contamination (e.g., caps touching and touching the inside of the cap with a glove), use a replacement cap.
12. Once all samples are barcoded, seal and vortex the plate.
13. Spin down the Smart-Seq2 plate and place into the thermocycler for the PCR amplification 2 (Nextera) with the following parameters: *3 min @ 72 °C, 30 s @ 95 °C, 10× of 10 s @95 °C, 30 s @55 °C, 30 s @72 °C, and then 5 min @72 °C.*
14. Aliquot 3 μL of each PCR sample into a 2 mL LoBind tube to pool them. Mix thoroughly.
15. Before starting purification of the Smart-Seq2 pool, let the Ampure XP beads equilibrate to room temperature (20 °C) for 30 min. In addition, prepare enough fresh 80% ethanol (EtOH) for the wash steps of the protocol.
16. Divide the Smart-Seq2 pool into a few equal sub-pools of about 25–50 μL by pipetting the determined amount into separate wells of a Costar 96-Well "U" Bottom Plate. Make sure that the initial pool is thoroughly mixed.
17. Purify each sub-pool with a 1:1 volume of magnetic bead using the same procedure as above (Day 2: 3′ Adapter Digestion, 5′ Adapter Ligation, Reverse Transcription and PCR Amplification (Small-Seq), Purification (Smart-Seq2), **steps 13–25**).
18. Once purified, combine each of the sub-pools in a new 2 mL DNA Lo-Bind tube. This is the final purified library pool, which along with the plate containing any remaining PCR product can be stored at −20 °C indefinitely.
19. Determine the concentration of the final purified library pool using the Qubit.
20. Run the final library pool on a bioanalyzer to see the profile and library size (bp). The final pool as well as the plate containing any remaining PCR product can be stored at −20 °C indefinitely.
21. Use the *Concentration* obtained with the Qubit and the *Average size* obtained on the Bioanalyzer to calculate the *Molarity* of the final library as follows (*see* **Note 9**):

$$\text{Molarity(nM)} = \frac{\text{Concentration(ng/}\mu\text{L)} \times 10}{\text{Average size(bp)} \times 660\,\text{(g/mol)}}$$

3.6 Data Analysis

The analysis will be divided into two parts, based on the type of dataset.

1. Various pipelines can be utilized for the Smart-Seq2 data analysis. Here we present a simplified example that was modified from the Smart-Seq2 pipeline available at https://github.com/nf-core/smartseq2 [19]. After quality control using FastQC [20], the raw fastq-files were mapped to the mm10 reference genome (v.2020-A, mm10), downloaded from the 10× Genomics website (https://support.10xgenomics.com/single-cell-gene-expression/software/release-notes/build#mm10_2020 A), using STAR with default settings. Raw read counts were further estimated using rsem-calculate-expression from RSEM tool with the option of "--single-cell-prior." We chose to use STAR for mapping and genome annotation from 10× genomics because it allows the possibility to compare datasets generated from 10× Chromium. High-quality cells from Smart-Seq2 datasets were required to have at least 2000 expressed genes and a percentage of mitochondrial genes less than 0.15. Read counts were further normalized to TP10K (transcripts per 10,000 transcripts) and MAST was utilized to calculate the differentially expressed genes (DEGs) between "Full cells" and "Half cells" using a FDR less than 0.05.

   Running scripts:

```
###  setting
STAR_REF=/STAR_INDEX_FOLDER/star      ### the genome index created by STAR tool
GTF=/GENOME_ANNOTATION_FOLDER/genes.gtf      ### the genome annotation
RSEM_REF=/RSEM_TRANSCRIPTS_FOLDER/rsem      ### prepared transcripts index for
RSEM


### reads mapping
STAR --runThreadN 4 --outFilterMultimapNmax 1 --genomeDir ${STAR_REF} --
readFilesIn ${SAMPLE}.fastq.gz --quantMode TranscriptomeSAM --outSAMstrand-
Field intronMotif --readFilesCommand zcat --outSAMtype BAM Unsorted --
outFileNamePrefix ${SAMPLE}.


###  feature counting
rsem-calculate-expression --no-bam-output --single-cell-prior --alignments -p
4 ${SAMPLE}.Aligned.toTranscriptome.out.bam ${RSEM_REF} ${SAMPLE}.rsem -q
```

2. The reads obtained from Small-seq were processed using a modified Small-Seq pipeline based on the original paper [21]. Briefly, the UMIs (8 bp associated with the unique molecular identifiers) were removed from sequence reads and appended to the reads' name using umi_tools. Then, cutadapt was used to remove adapter and polyA sequences (parameters: -e 0.1 -o 5 -m 18 -M 86 -u 2). The resulting clean reads were aligned to the mouse genome (mm10) using bowtie with the following parameters: "-a --best --strata -v 2 -m 50 -S -q -p 2." Reads with length less than 20 bp with one or more mismatches and reads with length between 20 bp and 40 bp with two mismatches were filtered out using custom Python script. PCR amplicons were collapsed, and RNA molecules were counted using UMI tools with the "dedup --method directional" option. Deduplicated reads were remapped onto the genome using the same procedure as above to recover multiple mapping information dampened by the "dedup" process. Annotation coordinates on the genome were then extracted from Mirbase V22 (miRNA) [22], Gencode V39 (snoRNA, snRNA, rRNA, and others) [23], and GtRNAdb v19 (tRNA) [24] to assign the reads to their corresponding categories. The reads were assigned to each category in a hierarchical manner based on their overlap with the above annotations. If a read mapped to more than one category, we assigned it to only one category based on the following order of priority: miRNA, snoRNA, snRNA, tRNA, rRNA, and other types of transcripts. If a read mapped to more than one feature belonging to the same category, the read would be split evenly among each overlapping feature. Normalization and differential expression analysis for miRNA expression between "Full cell" and "Half cells" were performed by edgeR [17]. Uniform Manifold Approximation and Projection (UMAP) [12] were calculated based on the all-expressed miRNA (at least in 2 cells) using umap function.

Running scripts:

```
### setting
N=86 ## 101-8-2-1 -4(adapter length 5)
ADAPTER_FILE=/DOC/cutadapt_3prime.fa  ### adapter sequence
RGB=/GENOME_BOWTIE_INDEX /bowtie      ### genome reference index
REF=/ GENOME_SEQUENCE /genome.fa      ### genome reference
ANDB=/DOC/small.anno.bed              ### Genome coordination of transcripts
annotation
ANOD=/DOC/annotation.report.order     ### categories assigning order
SRC=/SCRIPTS_FOLDER                   #### folder to store scripts
```

```
### removing adapter and getting UMI sequence
umi_tools extract --bc-pattern=NNNNNNNN -I ${SAMPLE}.fastq -S
${SAMPLE}.extract.fastq -L ${SAMPLE}.extract.log
cutadapt -a file:${ADAPTER_FILE}  -e 0.1 -o 5 -m 18 -M ${N} -u 2 -o
${SAMPLE}.extract.cutadp.fastq  ${SAMPLE}.extract.fastq  >${SAM-
PLE}.extract.cutadp.log


### mapping
bowtie  -a --best --strata -v 2 -m 50  -S -q -p 2 $RGB ${SAMPLE}.
extract.cutadp.fastq 2>bowie.log |python $SRC/filter_sam_mismatch.
py|samtools view -Sb -F 4 > ${SAMPLE}.bowtie.bam
samtools sort ${SAMPLE}.bowtie.bam  -o ${SAMPLE}.bowtie.sort.bam


### deduplication
samtools index ${SAMPLE}.bowtie.sort.bam
umi_tools dedup --method directional --output-stats dedup.log1  --
read-length -I ${SAMPLE}.bowtie.sort.bam -S ${SAMPLE}.bowtie.sort.
dedup.temp.bam --random-seed 123 > /dev/null
samtools  sort  ${SAMPLE}.bowtie.sort.dedup.temp.bam  -o  ${SAMPLE}.
bowtie.sort.dedup.bam
picard  SamToFastq  INPUT=${SAMPLE}.bowtie.sort.dedup.bam  FASTQ=
${SAMPLE}.bowtie.sort.dedup.bam.fastq.temp  2>> log.txt
seqkit  fx2tab  ${SAMPLE}.bowtie.sort.dedup.bam.fastq.temp|sort  |
uniq |seqkit tab2fx -w 0 > ${SAMPLE}.bowtie.sort.dedup.bam.fastq


### remapping
bowtie  -a --best --strata -v 2 -m 50  -S -q -p 2 $RGB ${SAMPLE}.
bowtie.sort.dedup.bam.fastq  2>bowie.dedup.log  |python  $SRC/fil-
ter_sam_mismatch.py|samtools  view  -Sb  -F  4  >  ${SAMPLE}.bowtie.
dedup.reMap.bam  ### recover the multiple mapping results
samtools sort ${SAMPLE}.bowtie.dedup.reMap.bam -o ${SAMPLE}.bowtie.
dedup.reMap.sort.bam


### feature counting
intersectBed  -a  <(bam2bed  <${SAMPLE}.bowtie.dedup.reMap.sort.bam|
awk -F "\t" '{OFS="\t";if (($3-$2) <  200) print $1,int(($2+$3)/2)-
1,int(($2+$3)/2),$4}') -b ${ANDB} -wa -wb |cut -f 4,11,12 |sort |
uniq > ${SAMPLE}_tmpCount.detail.out
python $SRC/count_smallRNA.ordered.exp.py -a ${ANOD} -c ${SAMPLE}
_tmpCount.detail.out -o ${SAMPLE}
```

(*see* **Note 10**)

**Above scripts are all included in https://github.com/zhaocheng3326/Pup_Savana_split_seq_protocol*

4 Notes

1. Plates can be used if using a cell sorting system.
2. This step (Day 1: Cell Lysis, 3′ Adapter Ligation (Small-Seq), Reverse Transcription and PCR Amplification (Smart-Seq2)) requires a 16 h incubation; therefore, it may be convenient to start in the afternoon or evening and incubate overnight.
3. The number of PCR cycles depends on the input amount of RNA. We have found that 18–22 cycles are sufficient to obtain an abundant amount of amplified cDNA from embryonic cells derived from various species when the lysate is split. The optimal number of cycles for your samples should be determined using the Qubit or Bioanalyzer after sample purification.
4. For purification, unfiltered tips (stacks) can be utilized. Work surfaces should be free of RNase and DNA (cleaned with DNA-OFF™, RNAse-AWAY™, and 70% ethanol) to minimize the risk of RNA degradation and cross-contamination.
5. Make sure the beads are homogeneous prior to adding. We recommend letting the stock of beads come to room temperature, mixing it thoroughly, and making 1.5–5 mL aliquots (in Eppendorf tubes) to make it easier for each purification.
6. If any wells appear to have excess EtOH remaining, you can use a pipette tip and carefully place it near the pellet. The capillary action will remove excess EtOH so that all wells dry around the same time.
7. It is possible there will be slightly less than 16 μL due to evaporation.
8. If this is the first time you are running the protocol or the quality/quantity of the samples is uncertain, it may be desirable to run a bioanalyzer after step 25 to see the cDNA profiles and concentrations from individual cells prior to tagmentation. This is not however required every time the protocol is run.
9. Check with the facility you will be sequencing with for their final pool requirements. Generally, they require a concentration of 2–10 nM.
10. Our maximum sequenced length for small-RNA-Seq was 101 bp. By requiring the removal of eight bases associated with UMIs and the "CA" bases as well as minimal 5 bp overlap of adapter or polyA sequence, read length less than 87 bp was possibly generated from small RNAs. That is how "-M 86" was set in cutadapt process.
11. It is helpful to make the 0.2% dilution of SDS and aliquot it into an eight well strip beforehand so the multichannel can be used to more quickly dispense.

5 Strengths and Limitations

The constituents of the Co-Seq methods, Small-Seq [8] and Smart-Seq2 [9], are validated, well-established protocols both of which have been utilized to examine biological systems such as the preimplantation embryo successfully in numerous prior studies [25–27]. For an in-depth explanation of the limitations of each individual method, see the original published protocols [8, 9]. Our method leverages the robustness of the two original methods but enables the simultaneous sequencing of both sncRNA and mRNA from an individual cell. This provides researchers the ability to leverage the mRNA cell type signatures and therefore study all the sncRNA biotypes in individual cells, such as in preimplantation embryos and blastoids, where cellular heterogeneity is present and no sncRNA markers of cellular lineage are yet established. In addition, associations can be drawn between changes to lineage-specific sncRNA profiles and those in the transcriptome, potentially providing insight into the regulatory role of specific sncRNAs on gene expression. Further, the protocol can be leveraged to analyze the effects of developmental insults on both the sncRNAs and gene expression simultaneously, providing a more complete picture of the molecular circuitries underlying phenotype.

To evaluate the mRNA and sncRNA libraries produced from our half-cell lysates, eight cell mouse embryos ($n = 11$ embryos (Small-Seq); $n = 14$ embryos (Smart-Seq2)) were dissociated and picked. Four cells from each embryo were subjected to the method described here, including the splitting of the lysate. For comparison, the other four cells from the same embryos were prepared for Smart-Seq2 or Small-Seq as full cell lysates. The quantity and quality of the cDNA and small RNA libraries were assessed using an Agilent 2100 BioAnalyzer, and libraries were sequenced using the Illumina NextSeq 500. Bioinformatics analysis using established in-house pipelines compared half cells to whole cells from the same embryo with both protocols to determine if any genes or non-coding RNA families are differentially detected as a result of the split-lysate procedure. Compared to full cells, the number of expressed genes (Fig. 2a) was decreased in half cells. However, the high correlation between the gene expression profiles of the two half-lysis samples (Fig. 2b) indicates that there were only minor, tolerable information losses when profiling gene expression after splitting cells. We identified only six differentially expressed genes that were upregulated in full cells compared to the half-lysis samples (adjusted p-value <0.05). For sncRNA profiling, the number of expressed miRNAs was also decreased as expected, but we did not

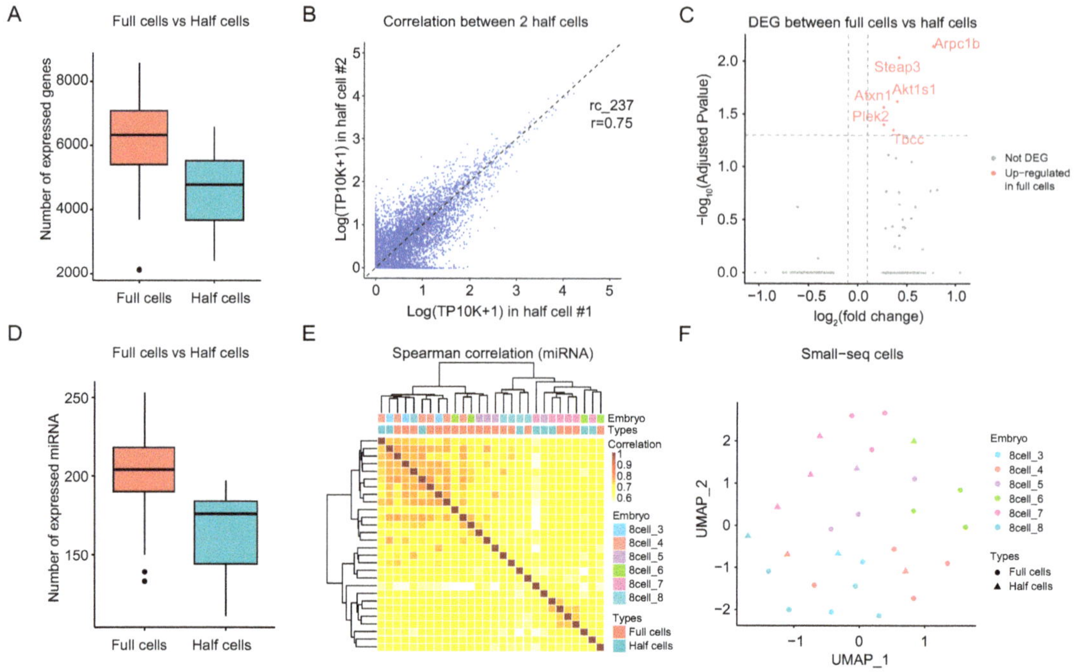

Fig. 2 Validation studies. (**a**) Comparison of the number of expressed genes detected in the Smart-seq2 data between full cells and half-cell lysates. (**b**) Representative correlation plot of a single cell (rc_237) which was lysed and divided into two halves. Smart-seq2 was performed on each half. A scatter plot was generated to show log-transformed gene expression levels (TP10K, transcripts per 10,000 transcripts) and the Spearman correlation coefficient between the two halves ($r = 0.75$). (**c**) A volcano plot was generated to visualize the expression fold change and the adjusted p-value between 35 full-cell and 95 half-cell lysates. (**d**) Comparison of the number of expressed miRNAs detected in the Small-seq data between full-cells and half-cell lysates. (**e**) A heatmap was generated to show the pairwise correlation of miRNA expression across all 28 samples included in the Small-seq dataset. The embryo ID or cell status (split or full) are indicated by colors above the heatmap. (**f**) Uniform Manifold Approximation and Projection (UMAP) plots were generated for the 28 samples, with shape indicating cell status (split or full) and color indicating the embryo ID, respectively

observe any significant differences due to splitting by checking the general correlation (Fig. 2e) among all samples and the UAMP results based on miRNA expression. This suggests that a significant amount of information was not lost due to lysate splitting and that the main factor driving differences is likely due to embryo origin. Furthermore, we did not detect any significantly differentially expressed miRNAs between full cells and half-lysis samples based on the FDR cut-off of less than 0.05, supporting the use of the split cell method as a practical protocol. Overall, Co-Seq is a promising method for simultaneous detection of mRNA and sncRNAs from a single cell. It is recommended that this protocol be used alongside Small-Seq on full cells to ensure important biological information is not lost.

Acknowledgments

This work was supported by grants from the Canadian Institutes of Health Research (PJT-178082), the Swedish Research Council (2016-01919), and Swedish Society for Medical Research (Dnr4-236-2107). SP holds the Canada Research Chair in Functional Genomics of Reproduction and Development (950-233204) and KV was granted a NSERC PGS D scholarship.

References

1. Biondic S, Canizo J, Vandal K et al (2023) Cross-species comparison of mouse and human preimplantation development with an Emphasis on Lineage Specification. Reproduction:R1–R14
2. Aalto AP, Pasquinelli AE (2012) Small non-coding RNAs mount a silent revolution in gene expression. Curr Opin Cell Biol 24: 333–340
3. Zhang Z, Zhuang L, Lin C-P (2019) Roles of MicroRNAs in establishing and modulating stem cell potential. Int J Mol Sci 20:3643
4. Gu K-L, Zhang Q, Yan Y et al (2016) Pluripotency-associated miR-290/302 family of microRNAs promote the dismantling of naive pluripotency. Cell Res 26:350–366
5. Viswanathan SR, Mermel CH, Lu J et al (2009) microRNA expression during trophectoderm specification. PLOS ONE 4:e6143
6. García-López J, Alonso L, Cárdenas DB et al (2015) Diversity and functional convergence of small noncoding RNAs in male germ cell differentiation and fertilization. RNA 21:946–962
7. Wang Y, Blelloch R (2009) Cell cycle regulation by MicroRNAs in embryonic stem cells. Cancer Res 69:4093–4096
8. Hagemann-Jensen M, Abdullayev I, Sandberg R et al (2018) Small-seq for single-cell small-RNA sequencing. Nat Protoc 13:2407–2424
9. Picelli S, Björklund ÅK, Faridani OR et al (2013) Smart-seq2 for sensitive full-length transcriptome profiling in single cells. Nat Methods 10:1096–1098
10. Dobin A, Davis CA, Schlesinger F et al (2013) STAR: ultrafast universal RNA-seq aligner. Bioinformatics 29:15–21
11. Li B, Dewey CN (2011) RSEM: accurate transcript quantification from RNA-Seq data with or without a reference genome. BMC Bioinform 12:323
12. Smith T, Heger A, Sudbery I (2017) UMI-tools: modeling sequencing errors in Unique Molecular Identifiers to improve quantification accuracy. Genome Res 27:491–499
13. Martin M (2011) Cutadapt removes adapter sequences from high-throughput sequencing reads. EMBnetjournal 17:10–12
14. Langmead B, Trapnell C, Pop M et al (2009) Ultrafast and memory-efficient alignment of short DNA sequences to the human genome. Genome Biol 10:R25
15. Shen W, Le S, Li Y et al (2016) SeqKit: a cross-platform and ultrafast toolkit for FASTA/Q file manipulation. PLoS One 11:e0163962
16. Quinlan AR, Hall IM (2010) BEDTools: a flexible suite of utilities for comparing genomic features. Bioinforma Oxf Engl 26:841–842
17. Robinson MD, McCarthy DJ, Smyth GK (2010) edgeR: a Bioconductor package for differential expression analysis of digital gene expression data. Bioinformatics 26:139–140
18. Finak G, McDavid A, Yajima M et al (2015) MAST: a flexible statistical framework for assessing transcriptional changes and characterizing heterogeneity in single-cell RNA sequencing data. Genome Biol 16:278
19. Ewels PA, Peltzer A, Fillinger S et al (2020) The nf-core framework for community-curated bioinformatics pipelines. Nat Biotechnol 38: 276–278
20. Andrews S FastQC A Quality Control tool for High Throughput Sequence Data. https://www.bioinformatics.babraham.ac.uk/projects/fastqc/
21. Faridani OR, Abdullayev I, Hagemann-Jensen M et al (2016) Single-cell sequencing of the small-RNA transcriptome. Nat Biotechnol 34: 1264–1266
22. Kozomara A, Birgaoanu M, Griffiths-Jones S (2019) miRBase: from microRNA sequences to function. Nucleic Acids Res 47:D155–D162
23. Harrow J, Frankish A, Gonzalez JM et al (2012) GENCODE: the reference human genome annotation for the ENCODE Project. Genome Res 22:1760–1774

24. Chan PP, Lowe TM (2016) GtRNAdb 2.0: an expanded database of transfer RNA genes identified in complete and draft genomes. Nucleic Acids Res 44:D184–D189
25. Petropoulos S, Edsgärd D, Reinius B et al (2016) Single-cell RNA-Seq reveals lineage and X chromosome dynamics in human preimplantation embryos. Cell 165:1012–1026
26. Posfai E, Petropoulos S, de Barros FRO et al (2017) Position- and Hippo signaling-dependent plasticity during lineage segregation in the early mouse embryo. elife 6:e22906
27. Zhao C, Biondic S, Vandal K et al (2022) Single-cell multi-omics of human preimplantation embryos demonstrates susceptibility to glucocorticoids. Genome Res gr.276665.122

Methods in Molecular Biology (2024) 2767: 213–250
DOI 10.1007/7651_2023_492

Published online: 24 June 2023

Evaluation of Stem-Cell Embryo Models by Integration with a Human Embryo Single-Cell Transcriptome Atlas

San Kit To, Bradley Balaton, and Vincent Pasque

Abstract

Single-cell RNA sequencing (scRNA-seq) revolutionized our understanding of the molecular processes of early development and provided us with the means to capture biological heterogeneity and assess the cellular composition in early embryos. Comparative analysis of the transcriptional landscapes of embryos with single-cell resolution allows us to better understand and improve stem-cell-based embryo models. However, proper comparison between different single-cell datasets acquired by different laboratories and through different technologies is imperative for adequate analysis and findings. In this chapter, we focus on the analysis of human blastoids, which model the blastocyst, and their integrative analysis with human embryo datasets and a 2D in vitro early development model system dataset, which models epiblast, extraembryonic mesoderm, and trophoblast cells.

Key words Bioinformatics, Blastoids, Human embryo, Integration, Seurat, Single-cell RNA sequencing

1 Introduction

The sudden surge in the number of single-cell studies in the last decade has revolutionized our understanding of the molecular processes of normal embryo development [2, 14, 27, 30, 32, 38]. Single-cell RNA sequencing (scRNA-seq) enables the profiling of transcriptome-wide gene expression at single-cell resolution, allows the capture of heterogeneity among cell types, and increases our ability to discover novel cell types [15, 26] and construct cell atlases [2, 4, 5, 18, 28]. scRNA-seq has been heavily applied to understand health and diseases [13, 31]. Due to ethical and legal reasons, it is difficult to study human embryogenesis [11]. In vitro model systems provide the ability to model and better understand the mechanism and pathology of early human development (reviewed in ([20]). However, in order to ensure that in vitro model systems accurately model embryo development, it is of great importance to profile the transcriptional landscape of stem-cell-based model systems and perform comparative analysis with published embryo single-cell transcriptomes. Blastoids are a recently developed system which cultures naive pluripotent stem cells in specific conditions to model the early blastocyst in 3D, and integration is often used to compare blastoid scRNA-seq data to that of the embryo [3, 9, 10, 19, 22, 33, 34, 37]. In this chapter, we describe how to acquire and analyze a published scRNA-seq dataset, in our case a dataset of an in vitro 3D model system representing the early blastocyst [9]. We further provide a guideline to perform integrative analysis with other published datasets including two embryo datasets [14, 27] and an in vitro model system [15].

2 Materials

2.1 Library

For this analysis, we used the R programming language, with the following packages:

```
library(R.utils)
library(umap)
library(Seurat)
library(reshape2)
library(plyr)
library(dplyr)

#For plotting graphs
library(ggplot2)
library(ggpubr)
library(ggrepel)

#For color schemes, optional
library(RColorBrewer)
```

2.2 Download Human Blastoid Data

Here, scRNA-seq data from human blastoids generated by [9] will be used, which is available from the following GEO accession number GSE177689 (https://www.ncbi.nlm.nih.gov/geo/query/acc.cgi?acc=GSE177689). This dataset was generated using Smart-seq2 technology [16, 17]. First, download the count matrix and the corresponding metadata. The values in the count matrix represent the number of molecules for each feature (row), for instance, genes, detected for each cell (column). The dataset is a dataframe with the dimension of 60,666 rows (genes) and 2,715 columns (cells) with 1 column representing the gene id. The corresponding metadata has the dimension of 2,715 rows (cells) and 9 columns (variables).

```
# set the URL
DataURL <- "https://ftp.ncbi.nlm.nih.gov/geo/series/GSE177nnn/GSE177689/suppl/
GSE177689_blastoid.H9.okae_bts5__scRNAseq.unfiltered__counts.RDS.gz"
MetaURL <- "https://ftp.ncbi.nlm.nih.gov/geo/series/GSE177nnn/GSE177689/suppl/
GSE177689_blastoid.H9.okae_bts5__scRNAseq.unfiltered__metadata.tsv.gz"

## set the destination pathway of your folder ###
DataFolder <- "/staging/leuven/stg_00041/Published_Data/"

# set the local file name to save the downloaded file
Data <- "kagawaData.RDS.gz"
Meta<- "kagawaMetadata.tsv.gz"

# download the file
download.file(DataURL , destfile = paste0(DataFolder,Data), method = "auto")
download.file(MetaURL , destfile = paste0(DataFolder,Meta), method = "auto")

# decompress data
gunzip(paste0(DataFolder,Data), remove=TRUE)
gunzip(paste0(DataFolder,Meta), remove=TRUE)

# read data and its corresponding metadata
CountMatrix <- readRDS(paste0(DataFolder,"kagawaData.RDS"))
MetaData <- read.table(paste0(DataFolder,"kagawaMetadata.tsv"), header = TRUE)

### Set your own figure output folder ###
FigureOutputFolder <- '/lustre1/project/stg_00041/SanKit/SKT022_BlastoidMethods
Paper/Figure/'

### Set your own data output folder ###
DataOutputFolder <- '/lustre1/project/stg_00041/SanKit/SKT022_BlastoidMethodsPaper/
DataOutput/'
```

3 Methods

3.1 Analyze Blastoid Data

3.1.1 Create Seurat Object

Since the establishment of scRNA-seq [25], along with the exponential rise in the numbers of cells being profiled [24], a plethora of scRNA-seq analysis tools written in a diversity of programming languages, most prevalently R and Python, have been developed [35, 36]. The preference of programming language is often determined by the choice between analysis tools. Widely used platforms are Seurat in R [1] and Scanpy in Python [29]. Here we employ Seurat as a toolkit for quality control, integration, analysis, and exploration of scRNA-seq data [1, 21, 23]. The website of the Satija laboratory provides comprehensive tutorials on the use of Seurat and is actively well maintained and updated (https://

satijalab.org/seurat/) [1, 6, 7, 21, 23]. However, we describe how to interact with the Seurat object, and insert and access data from it specifically with data from human embryos, and in vitro models such as blastoids. The count matrix is used as input to create a Seurat object. The object is able to store both data (e.g., count matrix, metadata) and analysis (e.g., principal component analysis (PCA), clustering results) for single-cell datasets. First, the Seurat object is created with a non-normalized, raw count matrix, and optionally the corresponding metadata is included in the object. The full content of the count matrix is loaded in this object; however, only a subset of particular cells of interest can be included (here illustrated as an example in the script below). Both the count matrix and the metadata should be a dataframe, as mentioned in the previous section. It is of great importance that the cell names in the row names of the metadata match with the column names of the count matrix. A good standard is to always include gene id in the count matrix, optional but appreciated is to include gene symbol. The authors of this dataset have included both gene id and gene symbol. Here we are only interested in the gene symbol, as the other datasets we will integrate with only include gene symbols.

```
# Split gene_id based on the hyphen as separator
Split <- separate(CountMatrix ,
            gene_id, into = c("GENEID", "SYMBOL"),  sep = "-", extra = "merge")

# Row sum value for each gene of every cell
Sum <- rowSums(Split[,-1:-2])

# Prepare count matrix as input for Seurat object
InputCountmatrix <-
cbind(Split , Sum) %>% # add the sum values
group_by(SYMBOL) %>% # group data based on gene symbol
slice(which.max(Sum)) %>%  # remove duplicated genes with lower values
subset(select = -Sum) %>% # remove the column with the sum values
column_to_rownames("SYMBOL") # have the gene symbols as rownames

# Extract cell type annotation
MetaData$type <-  sapply(strsplit(MetaData$samplename,  "-"),  function(x)  paste(x
[-length(x)], collapse = "-"))

# Prepare metadata as input for Seurat object
MetaData <- column_to_rownames(MetaData, "sampleid")

# Load the full count matrix
SeuratObject<-  CreateSeuratObject(counts=InputCountmatrix,  project="kagawaData",
min.cells=3, min.features=200, meta.data = MetaData)
```

```
# EXAMPLE: only include cells of interest:  96h blastoids, primed and naive H9 cells
SeuratObject<- CreateSeuratObject(counts=InputCountmatrix[,which(MetaData$type%in%c(
"primed_H9", "naive_H9", "blastoid_96h_TROP2_pl", "blastoid_96h_PDGFRa_pl", "blas-
toid_96h_DN"))], project="kagawaData", min.cells=3, min.features=200)
```

3.1.2 Quality Control of Single-Cell Data

To retain high-quality cells, quality control (QC) metrics are employed to filter cells based on user-defined criteria. The following QC metrics commonly applied [8] include the following:

1. The number of unique genes detected in each cell: low-quality cells will have very few genes and cell doublets or multiplets may display high values of gene counts (labeled in Seurat as nFeature_RNA).
2. Total number of detected molecules within a cell (labeled in Seurat as nCount_RNA).
3. The percentage of reads that map to the mitochondrial genome; a high percentage of mitochondrial reads indicates low-quality/dying cells (labeled in Seurat as percent.mt).

The percentage of reads that map to the mitochondrial genome is calculated with the PercentageFeatureSet () function by searching for genes indicated by the prefix "^MT-". Be aware of using non-human samples, the prefix for mitochondrial genes will be without capital letters "^mt-". To better estimate the threshold for each QC metric, it is recommended to visualize the distribution of genes (features), number of detected transcripts, and percentage of mitochondrial reads per cell (Fig. 1a). Although there is no clearly defined cutoff threshold, an initial threshold above the first quartile for number of genes and transcript, and below the last quartile of percentage of mitochondrial reads can help your threshold estimation. Different thresholds can be tested to assess to what degree is lost at a given cutoff and can help you select your desired threshold. However, it is useful to consider the biological context of your samples, such as if you expect cells with low transcriptional activity, or high mitochondrial activity due to highly active respiratory mechanisms, these cells can be considered to be included (*see* **Note 1**).

```
# Percentage calculation of mitochondrial genes present in each individual cell
SeuratObject[["percent.mt"]] <- PercentageFeatureSet(SeuratObject, pattern="^MT-")
# Plot the variables of which we use for quality control (Fig. 1a)
VlnPlot(SeuratObject, features = c("nFeature_RNA", "nCount_RNA", "percent.mt"), pt.
size=0.01, ncol = 3)
```

Low complexity of genes together with high mitochondrial percentage is a good indicator to define low-quality cells with

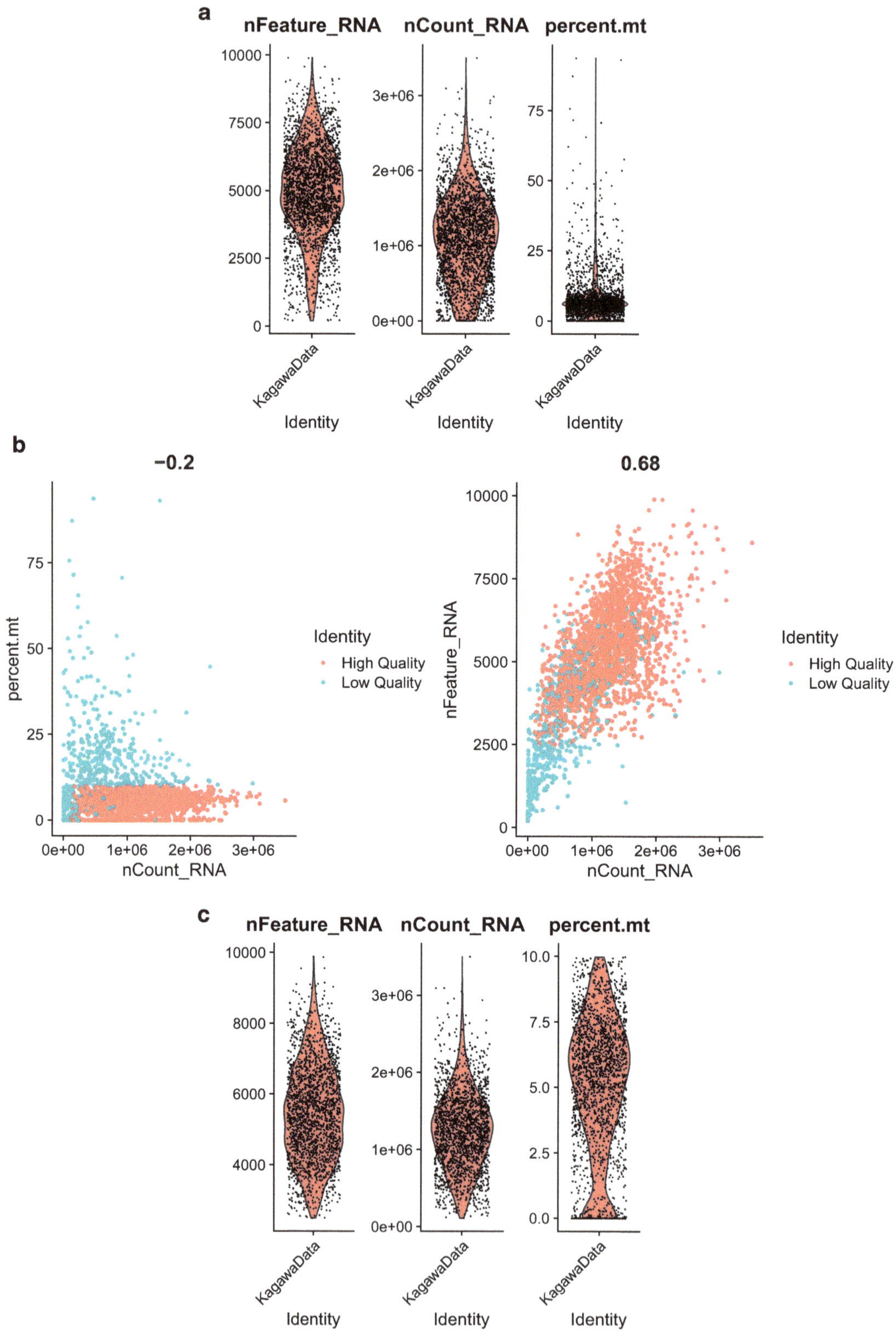

Fig. 1 Quality control metrics applied to the blastoid dataset from [9]. (**a**) Left: The number of genes (features) expressed per cell. Middle: The number of reads counted per cell. Right: The percentage of reads mapping to

higher confidence. While high complexity of genes with high number of transcripts gives us more confidence in the presence of doublets or multiplets in droplet-based sequencing [8, 12]. Therefore, it is recommended to visualize your low-quality cells after setting a QC threshold to assess with confidence whether your threshold is appropriate. Here we use scatter plots to assess the relationship between number of transcripts and mitochondrial reads, and the relationship between the number of transcripts and gene complexity (Fig. 1b). Low-quality/dying cells are filtered out by user-defined criteria for nFeature_RNA, nCount_RNA, and percent.mt and we assess the distribution of those variables again to assess whether the defined threshold is appropriate (Fig. 1c).

```
# Assess which cells would be removed based on the user-defined QC thresholds on
scatter plot (Fig. 1b)
SeuratObject@meta.data  <- SeuratObject@meta.data %>% mutate(
                             Quality = case_when(
     (nFeature_RNA >= 2500 & nCount_RNA >= 100000 &
                             percent.mt <= 10) ~ "High Quality",
     TRUE ~ "Low Quality"))
plot1 <- FeatureScatter(SeuratObject, feature1 = "nCount_RNA", feature2 = "percent.mt",
group.by = "Quality")
plot2 <- FeatureScatter(SeuratObject, feature1 = "nCount_RNA", feature2 = "nFeature_RNA",
             group.by = "Quality")
plot1 + plot2

# Filter low-quality cells
filter_check <- subset(SeuratObject,
subset = nFeature_RNA > 2500 & nCount_RNA >100000 & percent.mt < 10)
# Plot the filtered dataset (Fig. 1c)
VlnPlot(filter_check, features = c("nFeature_RNA", "nCount_RNA", "percent.mt"), pt.
size=0.01, ncol = 3)
```

3.1.3 Selection of Highly Variable Genes

Next, we aim to identify a collection of genes exhibiting high variability across cells that represent the heterogeneity in the data. Outliers on the mean-variance plot represent genes highly variable across cells (Fig. 2). Those features will be used in downstream

Fig. 1 (continued) mitochondrial genes, per cell. (**b**) Left: Relationship between number of transcripts (nCount_RNA) and percentage of mitochondrial reads (percent.mt). Right: Relationship between number of transcripts (nCount_RNA) and the number of genes (nFeature_RNA). Cells are labeled as high quality if they pass our quality control thresholds (cells with >2500 genes, >100,000 transcripts and <10% mitochondrial reads) or low quality (cells that do not conform to the threshold). (**c**) Quality control metrics as a, after filtering out low-quality cells and including only cells with over 2500 expressed genes, over 100,000 reads, and less than 10% of reads mapping to the mitochondria

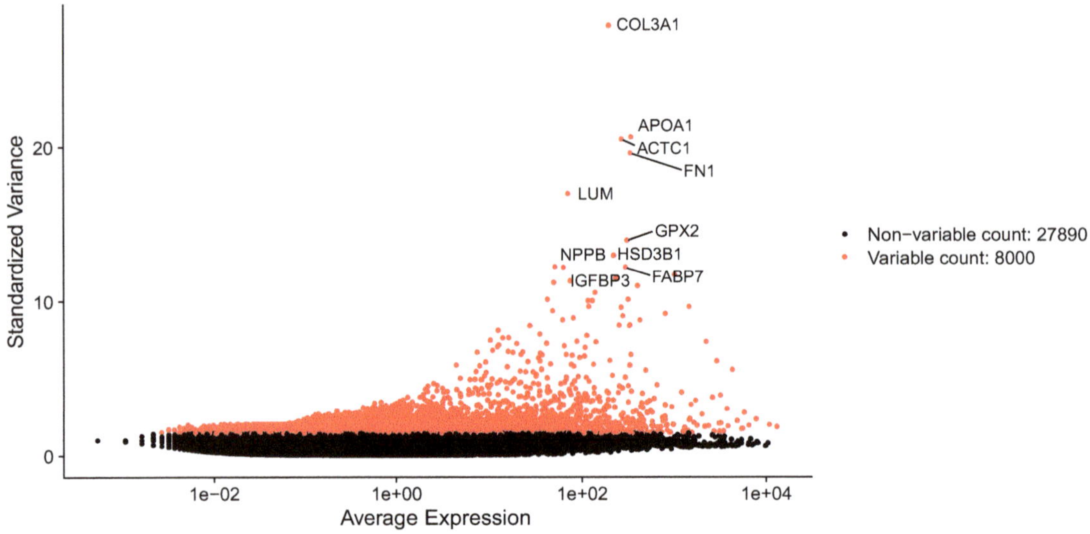

Fig. 2 A variable feature plot showing standardized variance versus gene expression per gene. The top 10 most variable genes are labeled. The top 8000 features are shown in red.

analysis such as PCA. Here, 8000 features instead of the standard 2000 features defined by Seurat are returned.

```
SeuratObject <- subset(SeuratObject,
subset = nFeature_RNA > 2500 & nCount_RNA > 100000  & percent.mt < 10)
SeuratObject<- FindVariableFeatures(SeuratObject, selection.method = "vst",
nfeatures = 8000)

# Identify the 10 most highly variable genes (Fig. 2)
top10 <- head(VariableFeatures(SeuratObject), 10)
plot1 <- VariableFeaturePlot(SeuratObject)
plot2 <- LabelPoints(plot = plot1, points = top10, repel = TRUE)
plot2
```

3.1.4 Data Normalization and Scaling, and Linear Dimension Reduction

We normalize the data by employing the global-scaling normalization method “LogNormalize”. Feature expression measurements for each cell are normalized by the total expression, multiplied by a scale factor of 10,000 and thereafter, log-transformed. Normalized values are stored in SeuratObject[[“RNA”]]@data or SeuratObject@data$RNA. Highly expressed genes can dominate downstream analysis; therefore, a linear transformation is applied to scale the data followed by using the dimensional reduction technique PCA. For each gene, the mean expression is centered across cells to 0, and the expression is scaled to achieve a variance of 1 across cells. Finally, PCA is run on the scaled data.

```
all.genes <- rownames(SeuratObject)
SeuratObject<- NormalizeData(SeuratObject, assay = "RNA")
SeuratObject<- ScaleData(SeuratObject, features = all.genes)
SeuratObject<- RunPCA(SeuratObject, features = VariableFeatures(object = Seurat
Object))
```

To overcome the technical noise characteristic to scRNA-seq data, Seurat clusters cells based on their PCA values. Each principal component (PC) represents a metafeature that combines information across a correlated feature set. The top principal components represent a robust compression of the dataset and display closely the nature of the dataset. Therefore, it is imperative to assess how many components are sufficient to explain the data, without including too many components which would cause technical noise. Here we describe three approaches to determine the dimensionality of the datasets. The first one is a supervised method where we explore the PCs to determine relevant sources of heterogeneity in our dataset. For instance, cells can be positioned in a 2D scatter plot based on their PCA score to assess different groups of closely related cells based on the two PCs that describe the variation in the data (Fig. 3a). Furthermore, we can assess the meaning of PCs from a biological standpoint by analyzing which genes contribute the most to each component (Fig. 3b). Here the knowledge of the biological relevance of the genes is crucial and gene set enrichment analysis can be complemented. Finally, we aim to estimate how many PCs to include for our analysis by plotting heatmaps visualizing scaled gene expression of the top contributing genes in each component for the top 500 cells (Fig. 3c). Here, we can assess to what extent each PC is able to explain the variation or heterogeneity of the data. If the PC is able to clearly divide cells into distinct groups, the borders between the groups are clearly defined by the scaled gene expression. A good indication for a cutoff of the number of dimensions is when the distinction bordering different groups becomes less clear. In practice, plotting as many PCs as possible is recommended (Seurat computes and stores 50 PCs by default, this can be increased by altering the parameters) to have the full overview, however, here we only plot 2 PCs due to space limitation.

```
options(repr.plot.width = 10, repr.plot.height = 5)
# 2D scatter plot of cells positioned based on the cell embeddings determined by PCA
(Fig. 3a)
DimPlot(object = SeuratObject, reduction = "pca")
# Visualize top genes that explain the principal components (Fig. 3b)
VizDimLoadings(SeuratObject, dims = 1:2, reduction = "pca")
# Heatmap where cells and genes are sorted based on their principal component values
(Fig. 3c)
p1 <- DimHeatmap(SeuratObject, dims = 1, cells = 500, balanced = TRUE, fast=FALSE ) &
```

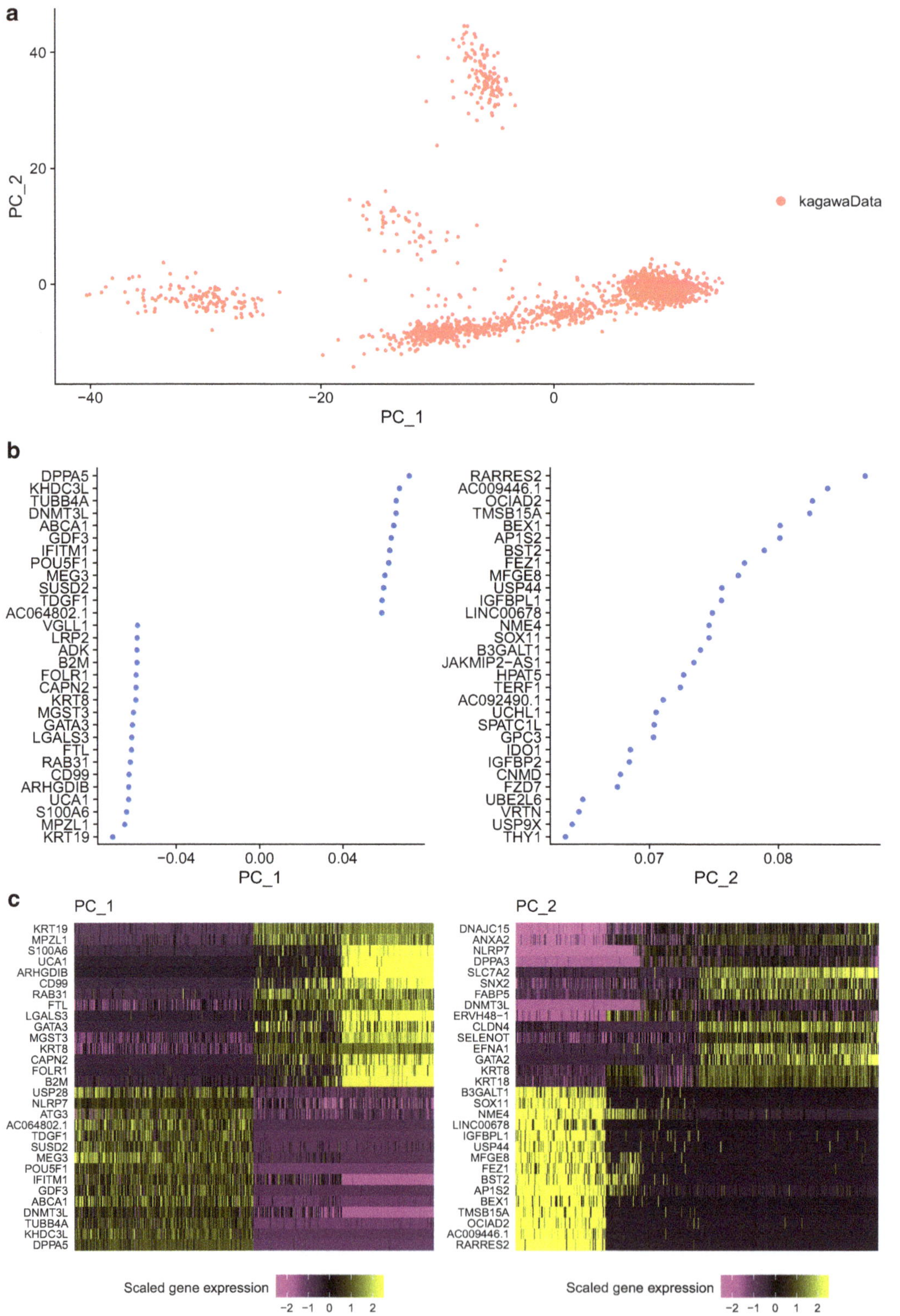

Fig. 3 Supervised method to explore the principal components to determine the dimensionality of the datasets. (**a**) A scatter plot showing the first two principal components of the blastoid dataset from [9]. (**b**) A figure showing the top genes correlated with the first two principal components. (**c**) Heatmaps of the top 2 principal components showing scaled expression of the genes shown in B, for the top 500 cells

```
ggtitle("PC_1") & theme(legend.direction="horizontal",legend.position="bottom") &
scale_fill_gradient2(low = "magenta", mid = "#000000", high = "yellow", name =
"Scaled gene expression\n")
p2 <- DimHeatmap(SeuratObject, dims = 2, cells = 500, balanced = TRUE, fast=FALSE) &
ggtitle("PC_2") & theme(legend.direction = "horizontal",legend.position="bottom") &
scale_fill_gradient2(low = "magenta", mid = "#000000", high = "yellow", name =
"Scaled gene expression\n")
p1|p2
```

The second method is through implementation of a statistical test based on a random null model. Here Seurat applies a resampling test inspired by the JackStraw procedure (Fig. 4). This procedure randomly creates a subset of the data (1% by default) and reruns PCA on these genes to compare those scores with the original PCA scores to determine the statistical significance for each gene associated with each PC. This procedure is repeated (100x in our case). Statistical significance for each PCs is determined based on a proportion test that compares the number of features with a p-value below a defined threshold with the proportion of features expected under a uniform distribution of p-values. Seurat provides a function to calculate and visualize the p-value distribution for each PC compared with a uniform distribution (dashed line in the plot). The plot presents the distribution of p-values for all genes across each PC and the overall significance of each PC. Significant PCs show strong enrichment of features with low p-values (solid curve above the dashed line.) and a p-value distribution strongly skewed to the left compared to the null distribution. The downside of this method is high time-consumption for large datasets and possible unclear PC cutoff.

```
# Perform significance test on PCs
SeuratObject <- JackStraw(SeuratObject, num.replicate = 100)
SeuratObject <- ScoreJackStraw(SeuratObject, dims = 1:20)
# Visualize the p-value distribution and the overall significance of each PC (Fig. 4)
JackStrawPlot(SeuratObject, dims = 1:20)
```

The third method is by generating an elbow plot that ranks principal components based on the percentage of variance explained by each principal component (Fig. 5). The elbow plot corresponds well with the significant PC and is much faster to calculate than JackStraw.

```
# Construct an elbow plot (Fig 5.)
ElbowPlot(SeuratObject)
```

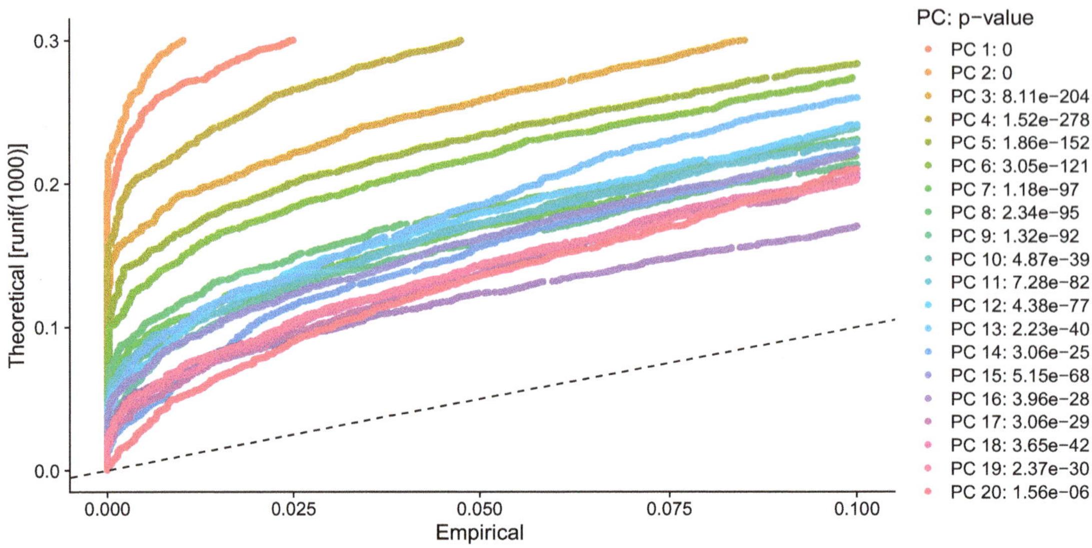

Fig. 4 A JackStraw plot showing the significance of each principal component.

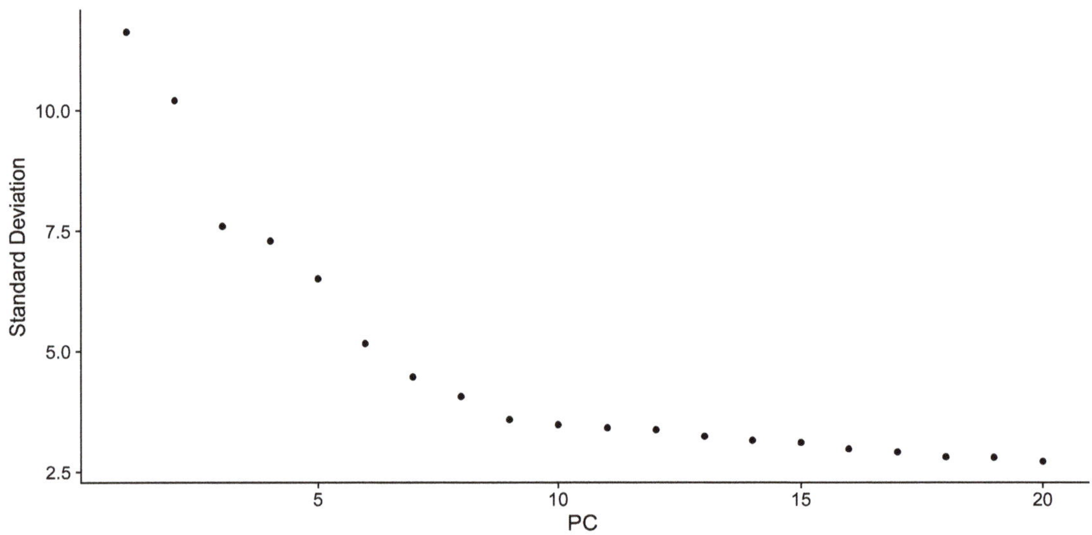

Fig. 5 An elbow plot showing the variance explained by each principal component

Next, Seurat calculates a k-nearest neighbors (KNN) graph based on the Euclidean distance in the PCA space, and the edge weights are refined between any two cells based on the shared overlap in their local neighborhoods (Jaccard similarity) (macosko et al.). This step is performed using the FindNeigbors() function with the previously defined number of PC dimensions. To cluster the cells, modularity optimization techniques such as the Louvain algorithm are applied to group cells together. The FindClusters() function implements this procedure. The resolution parameter is adjustable, with increased values leading to a higher number of clusters. Setting the resolution parameter between 0.4 and 1.2 is recommended for 3,000 cells. For larger datasets, increasing the resolution is recommended. It is encouraged to perform downstream analyses with different numbers of PCs for comparison, although the results often do not differ as the number of PCs gets higher. However, it is recommended to choose the higher number for this parameter.

```
SeuratObject<- FindNeighbors(SeuratObject, dims = 1:10)
SeuratObject<- FindClusters(SeuratObject, resolution = 0.8)
```

3.1.5 Non-linear Dimensional Reduction

Here, UMAP and t-SNE analysis is performed to learn the underlying manifold of the data to place similar cells together in a low-dimensional space (Fig. 6a–d). Cells within the graph-based clusters determined above should co-localize on these dimension-reduction plots.

```
SeuratObject<- RunTSNE(SeuratObject, dims = 1:10, seed.use = 1234, perplexity = 20)
SeuratObject<- RunUMAP(SeuratObject, dims=1:10)
# Plot UMAP and T-SNE (Fig. 6a, 6b, 6c, 6d)
DimPlot(SeuratObject, reduction = "umap", label = T)
DimPlot(SeuratObject, reduction = "umap", group.by = "type", label = T)
DimPlot(SeuratObject, reduction = "tsne", label = T)
DimPlot(SeuratObject, reduction = "tsne", group.by = "type", label = T)
```

3.1.6 Annotate Clusters by Marker Gene Expression

Next, we manually annotate cell clusters by expression of known marker genes for various cell types which we expect to be found in blastoids, or which we think may be off-target cells (Fig. 7). Heatmaps can visualize which marker genes are expressed in each cluster.

```
# Define the genes of interest
feats.core <- c("POU5F1","SOX2","NANOG")
feats.primed <- c("OTX2", "ZIC2", "CD24", "DUSP6", "TCF4")
feats.naive <- c("KLF17", "KLF4", "SUSD2", "DNMT3L", "DPPA5", "TFCP2L1")
feats.TE <- c("GATA2", "GATA3", "KRT7", "KRT18", "HAND1", "NR2F2")
feats.EXMC <- c("LUM","NID2", "FOXF1", "VIM", "POSTN", "ANXA1",  "PITX1")
feats.PrE <- c("SOX17", "GATA4", "GATA6", "FOXA2", "PDGFRA")
feats.EVT <- c("HLA-G", "MMP2")
feats.STB <- c("CGA",  "SDC1")
feats.amnion<- c("WNT6", "GABRP", "ISL1", "HEY1", "CDH10", "CTSV", "TPM1")
feats.mesoderm <- c("MIXL1", "MESP1", "EPHA4", "ZIC3", "GSC", "TBXT", "CDX1", "CDX2",
"CDX4", "EOMES")

combinedFeats <- c(feats.core, feats.primed, feats.naive, feats.TE, feats.EXMC,
feats.PrE, feats.EVT, feats.STB, feats.amnion, feats.mesoderm)

# Plot heatmap (Fig. 7)
```

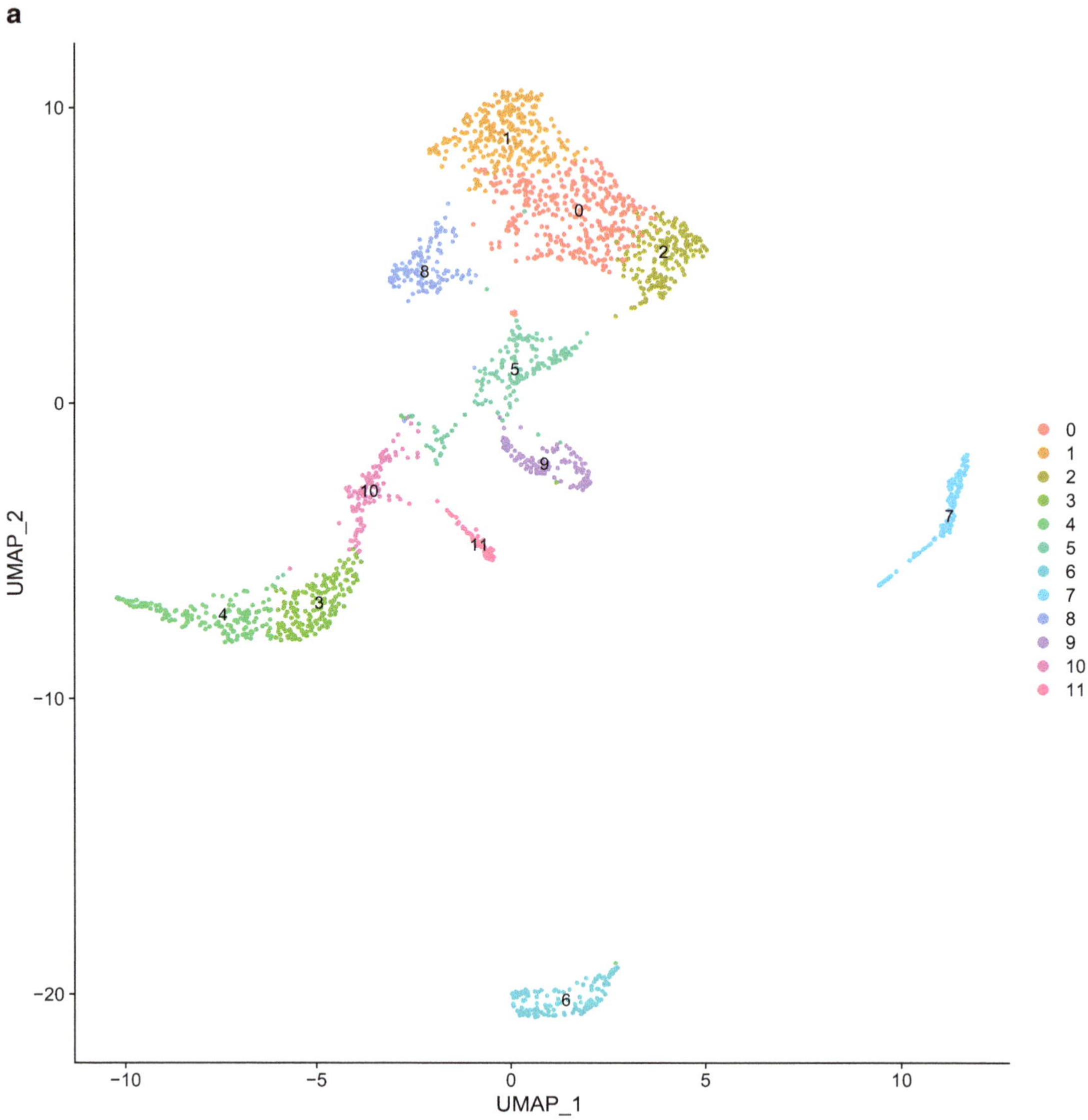

Fig. 6 Non-linear dimensional reduction of blastoid cells (**a**) UMAP labeled by clusters. (**b**) UMAP labeled by annotation from [9]. (**c**) t-SNE labeled by clusters. (**d**) t-SNE labeled by annotations from [9]

```
DoHeatmap(SeuratObject, features = combinedFeats) &
theme(legend.direction = "horizontal",legend.position="bottom") &
scale_fill_gradient2(low = "magenta", mid = "#000000", high = "yellow",
name = "Scaled gene expression\n")
```

In the heatmap, expected cell types in blastoid can be found: epiblast (with naive and core pluripotency markers), trophectoderm (with TE marker genes), and primitive endoderm (with PrE

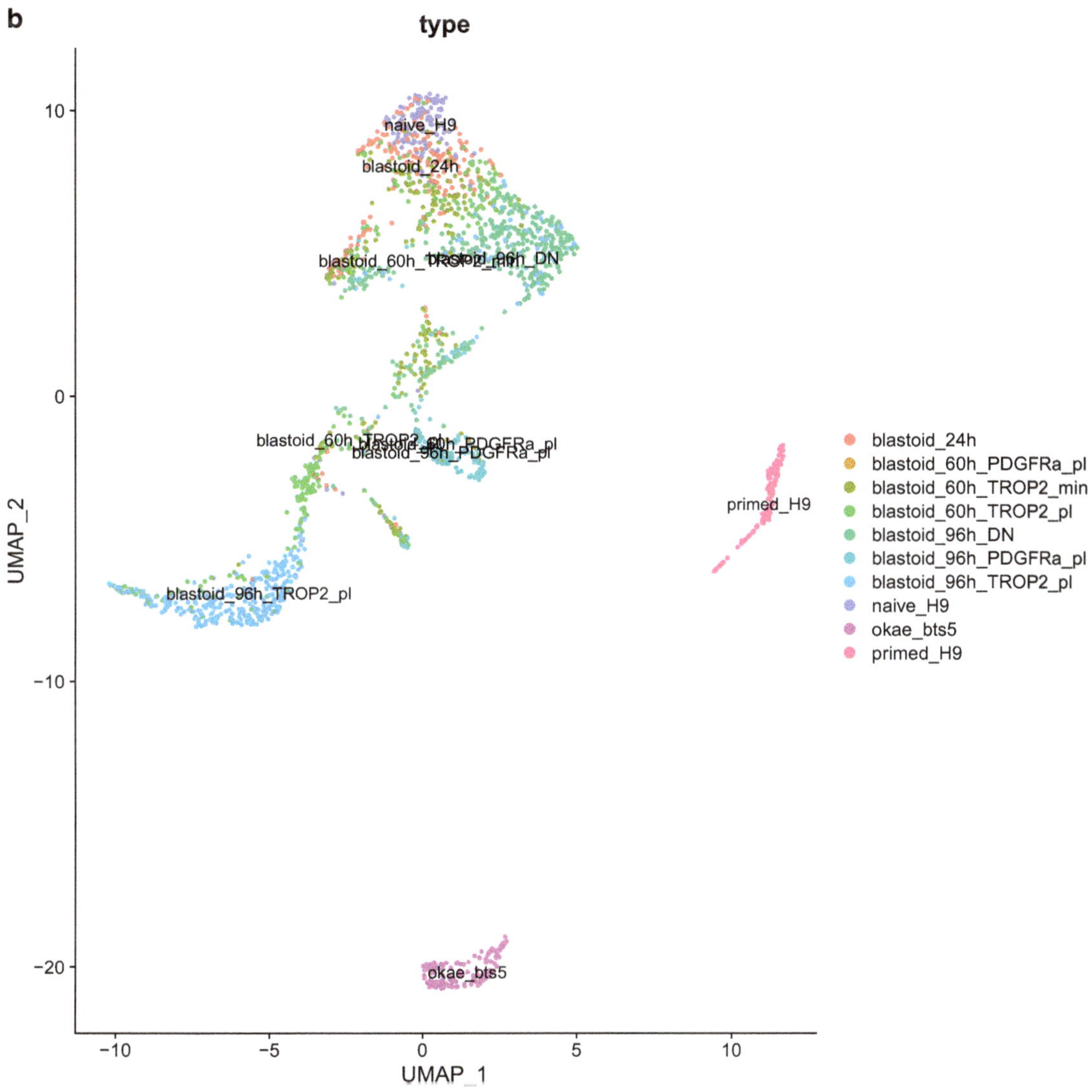

Fig. 6 (continued)

marker genes). Additionally, we see one cluster expressing EXMC marker genes which we annotated as extraembryonic mesoderm [15]. This is an example of what information can be extracted with this analysis. Including known marker genes for more cell types will increase your ability to identify populations in a dataset. Here we assign cell type identity to clusters based on the results of the earlier heatmap analysis.

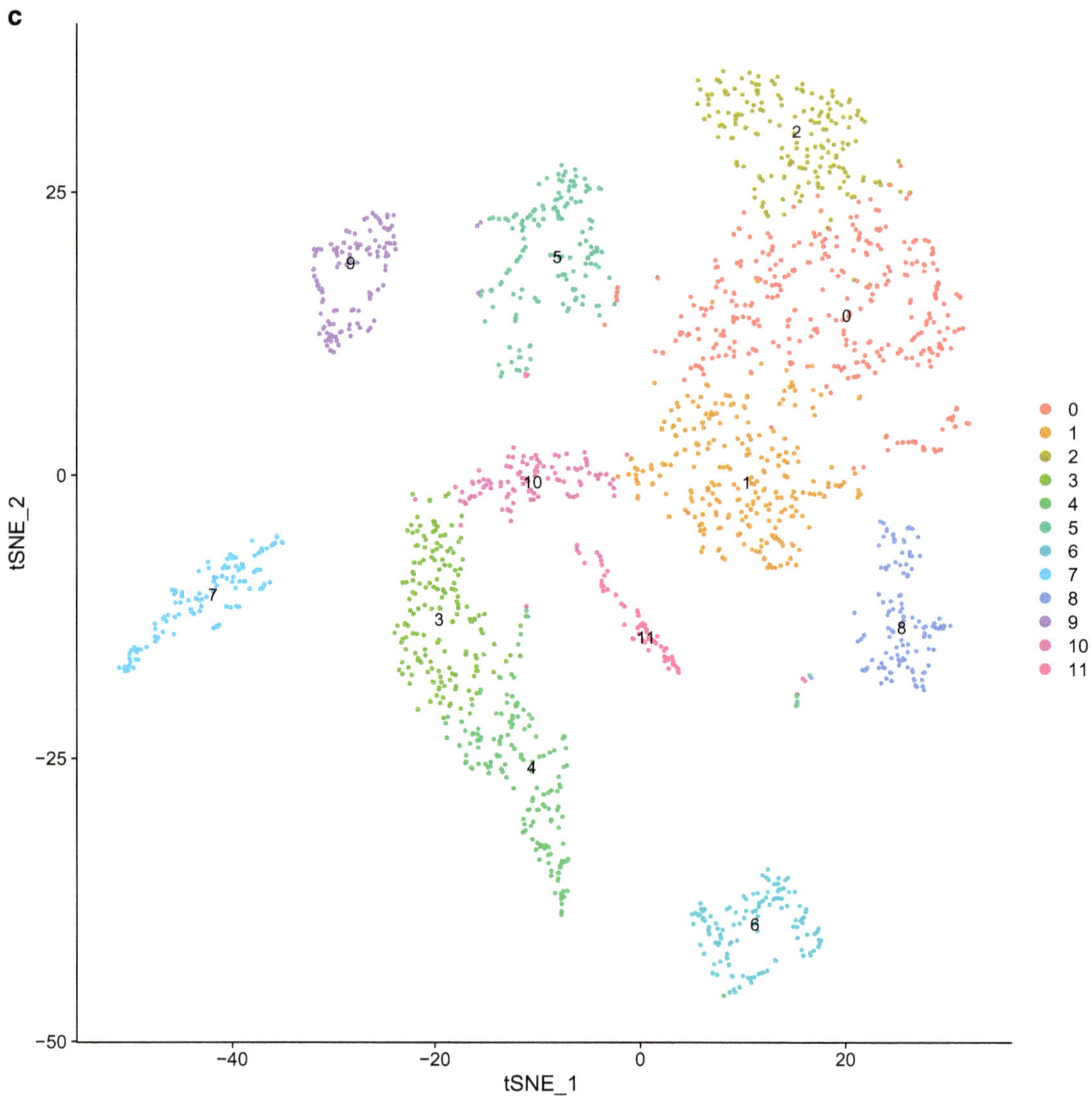

Fig. 6 (continued)

```
SeuratObject@meta.data$manualClusters <- NA
SeuratObject@meta.data$manualClusters[which(
SeuratObject@meta.data$seurat_clusters%in%c(0))] <- "Epi"
SeuratObject@meta.data$manualClusters[which(
SeuratObject@meta.data$seurat_clusters%in%c(1))] <- "Epi"
SeuratObject@meta.data$manualClusters[which(
SeuratObject@meta.data$seurat_clusters%in%c(2))] <- "Epi"
SeuratObject@meta.data$manualClusters[which(
SeuratObject@meta.data$seurat_clusters%in%c(3))] <- "TE"
SeuratObject@meta.data$manualClusters[which(
SeuratObject@meta.data$seurat_clusters%in%c(4))] <- "TE"
```

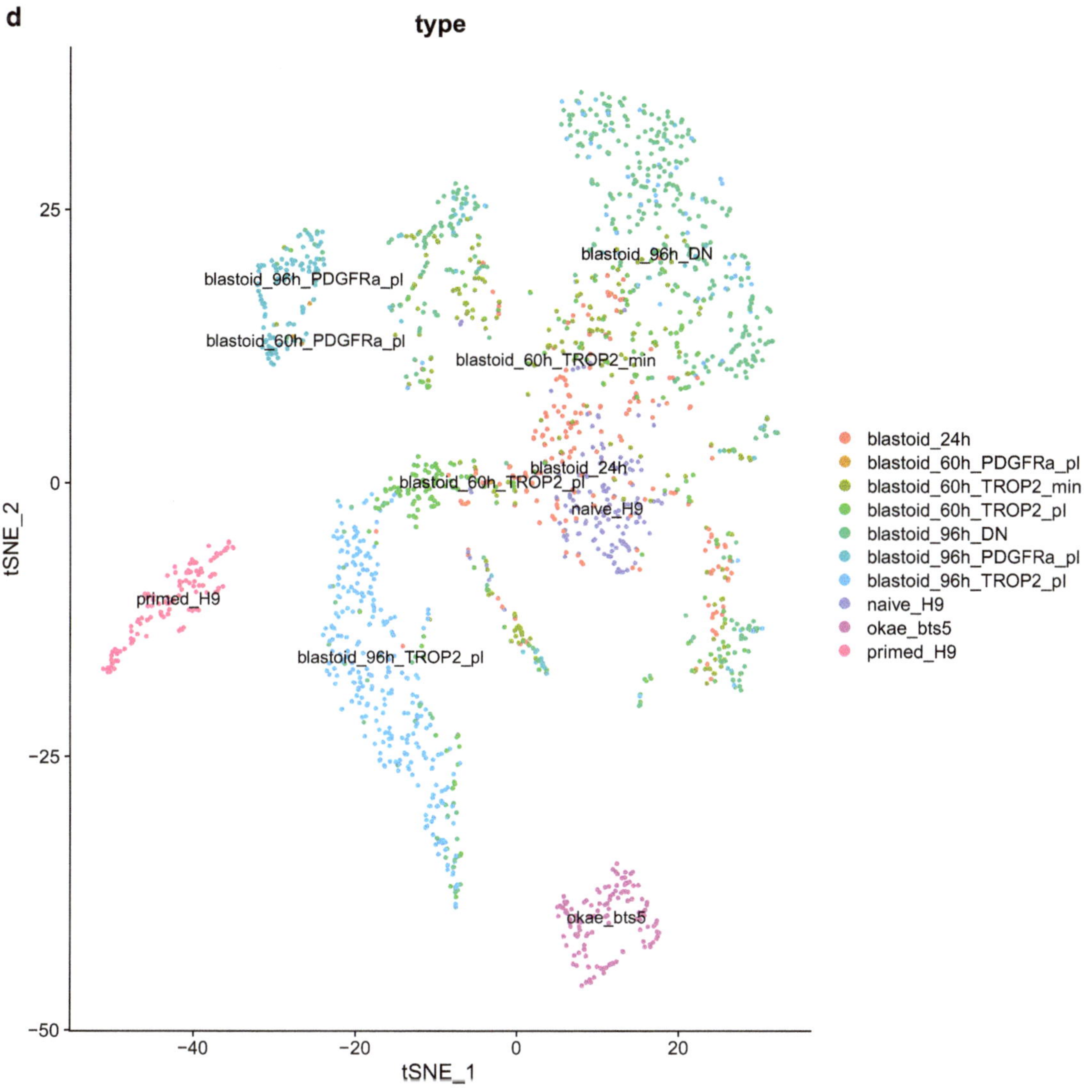

Fig. 6 (continued)

```
SeuratObject@meta.data$manualClusters[which(
SeuratObject@meta.data$seurat_clusters%in%c(5))] <- "Unknown"
SeuratObject@meta.data$manualClusters[which(
SeuratObject@meta.data$seurat_clusters%in%c(6))] <- "Okae"
SeuratObject@meta.data$manualClusters[which(
SeuratObject@meta.data$seurat_clusters%in%c(7))] <- "Primed"
SeuratObject@meta.data$manualClusters[which(
SeuratObject@meta.data$seurat_clusters%in%c(8))] <- "Epi"
SeuratObject@meta.data$manualClusters[which(
SeuratObject@meta.data$seurat_clusters%in%c(9))] <- "PE"
SeuratObject@meta.data$manualClusters[which(
SeuratObject@meta.data$seurat_clusters%in%c(10))] <- "TE"
```

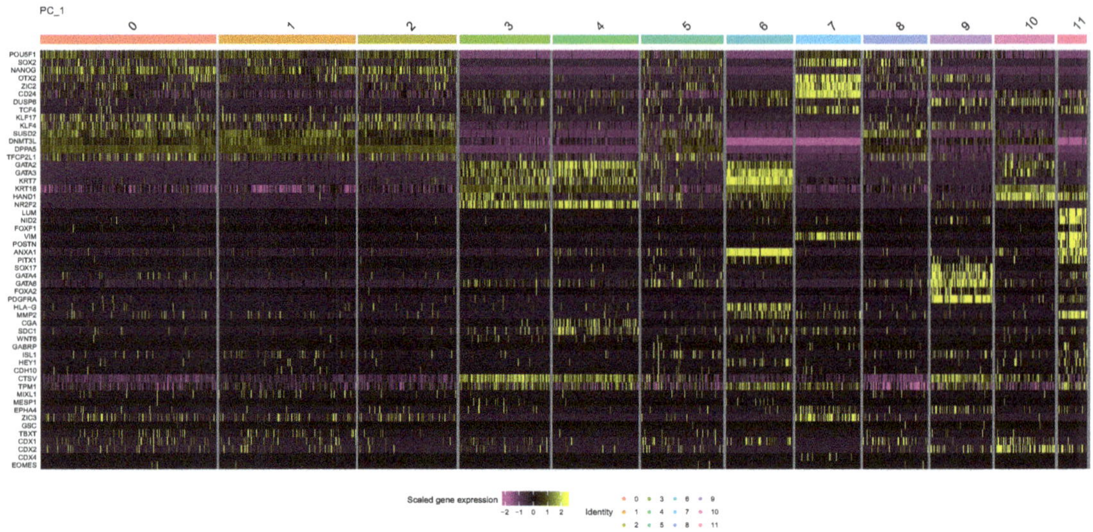

Fig. 7 Gene expression heatmap for lineage marker genes, with cells sorted by cluster

```
SeuratObject@meta.data$manualClusters[which(
SeuratObject@meta.data$seurat_clusters%in%c(11))] <- "EXM"
SeuratObject@meta.data$manualClusters <- factor(
SeuratObject@meta.data$manualClusters, levels=c("Epi", "TE", "PE", "EXM", "Primed",
"Okae", "Unknown"))
```

Finally, we plot the UMAP with newly curated labels, to ensure that cells of the same cell type cluster together (Fig. 8). It is recommended to frequently save your work at different stages of your analysis, especially after compute-intensive processes (*see* **Note 2**).

```
# Plot UMAP with newly curated labels  (Fig. 8)
DimPlot(SeuratObject, reduction = "umap", group.by="manualClusters", label=T, ) +
  NoLegend() + labs(title = NULL)

saveRDS(SeuratObject, paste0(DataOutputFolder ,"RivronResults.rds"))
```

3.2 Format Data for Integration

Here we introduce joint analysis of multiple published datasets. In addition to the blastoid data acquired from [9], we have collected 2 embryo datasets and 1 in vitro dataset [14, 15, 27]. Seurat possesses a collection of methods to align shared cell populations across datasets [23]. The aim of integration is to perform comparative scRNA-seq analysis across experimental conditions. Therefore, we need to identify cells across datasets that match or are similar in biological states ("anchors"). First, count data is collected from the different publications mentioned above. The challenge here is the absence of a consensus on how the count matrix and the meta data is formatted. As the data format is unique to each author, each

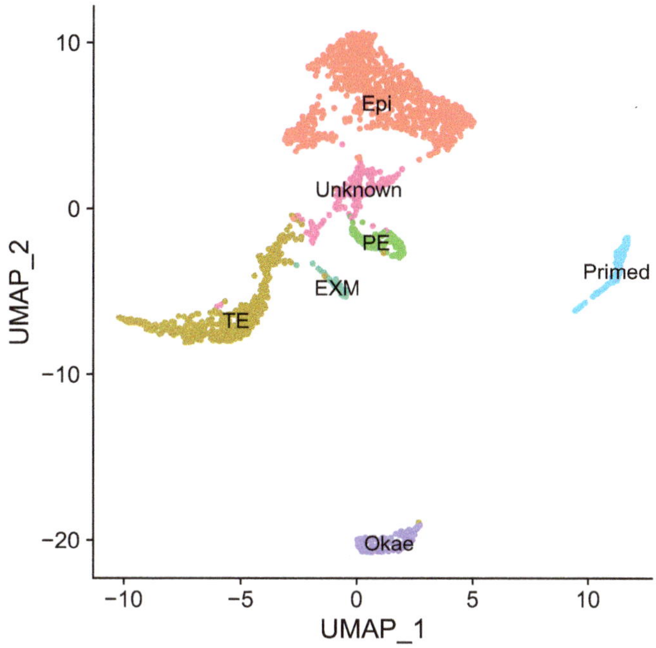

Fig. 8 UMAP of the blastoid cells, labeled by our cell type annotations

count matrix and the metadata need to be adjusted to have a uniform format compatible as input to create a Seurat object. To follow the methods, we have provided count matrices and metadata files from the authors mentioned above publicly available in our github website: https://github.com/pasquelab/BlastoidIntegration (*see* **Note 3**; Table 1).

```
# Set the destination pathway of your folder
DataFolder <- "/staging/leuven/stg_00041/Published_Data/"

# Kagawa
kagawa <- read.table(paste0(DataFolder, "kagawa_count.tsv"), h = T, sep = "\t")
kagawaCellMeta <- read.table(paste0(DataFolder, "kagawa_metadata.tsv"),
h = T, sep = "\t")
kagawaCellMeta$type <- NA
for(i in 1:nrow(kagawaCellMeta)){
kagawaCellMeta$type[i]<- strsplit(as.character(
kagawaCellMeta$title[i]), split="-")[[1]][1]}

# Petropoulos
petropoulos <- read.table(paste0(DataFolder, "petropoulos_count.tsv"),c
h=T, row.names = 1,  sep=",", row.names=1)
# We are using a reannotated version of the cell type labels,
from DOI: 10.1016/j.celrep.2020.108419
petropoulosCellMeta  <-  read.table(paste0(DataFolder,  "petropoulos_metadata.tsv"),
```

```
h=T, sep="," )
# Match the order of cells in the metadata to the order of cells in the counts matrix
petropouosCellMeta <- merge(data.frame(colnames(petropoulos)), petropoulosCellMeta,
by.x=1, by.y=8)

# Tyser
tyser <- read.table(paste0(DataFolder,  "tyser_count.tsv"), h=T, row.names = 1)
tyser <- t(tyser)
tyserCellMeta<- read.table(paste0(DataFolder, "tyser_metadata.tsv"), h=T, sep="\t")

# Pham
pham<- read.table(paste0(DataFolder, "pham_count.csv", h=T, sep=",", row.names = 1)
phamCellMeta <- read.table((paste0(DataFolder, "pham_metadata.tsv", h=T, sep="\t")

#example of what one of our tables looks like (Table 1)
#columns are individual cells and rows are individual genes
pham[1:10,1:4]
```

As the metadata for each dataset is formatted differently, unique to each respective publication, a new dataframe is created with only the specific variables of interest from each dataset. The following variables are included in the newly generated metadata: study (the origin of the dataset), sampleID (the origin of sample of which the cells are derived from), time (author-defined biological time point the cells represent), cell (individual cell identity), and cellLineageOriginal (author-defined lineage origin of the cells).

```
# Combine metadata
combinedMeta <- data.frame(
study=c(rep.int("kagawa", times=ncol(kagawa)), rep.int("petropoulos", times=ncol
(petropoulos)),
        rep.int("tyser", times=ncol(tyser)), rep.int("pham", times=ncol(pham))),
sampleID=c(kagawaCellMeta$type, petropoulosCellMeta$Embryo_ID, tyserCellMeta$ori-
gin,
        rep.int("pham", times=ncol(pham))),
time=c(kagawaCellMeta$characteristics..treatment, petropoulosCellMeta$Stage,
        rep.int("D16-19", times=ncol(tyser)), rep.int(NA, times=ncol(pham))),
cell=c( colnames(kagawa), colnames(petropoulos), tyserCellMeta$rownames.dat.,
        colnames(pham)),
cellLineageOriginal=c(kagawaCellMeta$type, petropoulosCellMeta$Lineage,
tyserCellMeta$cluster, phamCellMeta$cellType))
```

Table 1
Example of a count matrix. The columns represent cells and rows represent genes

	AAACGAAAGGTTGACG.1 <int>	AAACGCTAGATGGTAT.1 <int>	AAACGCTAGGTCCTGC.1 <int>	AAACGCTAGGTGATAT.1 <int>
DNAJC11	0	0	1	0
CDK11A	0	0	0	0
NADK	0	0	2	0
PAX7	0	0	0	0
MASP2	0	0	0	0
CLCN6	0	0	1	0
TNFRSF1B	0	0	0	0
FBXO42	0	0	0	0
SLC66A1	0	0	2	0

Table 2
The first five rows of our combined metadata file

	Study <chr>	sampleID <chr>	time <chr>	cell <chr>	cellLineageOriginal <chr>	cellLineage <chr>
1	rivron	primed_H9	E8	primed_H9.150182	primed_H9	primed
2	rivron	primed_H9	E8	primed_H9.150183	primed_H9	primed
3	rivron	primed_H9	E8	primed_H9.150185	primed_H9	primed
4	rivron	primed_H9	E8	primed_H9.150187	primed_H9	primed
5	rivron	primed_H9	E8	primed_H9.150188	primed_H9	primed
6	rivron	primed_H9	E8	primed_H9.150189	primed_H9	primed

The challenge of combining metadata from different sources is the annotations given by the authors. In the majority of cases, the same cell lineage labels are written differently by each author. The extent of detailed cell type annotation can differ between multiple datasets, and reannotating the cells to your own preferences will be inevitable. Here a new variable is generated to accommodate user-defined reannotation (Table 2).

```
# Reannotate cell lineage labels to match cell types or lineages across datasets
combinedMeta$cellLineage <- combinedMeta$cellLineageOriginal
combinedMeta$cellLineage[which(combinedMeta$cellLineage%in%c(
"Trophectoderm", "TE", "Early Trophectoderm", "early_TE", "late_TE", "medium_TE",
"ysTE"))] <- "TE"
combinedMeta$cellLineage[which(combinedMeta$cellLineage%in%c(
"EPI", "EPI_PE", "Epiblast", "PSA-EPI"))] <- "EPI"
combinedMeta$cellLineage[which(combinedMeta$cellLineage%in%c(
"ICM", "Inner Cell Mass", "Inner Cell Mass-Trophectoderm Transition"))] <- "ICM"
combinedMeta$cellLineage[which(combinedMeta$cellLineage%in%c(
"blastoid_24h", "blastoid_60h_PDGFRa_pl", "blastoid_60h_TROP2_min",
"blastoid_60h_TROP2_pl", "blastoid_96h_DN", "blastoid_96h_PDGFRa_pl",
"blastoid_96h_TROP2_pl", "XMU_MP_1", "SC144"))] <- "blastoid"
combinedMeta$cellLineage[which(combinedMeta$cellLineage%in%c(
"naiveWIBR3", "naive_H9"))] <- "naive"
combinedMeta$cellLineage[which(combinedMeta$cellLineage%in%c(
"primed_H9" ))] <- "primed"
combinedMeta$cellLineage[which(combinedMeta$cellLineage%in%c(
"Undefined", "unknown", "Unknown","Transitioning", "MIX", "NA", NA))] <- "unknown"
combinedMeta$cellLineage[which(combinedMeta$cellLineage%in%c(
"Hypoblast"))] <- "hypo"
combinedMeta$cellLineage[which(combinedMeta$cellLineage%in%c(
"Advanced Mesoderm", "Axial Mesoderm", "Emergent Mesoderm", "Nascent Mesoderm", "YS
Mesoderm", "EXMC"))] <- "mesoderm"
combinedMeta$cellLineage[which(combinedMeta$cellLineage%in%c(
```

```
"TB.apoptosis", "TB.early", "TB.late", "TB.medium1", "TB.medium2", "TB.medium3" ,
"TS", "TSC", "okae_bts5", "okae_bts5_3D"))] <- "TB"
combinedMeta$cellLineage[which(combinedMeta$cellLineage%in%c(
"EVT", "EVTs", "Pre.EVT"))] <- "EVT"
combinedMeta$cellLineage[which(combinedMeta$cellLineage%in%c(
"Erythrocytes", "Hemogenic Endothelial Progenitors"))]<- "erythrocytes"
combinedMeta$cellLineage[which(combinedMeta$cellLineage%in%c(
"ST", "STBs", "Pre.ST")) ] <- "STB"
combinedMeta$cellLineage[which(combinedMeta$cellLineage%in%c(
"Endoderm")) ] <- "PrE"

table(combinedMeta$cellLineage, combinedMeta$study, useNA="always")
saveRDS(combinedMeta, paste0(DataOutputFolder ,"CombinedMeta.rds"))

# Example of the format of our combined metadata (Table 2)
head(combinedMeta)


# Construct a combined R-object with all datasets in a standardized format
combinedCounts <-list()

# Kagawa
combinedCounts$kagawa <- list(counts = kagawa,
annotation = combinedMeta[which(combinedMeta$study=="kagawa"),])
# Petropoulos
combinedCounts$kagawa <- list(counts = petropoulos,
annotation = combinedMeta[which(combinedMeta$study=="petropoulos"),])
#Tyser
combinedCounts$kagawa <- list(counts = tyser,
annotation = combinedMeta[which(combinedMeta$study=="tyser"),])
# Pham
combinedCounts$pham <- list(counts=pham, annotation=combinedMeta[which(combinedMeta
$study=="pham"),])

saveRDS(combinedCounts, paste0(DataOutputFolder ,"CombinedCounts.rds"))
```

3.3 Integration of Blastoid and Embryo Data

After formatting datasets, we load our combined datasets and metadata to generate an integrated Seurat object. Combining multiple large datasets can strain heavily on memory storage; therefore, we constructed a function to perform QC filtering on cells before merging the count matrices to generate a Seurat object. Loading fewer cells after QC can make a difference in speed and memory storage as the number of included datasets increases. It is recommended to load only what is deemed to be required for each step (*see* **Note 4**). Therefore, we reset our workspace between sections. As described above, cells are filtered with a high percentage of mitochondria and low number of RNA counts.

```
combinedCounts <- readRDS(paste0(DataOutputFolder ,"CombinedCounts.rds"))
combinedMeta <- readRDS(paste0(DataOutputFolder ,"CombinedMeta.rds"))
```

To filter cells with a high percentage of mitochondria, we provided a function that is easy to apply across count matrix data frames to calculate the percentage of mitochondria.

```
# Function to calculate the percentage of mitochondria
manualCalculatePercentMT <- function(cellGeneCounts, geneNames){
    geneNames <- as.character(geneNames)
    return((sum(cellGeneCounts[grep(pattern="^MT-", x = geneNames)])
          /sum(cellGeneCounts))
          *100) }
```

We generate two sublists that represent the variables RNA count and mitochondrial percentage.

```
# Generate new column with sum values of the total RNA count for each cell
for(i in 1:length(combinedCounts)){
combinedCounts[[i]]$RNAcounts <- apply(combinedCounts[[i]]$counts[,
which(colnames(combinedCounts[[i]]$counts)!=\"gene\")], 2, sum)}

# Generate new column with the mitochondrial percentage for each cell
for(i in 1:length(combinedCounts)){
combinedCounts[[i]]$percentMT <- apply(combinedCounts[[i]]$counts[,
which(colnames(combinedCounts[[i]]$counts)!=\"gene\")], 2, manualCalculatePer-
centMT,
geneNames=rownames(combinedCounts[[i]]$counts))}
```

Next, we constructed a QC filter function that is able to filter data sets individually based on the variables "RNAcount" and "percentMT".

```
# Filter for MT % and read counts
applyFilters <- function(combinedCountsPiece){
  filtered <-list()
  survivesFilter <- which(
combinedCountsPiece$RNAcounts > # define your threshold
# we defined our threshold in the line below
10^(log10(summary(combinedCountsPiece$RNAcounts)[2])-
IQR(log10(combinedCountsPiece$RNAcounts)))
& combinedCountsPiece$percentMT < 10) # define your threshold for MT %
# we set up less than 10%

  filtered$counts <- combinedCountsPiece$counts[,survivesFilter]
  filtered$annotation <- combinedCountsPiece$annotation[survivesFilter,]
  filtered$RNAcounts <- combinedCountsPiece$RNAcounts[survivesFilter]
  filtered$percentMT <- combinedCountsPiece$percentMT[survivesFilter]
```

```
  filtered$counts <- as.data.frame(filtered$counts)
  filtered$counts$gene <- row.names(filtered$counts)
  return(filtered)
}
# The threshold for minimum RNA read count we use here is the first quartile
# minus 1.5*the interquartile range, log10 transformed to achieve a normal distribu-
tion of the dataset.
# NOTE: You may need to use a different filter depending on how your datasets look
like.
```

Here, we generate a function to merge multiple count matrices in our list after individually applying the QC filter function on each data set and only include genes that are present in all datasets.

```
allMerge <- function(studies){
              combinedCountsMerge <- merge(
                   applyFilters(combinedCounts[[studies[1]]])$counts,
                   applyFilters(combinedCounts[[studies[2]]])$counts,
                   by="gene", all=F)
          if(length(studies)>2){for(i in 3:length(studies)){
              combinedCountsMerge <- merge(
                  combinedCountsMerge,
                  applyFilters(combinedCounts[[studies[i]]])$counts,
                  by="gene", all=F)}}
          return(combinedCountsMerge)}
```

The function below includes multiple processing and analysis steps such as count matrix merging after QC, Seurat object creation, data normalization and variable feature identification, and data integration. After creating a Seurat object from the count matrix merged from different datasets, the count data is normalized, and variable genes (features) are identified for each dataset independently. The next step is to identify cell populations that are expressed similarly across multiple datasets. Seurat provides a set of methods to realize this. First, SelectIntegrationFeatures() function selects top scoring features that are variable across multiple datasets, ranked by the number of datasets they are deemed variable in. The FindIntegrationAnchors() function then identifies anchors across a user-defined list of Seurat objects. This is realized by performing a dimensional reduction technique, canonical correlation analysis, followed by L2 normalization of the canonical correlation vectors. This enables the projection of a shared correlation space where pairs of cells across datasets are within each other's neighborhood (mutual nearest neighbors). Those pairs represent a shared biological state across datasets and serve as anchors to guide dataset integration. Low confidence anchors that are not consistent in their local neighborhoods are filtered out. Finally, an anchor score is defined by the shared neighbor (shared nearest neighbor) overlap between the k-score (K-nearest neighbor) of each anchor within its own dataset. To enable joint analysis of multiple datasets, the IntegrateData() function integrates the datasets together. This is realized by subtracting weighted correction vectors from the query gene expression matrix enabling the joint analysis with the original reference gene expression matrix. The weighted corrections are determined by the cell similarity score (distance between each query cell and its k-nearest anchors in principal component space) and the anchor score determined previously. For multiple datasets, iterative pairwise integration will be performed where the order of the integration is based on the similarity between all pairs of datasets. A comprehensive description of integration methodology can be found in this paper [23]. After data integration, we perform the standard downstream analysis described already in section 4.

```
integrateAsFunction <- function(studyNums){# This variable allows us to select which
studies to integrate
                        #This can be useful to troubleshoot integrating many datasets

# Merge multiple count matrices
mergedDf<- allMerge(studyNums)
rownames(mergedDf) <- mergedDf$gene

# Remove genes that are not expressed
mergedDf<- mergedDf[apply(mergedDf[, which(colnames(mergedDf)!="gene")], 1,function
(x){!all(x==0)}),]

# Include mitochondrial percentage
percentMT <- c()
for(i in studyNums){percentMT <- c(percentMT, mergedDf[[i]]$percentMT)}
percentMT <- data.frame(percentMT, cell=names(percentMT))

# Create seurat object from merged count data and metadata
SeuratObject<- CreateSeuratObject(
                      counts = mergedDf[,which(colnames(mergedDf)!="gene")],
                      project = "combinedCountsSeurat", min.cells = 0, min.features = 0)
SeuratObject@meta.data$cell <- rownames(SeuratObject@meta.data)
SeuratObject@meta.data <- join(SeuratObject@meta.data, combinedMeta[,], by="cell",
type="left")
SeuratObject@meta.data <- join(SeuratObject@meta.data, percentMT, by="cell", type="-
left")
rownames(SeuratObject@meta.data) <- SeuratObject@meta.data$cell

# Normalize and identify variable features for each dataset independently
SeuratObjectSplit <- SplitObject(SeuratObject, split.by="study")
SeuratObjectSplit <- lapply(SeuratObjectSplit , function(x){
                      x <- NormalizeData(x)
```

```
                    x <- FindVariableFeatures(x, selection.method="vst", nfeatures
                    = 2000)
                                                             })
# Select features that are variable across datasets for integration
features <- SelectIntegrationFeatures(object.list=SeuratObjectSplit)
# Perform integration and integrated analysis
anchors <- FindIntegrationAnchors(SeuratObjectSplit, anchor.features=features)
SeuratIntegration<- IntegrateData(anchorset=anchors, dims=1:50)
SeuratIntegration<- ScaleData(SeuratIntegration , verbose=FALSE)
SeuratIntegration<- RunPCA(SeuratIntegration , npcs=30, verbose=FALSE)
SeuratIntegration<- FindNeighbors(SeuratIntegration , reduction="pca", dims=1:30)
SeuratIntegration<- RunUMAP(SeuratIntegration, reduction="pca", dims=1:30, seed.use
= 123)
  return(SeuratIntegration )
}


# Run the function and save the output:
IntegratedSeuratObject <- integrateAsFunction(c(1:4))
saveRDS(IntegratedSeuratObject , paste0(DataOutputFolder ,"SeuratIntegration.rds"))
```

3.4 Analyze Blastoids Integrated with Embryos

```
### Set your own figure output folder ###
FigureOutputFolder <-
'/lustre1/project/stg_00041/SanKit/SKT022_BlastoidMethodsPa-
per/Figure/'
```

After integrating the data, you are able to perform downstream analysis as described in section 4. In addition, we have included a few scripts for specific cases.

```
IntegratedSeuratObject <- readRDS(paste0(DataOutputFolder ,"SeuratIntegration.rds"))
SeuratSplit<- SplitObject(IntegratedSeuratObject, split.by="study")
```

Here, a particular subset of cells of interest are included. It is important to know that the downstream analysis is performed again after subsetting data. Here we also reintegrate the data after subsetting the data, but this is optional. However, when subsetting to very few cells, this might cause some issues and thus, keeping the integration of the whole dataset is recommended.

```
# Subset Kagawa dataset to only include naive and  primed cells, and 96h blastoids
(the final time point)
SeuratSplit$kagawa <- subset(SeuratSplit$kagawa, subset = cellLineageOriginal%in%c(
"blastoid_96h_DN", "blastoid_96h_PDGFRa_pl", "blastoid_96h_TROP2_pl", "naive_H9",
"primed_H9"))

# Reintegrate
features <- SelectIntegrationFeatures(object.list=SeuratSplit)
```

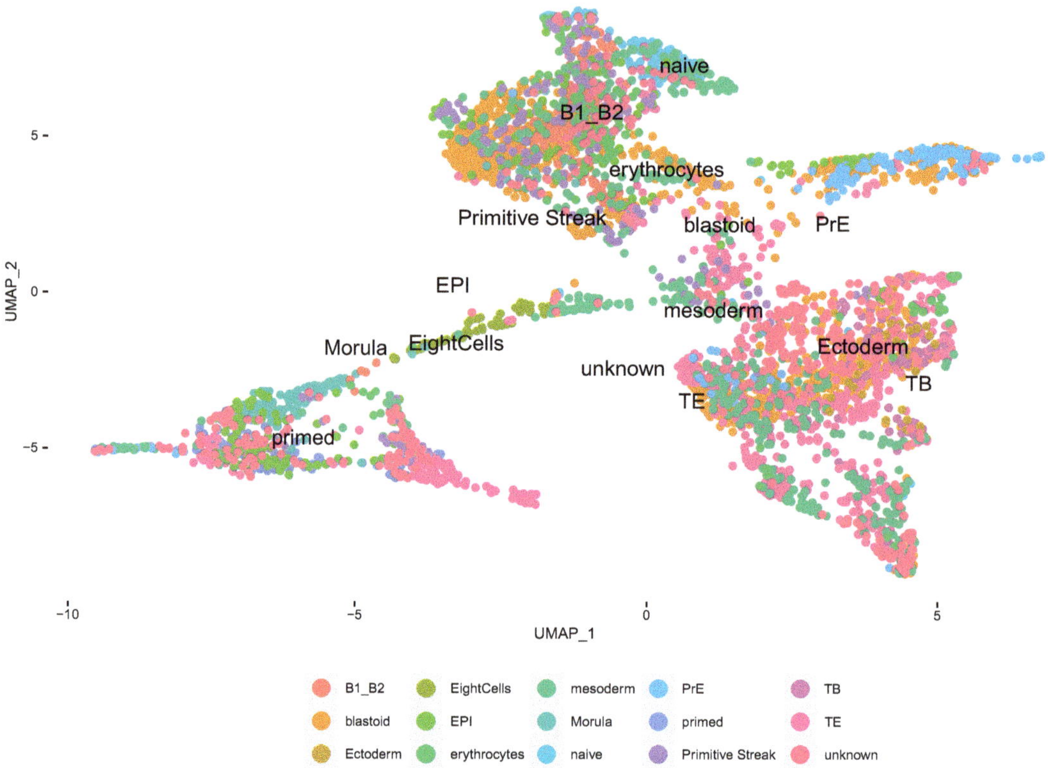

Fig. 9 UMAP of integrated scRNA-seq datasets of human blastoids, embryos, and human pluripotent stem cells colored by cell type

```
anchors <- FindIntegrationAnchors(SeuratSplit, anchor.features=features)
SeuratReintegrated<- IntegrateData(anchorset=anchors, dims=1:50)

SeuratReintegrated<- ScaleData(SeuratReintegrated, verbose=FALSE)
SeuratReintegrated<- RunPCA(SeuratReintegrated, npcs=30, verbose=FALSE)
SeuratReintegrated<- FindNeighbors(SeuratReintegrated, reduction="pca", dims=1:30)
SeuratReintegrated<- RunUMAP(SeuratReintegrated, reduction="pca", dims=1:30, seed.
use = 123)
```

3.4.1 Non-linear Dimensional Reduction of Integrated Data

For plotting a UMAP, the Seurat-provided DimPlot() can be used. Here we opt for an alternative using the ggplot 2 package to take advantage for its highly customizable characteristics. The difficulty of integrating multiple datasets, especially datasets with a high number of cell annotations, is the lack of an organized and clear vision of the labels on a UMAP or t-SNE. Here we have constructed a function to center the labels for each lineage and plotted this on the UMAP as an example (Fig. 9).

```
# Function to determine the mean x and y position on the UMAP for each cell type. This
will be used to plot labels and makes it easier to visualize where the populations
are.
centerLabel<- function(SeuratObject, colOfInterest="cellLineage"){
  outputFrame<- data.frame(
                    Label=unique(SeuratObject@meta.data[colOfInterest])[,1],
                    UMAP1=NA, UMAP2=NA)
  for(i in 1:nrow(outputFrame)){
      outputFrame$UMAP1[i] <- mean(
        SeuratObject@reductions$umap@cell.embeddings[which(
        SeuratObject@meta.data[colOfInterest]== outputFrame$Label[i]),
        1], na.rm=T)

        outputFrame$UMAP2[i] <- mean(
        SeuratObject@reductions$umap@cell.embeddings[which(
        SeuratObject@meta.data[colOfInterest]== outputFrame$Label[i]),
        2], na.rm=T)
             }
return( outputFrame)
 }

# Determine label coordinates
LabelCoordinates<- centerLabel(SeuratReintegrated, colOfInterest="cellLineage")
# Plot UMAP (Fig .9)
ggplot(data.frame(SeuratReintegrated@reductions$umap@cell.embeddings),
       aes(x=UMAP_1, y=UMAP_2, color=SeuratReintegrated@meta.data$cellLineage)) +
geom_point() +
theme(legend.position="bottom") +
geom_text_repel(data=LabelCoordinates , aes(x=UMAP1, y=UMAP2, label=Label),
         face="bold", size=4, color="black") +
guides(colour = guide_legend(override.aes = list(size=4), title ="")) + theme
(text=element_text(size=8), panel.background=element_blank())
```

Next, cells are clustered to assess the cell distribution in each cluster and which cells across different datasets cluster together (Fig. 10).

```
SeuratReintegrated<- FindNeighbors(SeuratReintegrated , dims = 1:20)
# Change resolution in FindClusters if you want more or less clusters
SeuratReintegrated<- FindClusters(SeuratReintegrated , resolution = 0.3)
# Compare this resulting UMAP to the previous one, to help determine which clusters go
where
DimPlot(SeuratReintegrated, reduction = "umap", pt.size = 2, label = T)
```

3.4.2 Change Annotation

We then take the blastoid cell type annotations from section 4.6 and add them here. Additionally, we change all of the labels to specify that those cells are from blastoids.

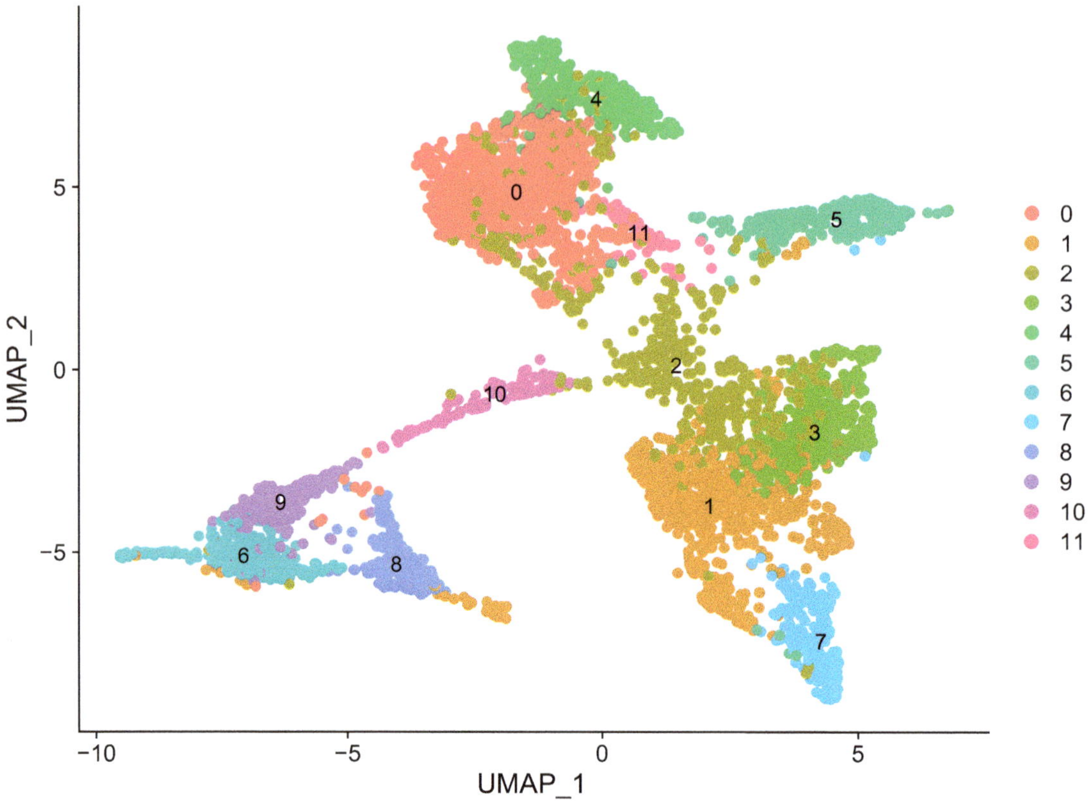

Fig. 10 UMAP of integrated scRNA-seq datasets of human blastoids, embryos, and human pluripotent stem cells colored by cluster

```
# Retrieve the annotations of the blastoid data from section 4.6
kagawaSeurat <- readRDS("RivronResults.rds")

# Seurat does not like hyphens in cell names, so it changed them to periods
# If your cell names do not match between your annotation and Seurat object, this is
something to check for (Note 5)
head(SeuratReintegrated@meta.data$cell)
head(rownames(kagawaSeurat@meta.data))
```

Here are the outputs from this command:

```
'primed_H9.150182''primed_H9.150183''primed_H9.150185''primed_H9.150187''pri-
med_H9.150188'

'primed_H9-150182''primed_H9-150183''primed_H9-150185''primed_H9-150187''primed_H9-
150188'
```

Note that one has periods and the other has hyphens. This will need to be changed to merge the metadata

Table 3
The first five rows of our table to change labels. "currLabels" refers to the current cell type label, and "FinalLabels" is a new combined label including study and cell type. This was generated manually in excel and imported here

	Study <chr>	currLabels <chr>	FinalLabels <chr>
1	rivron	primed_H9	Kagawa Primed
2	rivron	blastoid epiblast	Blastoid Epiblast
3	rivron	naive_H9	Kagawa Naive
4	rivron	blastoid EXM	Blastoid EXMC
5	rivron	blastoid TE	Blastoid TE

```
# Change the hyphens to periods to match annotations
kagawaSeurat@meta.data$cell <-unlist(lapply(as.list(rownames(kagawaSeurat@meta.
data)),
            function(nameWhyphen){return(gsub(nameWhyphen,pattern = "-",
            replacement = "."))}))
```

Here, we made a new variable copied from the existing variable "cellLineageOriginal". In this new variable, we will adjust the cell labels for blastoid cells where the annotations define their sample origin. We used the metadata from the Seurat object generated in section 4 to designate which cells will be annotated for particular labels. We are also changing all of our cell type labels to how we want them to appear in our final figures (Table 3).

```
# Change blastoid labels to match those defined in section 4
SeuratReintegrated@meta.data$finalLabels <-
        SeuratReintegrated@meta.data$cellLineageOriginal

SeuratReintegrated@meta.data$finalLabels[which(

SeuratReintegrated@meta.data$cell%in%kagawaSeurat@meta.data$cell[which(
        kagawaSeurat@meta.data$manualClusters=="epiblast")])] <-
        "blastoid epiblast"

SeuratReintegrated@meta.data$finalLabels[which(

SeuratReintegrated@meta.data$cell%in%kagawaSeurat@meta.data$cell[which(
        kagawaSeurat@meta.data$manualClusters=="primitive endoderm")])] <-
        "blastoid PE"

SeuratReintegrated@meta.data$finalLabels[which(

SeuratReintegrated@meta.data$cell%in%kagawaSeurat@meta.data$cell[which(
```

```
        kagawaSeurat@meta.data$manualClusters=="trophectoderm")])] <- "blastoid TE"

SeuratReintegrated@meta.data$finalLabels[which(
SeuratReintegrated@meta.data$cell%in%kagawaSeurat@meta.data$cell[which(
          kagawaSeurat@meta.data$manualClusters=="exta-embyronic mesoderm")])] <-
          "blastoid EXM"

# Manually generated table to compare all unique combination of the variable
"finalLabels" and "study"
# This data is available in our github page.
DataFolder <- "/staging/leuven/stg_00041/Published_Data/"

# Changing the cell labels to the new annotation
for(i in 1:nrow(changeLabels)){
      SeuratReintegrated@meta.data$finalLabels[which
(SeuratReintegrated@meta.data$finalLabels==changeLabels$currLabels[i] &
        SeuratReintegrated@meta.data$study==changeLabels$study[i])] <- changeLabels
        $FinalLabels[i]}

#This is how the table looks like (Table 3)
head(changeLabels)
```

After adjusting the annotation, we plot them in an UMAP as an example (Fig. 11). This UMAP shows the relationship between different embryo cell types with the cells from the in vitro data. The various blastoid cell types overlap with their corresponding cell types from the embryo.

```
# Rerun the function to put labels at the center of cell types, now with the new
labels
LabelCoordinates <-centerLabel(SeuratReintegrated, colOfInterest="finalLabels")

# Plot the UMAP with the final labels
ggplot(data.frame(SeuratReintegrated@reductions$umap@cell.embeddings), aes
(x=UMAP_1, y=UMAP_2)) +
    # Draw the points
geom_point(data=data.frame(SeuratReintegrated@reductions$umap@cell.embeddings,
                          lineage=SeuratReintegrated@meta.data$finalLabels,
                          study=SeuratReintegrated@meta.data$study),
                          aes(x=UMAP_1, y=UMAP_2, color=lineage), size=0.75) +
    # Draw the labels
     geom_text_repel(data=LabelCoordinates, aes(x=UMAP1, y=UMAP2,
                       label=Label), size=3, color="black") +
     # Change the aesthetics to make it easier to read the legend and removing
     background
     guides(colour = guide_legend(override.aes = list(size=2), ncol=1)) +
       theme(text=element_text(size=7), panel.grid = element_blank(), axis.ticks =
       element_blank(),
```

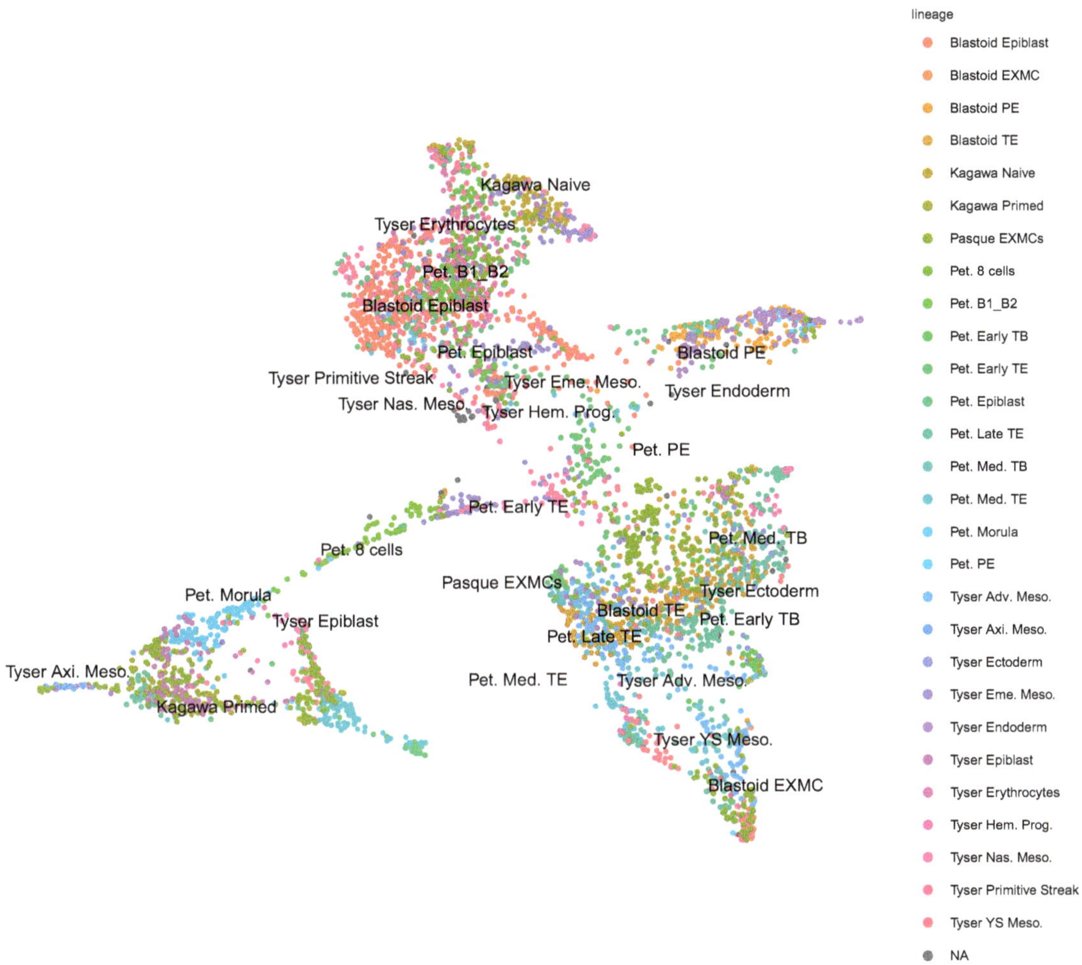

Fig. 11 UMAP of integrated scRNA-seq datasets of human blastoids, human embryos, and human pluripotent stem cells colored and labeled with our final labels

```
                    axis.text=element_blank(), axis.line = element_blank(), panel.
                    border = element_blank(),
                   legend.position="right") +
    theme_linedraw() + xlab("") + ylab("")
```

3.4.3 Highlight Specific Cells of Interest

Here we provided a function that enables us to highlight specific cells defined by a list of annotations from the metadata. This allows us to focus on specific cells of interest without other labels (Fig. 12). These graphs better emphasize the overlap of blastoid cells with their corresponding cell types from the embryo, and give us a clearer overview of individual cell types.

```
# Function to graph specific cell lineages over the rest of the UMAP. This makes it
easier to view specific cell types and how they compare and overlap between studies.
```

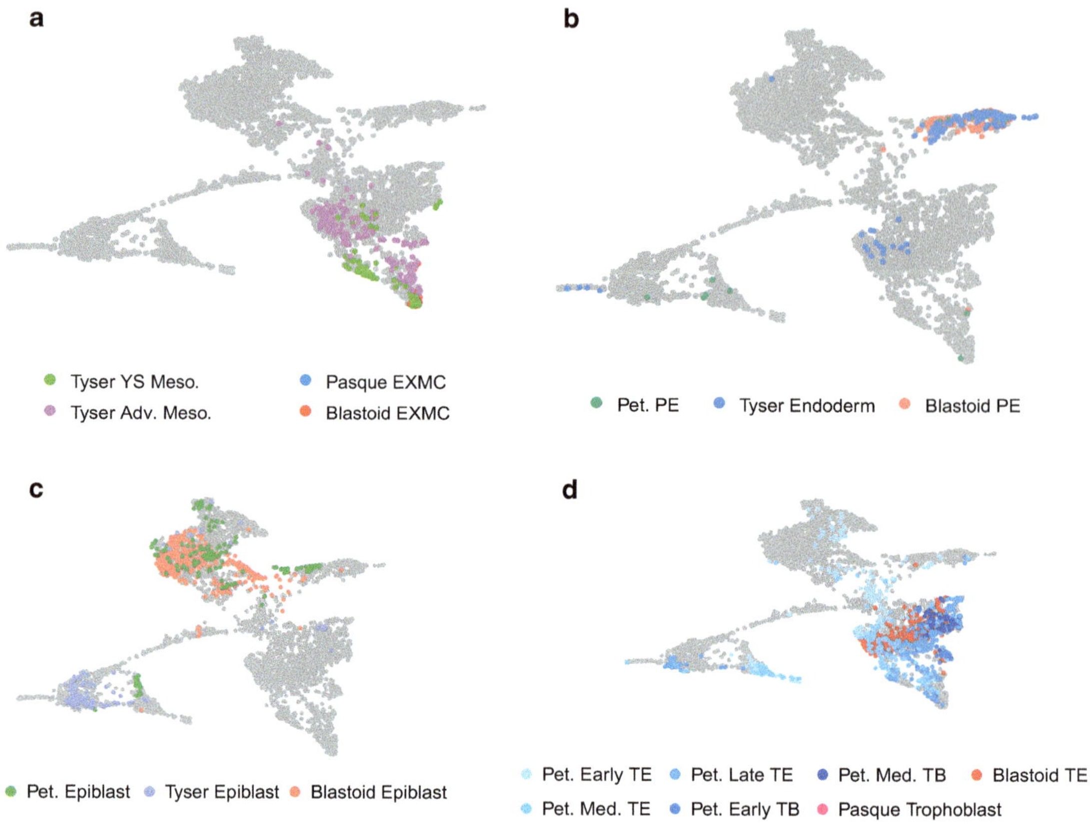

Fig. 12 UMAP of integrated scRNA-seq datasets of human blastoids, embryos, and human pluripotent stem cells showing individual cell lineages. (**a**) Extraembryonic mesoderm. (**b**) Primitive endoderm. (**c**) Epiblast. (**d**) Trophectoderm

```
graphSpecificLineage <- function(lineageToGraph, outname, custColor=col_vector, col-
orLabelRow=1){

   # Output graphs as pdfs. If you remove this, it will print to output instead.
    pdf(outname, width=2, height=2 )

   # Match the custom colors to the array of cell types
    names(custColor)<- lineageToGraph

    # Print all data points in gray
    print(
ggplot(data.frame(combinedCountsSeuratReintegrated@reductions$umap@cell.
embeddings),
     aes(x=UMAP_1, y=UMAP_2)) +
    geom_point(data=data.frame(
combinedCountsSeuratReintegrated@reductions$umap@cell.embeddings,
lineage=combinedCountsSeuratReintegrated@meta.data$finalLabels,
```

```
            study=combinedCountsSeuratReintegrated@meta.data$study),
                        aes(x=UMAP_1, y=UMAP_2),color="gray", size=0.01) +
# Color points/cell of interest with the custom colors
  geom_point(data=data.frame(
combinedCountsSeuratReintegrated@reductions$umap@cell.embeddings,
lineage=combinedCountsSeuratReintegrated@meta.data$finalLabels,
study=combinedCountsSeuratReintegrated@meta.data$study)[which(
combinedCountsSeuratReintegrated@meta.data$finalLabels%in%lineageToGraph),],
            aes(x=UMAP_1, y=UMAP_2, color=lineage), size=0.01) +
    # Adjustment of the graph
    theme_linedraw() +
    theme(legend.position="bottom", panel.grid.major = element_blank(), panel.grid.
    minor = element_blank()) +
    guides(colour = guide_legend(override.aes = list(size=1), nrow=colorLabelRow)) +
      theme(text=element_text(size=7), axis.line = element_blank(), axis.text =
      element_blank(),
    panel.border = element_blank(), axis.ticks = element_blank()) +
    scale_color_manual(values=custColor, name=NULL)+xlab(NULL)+ylab(NULL)
    )

   #This turns off the command to send the graph to pdf
    dev.off()
}

# Graph all of the cell types in each lineage together
graphSpecificLineage(outname = paste(c( FigureOutputFolder, "5.5a.integration.
EXMonly.pdf"), collapse=""), lineageToGraph = c("Tyser YS Meso.","Tyser Adv. Meso.",
"Pham EXMC", "Blastoid EXMC"), custColor=c("#66A61E","#BC80BD" ,
"#1F78B4","#E41A1C"))

graphSpecificLineage(outname = paste(c( FigureOutputFolder, "5.5b.integration.
PrEonly.pdf"), collapse=""), lineageToGraph = c("Pet. PE", "Tyser Endoderm",
"Blastoid PE"), custColor=c("#1B9E77", "#377EB8", "#FB8072"))

graphSpecificLineage(outname = paste(c( FigureOutputFolder , "5.5c.integration.
EPIonly.pdf"), collapse=""), lineageToGraph = c("Pet. Epiblast","Tyser Epiblast",
"Blastoid Epiblast"), custColor=c("#33A02C","#8DA0CB", "#FB8072" ))

graphSpecificLineage(outname = paste(c( FigureOutputFolder, "5.5d.integration.
TEonly.pdf"), collapse=""), lineageToGraph = c("Pet. Early TE", "Pet. Med. TE",
"Pet. Late TE", "Pet. Early TB","Pet. Med. TB", "Pham Trophoblast", "Blastoid TE"),
                  custColor=c(brewer.pal(n = 8, name = "Blues")[c(4:8)], "#E7298A",
"#E31A1C"), colorLabelRow=2)
```

4 Notes

1. When setting QC thresholds, it is useful to consider the biological context of your samples, such as if you expect cells with low transcriptional activity, or high mitochondrial activity due to highly active respiratory mechanisms, these cells can be considered to be included.
2. It is recommended to frequently save your work at different stages of your analysis, especially after compute-intensive processes.
3. To follow the methods, we have provided count matrices and metadata files from the authors mentioned above publicly available in our github website: https://github.com/pasquelab/BlastoidIntegration.
4. It is recommended to load only what is deemed to be required for each step. Therefore, we reset our workspace between sections. Remember to reload libraries and any variable used.
5. Seurat does not like hyphens in cell names, so it changed them to periods. If your cell names do not match between your annotation and Seurat object, this is something to check for.

Acknowledgments

We thank the Vlaams Supercomputer Center Leuven (https://www.vscentrum.be) which we used for all computing. Research in the Pasque lab is funded by The Research Foundation–Flanders (FWO; Odysseus Return Grant G0F7716N to V.P.; FWO grants G0C9320N and G0B4420N to V.P.; EOS project PANDAROME G0I7822N) and the KU Leuven Research Fund (C1 grant C14/21/19 to V.P.), with additional FWO scholarships to S.K.T. (1S75720N) and B.P.B. (1263323 N).

References

1. Butler A, Hoffman P, Smibert P, Papalexi E, Satija R (2018) Integrating single-cell transcriptomic data across different conditions, technologies, and species. Nat Biotechnol 36(5):411–420
2. Cao J, O'Day DR, Pliner HA, Kingsley PD, Deng M, Daza RM, Zager MA et al (2020) A human cell atlas of fetal gene expression. Science 370(6518):eaba7721
3. Fan Y, Min Z, Alsolami S, Zhenglai Ma E, Zhang WC, Zhong K et al (2021) Generation of human blastocyst-like structures from pluripotent stem cells. Cell Discover 7(1):1–14
4. Fei L, Chen H, Lifeng Ma E, Weigao RW, Fang X, Zhou Z et al (2022) Systematic identification of cell-fate regulatory programs using a single-cell atlas of mouse development. Nat Genet 54(7):1051–1061
5. Haniffa M, Taylor D, Linnarsson S, Aronow BJ, Bader GD, Barker RA, Camara PG et al (2021) A roadmap for the human developmental cell atlas. Nature 597(7875):196–205
6. Hao Y, Hao S, Andersen-Nissen E, Mauck WM 3rd, Zheng S, Butler A, Lee MJ et al (2021) Integrated analysis of multimodal single-cell data. Cell 184(13):3573–87.e29

7. Hao Y, Stuart T, Kowalski M, Choudhary S, Hoffman P, Hartman A, Srivastava A et al (2022) Dictionary learning for integrative, multimodal, and scalable single-cell analysis. bioRxiv. https://doi.org/10.1101/2022.02.24.481684
8. Ilicic T, Kim JK, Kolodziejczyk AA, Bagger FO, McCarthy DJ, Marioni JC, Teichmann SA (2016) Classification of low quality cells from single-cell RNA-Seq data. Genome Biol 17(February):29
9. Kagawa H, Javali A, Khoei HH, Sommer TM, Sestini G, Novatchkova M, Op Reimer YS et al (2022) Human blastoids model blastocyst development and implantation. Nature 601(7894):600–605
10. Liu X, Nefzger CM, Rossello FJ, Chen J, Knaupp AS, Firas J, Ford E et al (2017) Comprehensive characterization of distinct states of human naive pluripotency generated by reprogramming. Nat Methods 14(11):1055–1062
11. Lovell-Badge R, Anthony E, Barker RA, Bubela T, Brivanlou AH, Melissa Carpenter R, Charo A et al (2021) ISSCR guidelines for stem cell research and clinical translation: the 2021 update. Stem Cell Reports 16(6):1398–1408
12. Luecken MD, Theis FJ (2019) Current best practices in single-cell RNA-Seq analysis: a tutorial. Mol Syst Biol 15(6):e8746
13. Naydenov DD, Vashukova ES, Barbitoff YA, Nasykhova YA, Glotov AS (2023) Current status and prospects of the single-cell sequencing technologies for revealing the pathogenesis of pregnancy-associated disorders. Genes 14(3). https://doi.org/10.3390/genes14030756
14. Petropoulos S, Edsgärd D, Reinius B, Deng Q, Panula SP, Codeluppi S, Reyes AP, Linnarsson S, Sandberg R, Lanner F (2016) Single-cell RNA-Seq reveals lineage and X chromosome dynamics in human preimplantation embryos. Cell 165(4):1012–1026
15. Pham, Thi Xuan Ai, Amitesh Panda, Harunobu Kagawa, San Kit To, Cankat Ertekin, Grigorios Georgolopoulos, Sam S. F. A. van Knippenberg, et al. 2022. "Modeling human extraembryonic mesoderm cells using naive pluripotent stem cells." Cell Stem Cell 29 (9): 1346–65.e10.
16. Picelli S, Björklund ÅK, Faridani OR, Sagasser S, Winberg G, Sandberg R (2013) Smart-seq2 for sensitive full-length transcriptome profiling in single cells. Nat Methods 10(11):1096–1098
17. Picelli S, Faridani OR, Björklund AK, Winberg G, Sagasser S, Sandberg R (2014) Full-length RNA-Seq from single cells using smart-seq2. Nat Protoc 9(1):171–181
18. Regev A, Teichmann SA, Lander ES, Amit I, Benoist C, Birney E, Bodenmiller B et al (2017) The human cell atlas. eLife 6 (December). https://doi.org/10.7554/eLife.27041
19. Rivron NC, Frias-Aldeguer J, Vrij EJ, Boisset J-C, Korving J, Vivié J, Truckenmüller RK, van Oudenaarden A, van Blitterswijk CA, Geijsen N (2018) Blastocyst-like structures generated solely from stem cells. Nature 557(7703): 106–111
20. Rossant J, Tam PPL (2022) Early human embryonic development: blastocyst formation to gastrulation. Dev Cell 57:152–165. https://doi.org/10.1016/j.devcel.2021.12.022
21. Satija R, Farrell JA, Gennert D, Schier AF, Regev A (2015) Spatial reconstruction of single-cell gene expression data. Nat Biotechnol 33(5):495–502
22. Sozen B, Jorgensen V, Weatherbee BAT, Chen S, Zhu M, Zernicka-Goetz M (2021) Reconstructing aspects of human embryogenesis with pluripotent stem cells. Nat Commun 12(1):5550
23. Stuart T, Butler A, Hoffman P, Hafemeister C, Papalexi E, Mauck WM 3rd, Hao Y, Stoeckius M, Smibert P, Satija R (2019) Comprehensive integration of single-cell data. Cell 177(7):1888–1902.e21
24. Svensson V, Vento-Tormo R, Teichmann SA (2018) Exponential scaling of single-cell RNA-Seq in the past decade. Nat Protoc 13(4):599–604
25. Tang F, Barbacioru C, Wang Y, Nordman E, Lee C, Nanlan X, Wang X et al (2009) mRNA-Seq whole-transcriptome analysis of a single cell. Nat Methods 6(5):377–382
26. Taubenschmid-Stowers J, Rostovskaya M, Santos F, Ljung S, Argelaguet R, Krueger F, Nichols J, Reik W (2022) 8C-like cells capture the human zygotic genome activation program in vitro. Cell Stem Cell 29(3):449–59.e6
27. Tyser RCV, Mahammadov E, Nakanoh S, Vallier L, Scialdone A, Srinivas S (2021) Single-cell transcriptomic characterization of a gastrulating human embryo. Nature 600(7888):285–289
28. Wang R, Zhang P, Wang J, Lifeng Ma E, Weigao SS, Jiang M et al (2022) Construction of a cross-species cell landscape at single-cell level. Nucleic Acids Res 51(2):501–516
29. Wolf FA, Angerer P, Theis FJ (2018) SCANPY: large-scale single-cell gene expression data analysis. Genome Biol 19(1):15

30. Xiang L, Yin Y, Zheng Y, Ma Y, Li Y, Zhao Z, Guo J et al (2020) A developmental landscape of 3D-cultured human pre-gastrulation embryos. Nature 577(7791):537–542
31. Xu X, Hua X, Mo H, Hu S, Song J (2023a) Single-cell RNA sequencing to identify cellular heterogeneity and targets in cardiovascular diseases: from bench to bedside. Basic Res Cardiol 118(1). https://doi.org/10.1007/s00395-022-00972-1
32. Xu Y, Zhang T, Zhou Q, Mengzhu H, Qi Y, Xue Y, Nie Y, Wang L, Bao Z, Shi W (2023b) A single-cell transcriptome atlas profiles early organogenesis in human embryos. Nat Cell Biol 25(4):604–615
33. Yanagida A, Spindlow D, Nichols J, Dattani A, Smith A, Guo G (2021) Naive stem cell blastocyst model captures human embryo lineage segregation. Cell Stem Cell 28(6):1016–22.e4
34. Yu L, Wei Y, Duan J, Schmitz DA, Sakurai M, Wang L, Wang K, Zhao S, Hon GC, Jun W (2021) Blastocyst-like structures generated from human pluripotent stem cells. Nature 591(7851):620–626
35. Zappia L, Theis FJ (2021) Over 1000 tools reveal trends in the single-cell RNA-Seq analysis landscape. Genome Biol 22(1):301
36. Zappia L, Phipson B, Oshlack A (2018) Exploring the single-cell RNA-Seq analysis landscape with the scRNA-tools database. PLoS Comput Biol 14(6):e1006245
37. Zhao C, Reyes AP, Schell JP, Weltner J, Ortega NM, Zheng Y, Björklund ÅK et al (2021) Reprogrammed blastoids contain amnion-like cells but not trophectoderm. bioRxiv. https://doi.org/10.1101/2021.05.07.442980
38. Zhou F, Wang R, Yuan P, Ren Y, Mao Y, Li R, Lian Y et al (2019) Reconstituting the transcriptome and DNA methylome landscapes of human implantation. Nature 572(7771):660–664

Methods in Molecular Biology (2024) 2767: 251–262
DOI 10.1007/7651_2022_465

Published online: 16 Febuary 2023

Inferring Gene Regulatory Networks and Predicting the Effect of Gene Perturbations via IQCELL

Tiam Heydari and Peter W. Zandstra

Abstract

IQCELL is a platform that infers Boolean gene regulatory networks from single-cell RNA sequencing data. Boolean networks can be simulated under normal and perturbed conditions. In this chapter, we provide a detailed guideline for implementing IQCELL from a raw dataset. The steps include processing data, inferring informative genes, inferring gene regulatory network, and simulating the resulted network under normal and perturbed conditions.

Key words Gene regulation, Boolean GRNs, scRNA-seq, Gene perturbations, Developmental trajectory, Causal inference

1 Introduction

Studying development and uncovering the principles of gene regulatory networks (GRNs) are historically connected. The early discovery that the blueprint of development is located inside maternal and paternal chromosomes [3] along with the speculations that stem cell differentiation is governed by differential gene expression [10] set a foundation for conceptually relating the molecular basis of gene regulation to stem cell differentiation [4]. It has been shown that during development, cis-regulatory systems can perform simple combinatorial calculations similar to computer gates [14, 21], and these gates can be integrated into gene regulatory networks (GRNs) that process information [5]. An important hallmark of research was the elaboration of a logical (Boolean) model GRN based on cis-regulatory connections that has predictive power for endomesoderm specification of Echinoidea [12, 13]. While the discovery of regulatory systems is rooted in the study of model organisms, applying this methodology to the study of gene regulation during mammalian development has started to inform cell-based therapeutic applications [17]. An early and prominent example is reprogramming the embryonic and adult somatic cell's GRNs into the pluripotent state [16]. It has also been shown that the developmental trajectories of stem cell differentiation can be modeled as Boolean GRNs undergoing state transitions [20] and that simulation of GRN can identify control points in stem cell-based developmental systems [7, 9]. Ultimately, having a more accurate picture of regulatory network evolution during stem cell differentiation is providing new strategies to manipulate cell fate and could also lead to improved stem cell-based cell manufacturing processes [8].

Herein we demonstrate the use of IQCELL [6], a multi-paradigm method and a software package for reconstructing GRNs directly from single-cell RNA sequencing (scRNA-seq) data (Fig. 1a). We use a publicly available dataset of blood development [11] and demonstrate a detailed pipeline to infer a Boolean GRN for the system and to predict the effect of gene perturbations on differentiation trajectories. By following our methods section

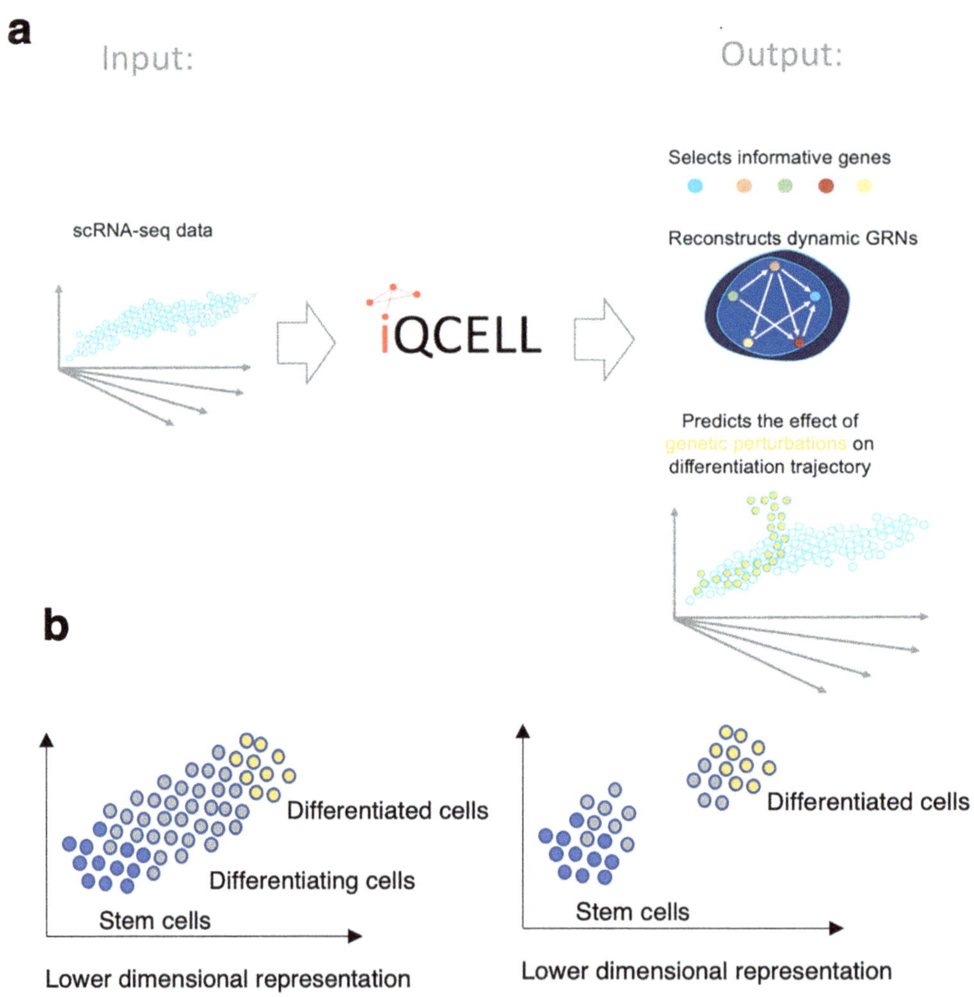

Fig. 1 General IQCELL workflow. (**a**) IQCELL infers executable GRNs from scRNA-seq data. (**b**) Example of a dataset that captures (left) and does not capture (right) the transitioning states of differentiating cells

and applying IQCELL into appropriate datasets of interest, users will be able to integrate IQCELL GRN inference pipeline into their research.

2 Materials

2.1 System Requirements

The following packages are required for implementing the IQCELL pipeline:

Python (v 3.7.3 or later), numpy, matplotlib, pandas (v 0.24.2), sklearn, scipy, seaborn, scprep, network (v 2.3), magic, pygraphviz, z3, booleannet3, ply, scanpy, slingshot (python version), pyscenic, and iqcell.

Follow the instructions of setting up the system (tested on Linux and Mac) for IQCELL package on this page: https://gitlab.com/stemcellbioengineering/iqcell.

2.2 Dataset Requirements

IQCELL infers GRNs from scRNA-seq data. An optimal scRNA-seq dataset for IQCELL analysis will have below characteristics:

1. Measures the mRNAs of a process in which the cells undergo state transition. This process should be reflected in the transcriptional state of the cells (e.g., stem cell differentiation).
2. The process should be mapped into a single pseudo-time order. In the case that there is bifurcation point, this pipeline should be run for each individual branch separately.
3. The dataset should have sufficient sequencing depth that allows capturing the activity of transcription factors (TFs) (*see* **Notes 1** and **2**).
4. The scRNA-seq dataset capture a sufficient time resolution of the differentiation process. This is subject to the system of interest and the underlying biology. Some differentiation processes are happening very fast (hours to couple of days), and some others are very slow (multiple weeks). The ideal dataset captures the transitioning state of cells and provides a smooth cloud of cells at different states (Fig. 1b).

2.3 Notebooks

All steps in the Methods section are implemented in four publicly accessible jupyter notebooks (Notebook1, Notebook2a, Notebook2b, Notebook3). Please find them in the link below:

https://gitlab.com/stemcellbioengineering/iqcell/example2.

3 Methods

3.1 Processing scRNA-seq Data (Notebook1)

Here we provide an example of processing the scRNA-seq data to generate inputs for IQCELL (Fig. 2a) from a publicly available dataset.

1. *Initial clustering and selecting cells*: The dataset that we use [11] provides a snapshot of blood development from blood *stem* cells into different lineages including erythrocytes (*Ery*), megakaryocytes (*Mk*), monocytes (*Mo*), neutrophils (*Neu*), and other blood cell types [19] (Fig. 2b). Here we will focus on the differentiation of stem cells toward monocytes. As is illustrated in Fig. 2b and based on the analysis performed in [19], cells are annotated into stem cells, *Mo*, *Ery*, *Mk*, *Neu*, etc. (Fig. 2b). First, we perform clustering on the dataset and select the clusters that are included in a continuous path between *stem* and progenitors. Using the Louvain clustering algorithm, we obtain 13 clusters, cluster 0 includes the *stem*, and cluster 7 includes *Mo* (Fig. 2c). We will include cluster 1 as the cluster that captures some transitioning cells as well; this is based on including all cells that make a continuous path between *stem* and *Mo* (Fig. 2d) (*see* **Note 3**).

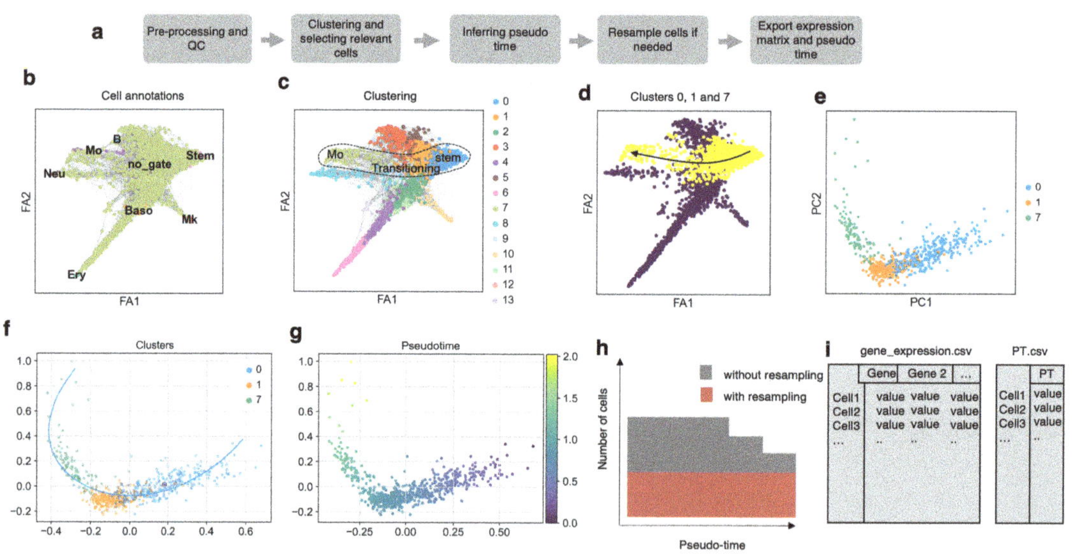

Fig. 2 Analyzing scRNA-seq data. (**a**) Overview of processing scRNA-seq data. (**b**) Lower-dimensional representation of scRNA-seq. Some cells are annotated with the appropriate cell types. (**c**) Clustering allows for selecting a smooth path between the cell states of interest. (**d**) The selected clusters show a connected path. (**e**) Re-performing PCA on the selected data. (**f**, **g**) Inferring pseudo-time order of the cells. (**h**) Re-sampling the cells along the pseudo-time removes bias from populated regions of pseudo-time. (**i**) Exporting the gene expression and pseudo-time datasets in appropriate format

2. *Inferring the pseudo-time order*: After cleaning the cells and selecting the cells of interest, we use the slingshot method [15] to infer pseudo-time order of the cells. Slingshot method fits a curve to the lower-dimensional representation of the cells (Fig. 2e) and, based on this, infers a pseudo-time order of differentiating cells (starting from $t = 0$) (*see* **Note 4**). We set the cluster 0 to be the starting point, and the algorithm automatically selects the cluster 7 to be the ending point (Fig. 2f, g).
3. *Exporting the results*: Finally, we export the gene expressions and pseudo-time order into two separate files (gene_expression.csv and PT.csv). The format that we will use in the expression file is the cell names as row names and gene names as column names. In the PT file, we will use cell names as row names and the pseudo-time (PT) as the column names (Fig. 2i).

3.2 Selecting Informative Genes (Notebook2a and Notebook2b)

Using the output of above, we will select a list of genes (Fig. 3a) to feed into the GRN inference module (Notebook3).

1. *Using pySCENIC to infer active regulons (Notebook2a)*: First we use the pySCENIC platform [1, 18] to infer active regulons. Regulons are sets of TFs and their downstream targets. Selecting Regulons instead of TFs helps with the robustness of the selection process of active TFs in the dataset. Notebook2a provides a premade template for this step in the both cases of human and mouse scRNA-seq datasets. The steps include:
   (a) Loading gene_expression.csv
   (b) Loading the list of TFs (for human: hs_hgnc_curated_tfs.txt, for mouse: mm_mgi_tfs.txt)
   (c) Running GRNboost module to find the co-expression matrix
   (d) Prune regulons modules based on cis-regulatory footprints
   (e) Regulon enrichment based on gene expression
   (f) Exporting the outputs

   The output of this step is list of active regulons (Fig. 3b) and is exported as "auc_mtx.csv."
2. *Selecting a set of TFs based on active regulons (Notebook2b)*: Next, we use active regulons to infer a set of genes to be used in GRN inference module. The criteria that we use for selecting TFs is based on their variability along the pseudo-time axis by using gene dynamic score (GDS) (Notebook2b). To obtain GDS, we first divide the pseudo-time into multiple fragments (N bins). Then for each TF, we calculate the difference between the mean expression value between all pairs of bins (i, j). The GDS of a TF will be the maximum of all differences (Fig. 3c),

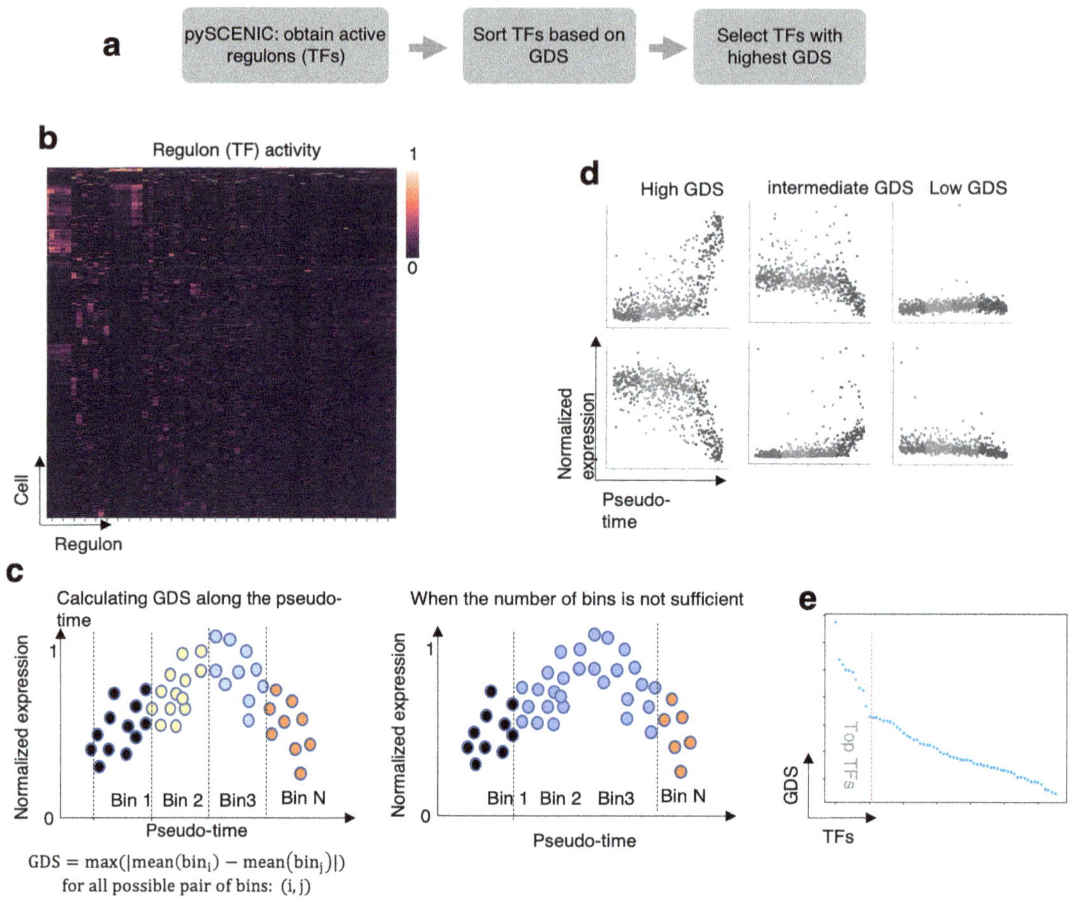

Fig. 3 IQCELL's gene selection pipeline. (**a**) Overview of the pipeline steps. (**b**) Regulon activity identifies active regulons and TFs along the dataset. (**c**) Calculating GDS based on binning the pseudo-time. (**e**) Selecting the top TFs based on GDS

and we assign a GDS to each TF (gene). Some TFs will have higher GDS than other TFs (Fig. 3d) (*see* **Note 5**). Finally, we select the top TFs based on their GDS (Fig. 3e) (*see* **Notes 6** and 7).

3.3 GRN Inference (Notebook3)

In this section, we explain how to infer GRN from the outputs of previous steps (gene expression profile, pseudo-time, and the list of genes) (Fig. 4a).

1. *Correcting for dropout effects.* After importing the iqcell library and loading the expression data, we correct for dropout effects (Fig. 4b) (*see* **Note 8**):

   From iqcell library: *run_magic_and_write_data*

2. *Calculating mutual information.* Next, after specifying the list of genes of interest, we calculate the correlation and mutual

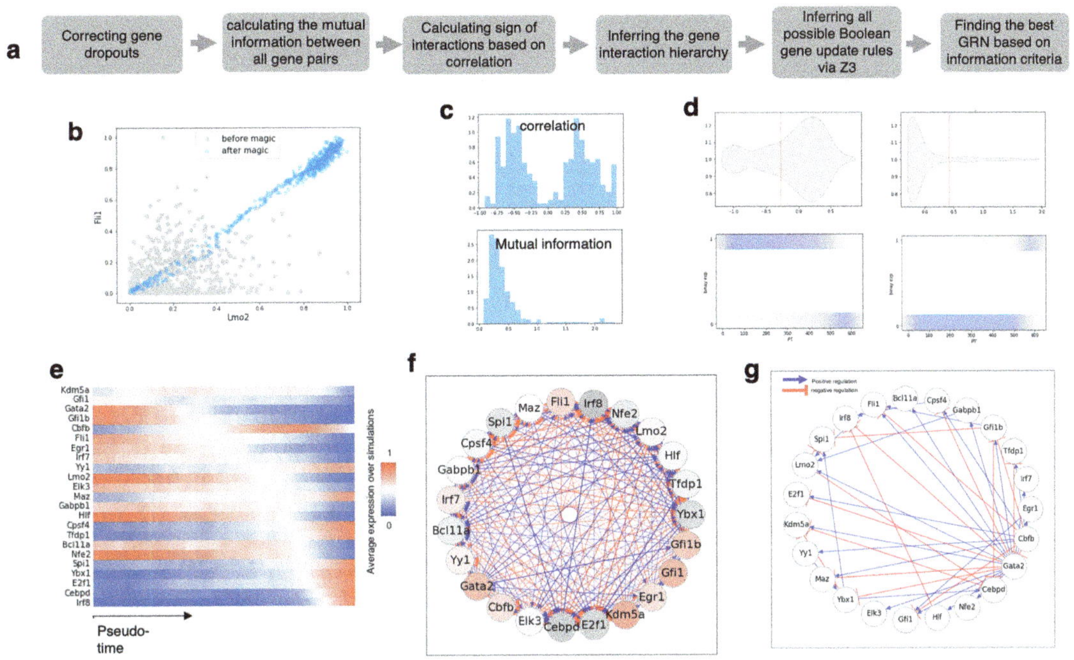

Fig. 4 IQCELL's GRN inference pipeline. (**a**) Overview of GRN inference pipeline. (**b**) Gene expression recovery recovers the dropouts. (**c**) The correlations and mutual information are calculated between all gene pairs. (**d**) Genes are binarized into two states: *on*/*off*. (**e**) The genes are sorted based on transitioning point. (**f**) Initial gene–gene interaction network. (**g**) The final Boolean GRN is inferred using Z3 engine and maximizing the mutual information

information between gene pairs. These will be used in the downstream applications (Fig. 4c).

From iqcell library: *export_selecet_genes*
From iqcell library: *calculate_the_gene_gene_interaction_in_three_ways*

3. *Binarizing genes.* At this step, we binarize the gene expression into *on*/*off*. There are two methods provided to binarize the gene expression, thresholding based on mean or k-means clustering. The default thresholding is set to the mean method (Fig. 4d).

   From iqcell library: *mean_thresholding_method_and_plot*
   From iqcell library: *K_mean_thresholding_method_and_plot*
   From iqcell library: *export_discretized_expression*

4. *Calculate gene hierarchy.* Next, we use the output of binarized gene expression to calculate the gene hierarchy. The hierarchy is made based on the transitioning points of genes from *on* to *off* or *off* to *on*. The genes that transition earlier in pseudo-time point will be on the higher level in gene hierarchy (Fig. 4e) (*see* **Note 9**).

From iqcell library: *generating_cell_order*
From iqcell library: *generating_expression_file*
From iqcell library: *calculate_jump_thresholds_from_data*
From iqcell library: *heatmap_avraging_expression_over_window_along_PT*

5. *Calculate gene-gene interaction network.* At this step, we calculate the initial gene-gene interaction network which makes the foundation for final GRN (Fig. 4f) (*see* **Note 10**).

   *sign_threshold = 0.65*pearson_std*
   From iqcell library: *z3_draw_propsed_graph*

6. *Determining update rules.* Using the results of the previous step, first we find the possible update rules for each gene using Z3 engine (*see* **Note 11**). Next, we find the best GRN based on mutual information criteria (the GRN with maximized mutual information) (Fig. 4g) (*see* **Note 12** and **13**).

   From iqcell library: *check_z3_interactions*
   From iqcell library: *Run_parallel_network_inference*
   From iqcell library: *convert_z3_to_simbool*
   From iqcell library: *select_model_GRN*
   From iqcell library: *convert_GRN_to_networkx*

3.4 Run Boolean Simulations and GRN Perturbations (Notebook3)

In this part, we will use the inferred Boolean GRN, to study its dynamic behavior under normal and perturbed conditions (*see* **Note 14**).

1. *Simulating a Boolean GRN.* First, we set up the simulations by finding the initial state of the system based on the pseudo-time and binarized gene expression; next we simulate the Boolean GRN and plot the outcomes (Fig. 5a, b).

   From iqcell library: find_and_export_initial_states
   From iqcell library: Check_GRN_for_ABN_simulation
   From iqcell library: simulate_Boolean_GRN
   From iqcell library: read_run_analysis
   From iqcell library: read_run_trajectory_analysis
   From iqcell library: plot_simunation_atractor_states
   From iqcell library: PCA_compare_sims_vs_exps_overlay_Refrences

2. *Simulating perturbations.* Next, we can simply simulate all gene perturbations (knock downs and over expression) (Fig 5c, d).

   From iqcell library: run_gene_perturbations_simulations
   From iqcell library: plot_simunation_atractor_states

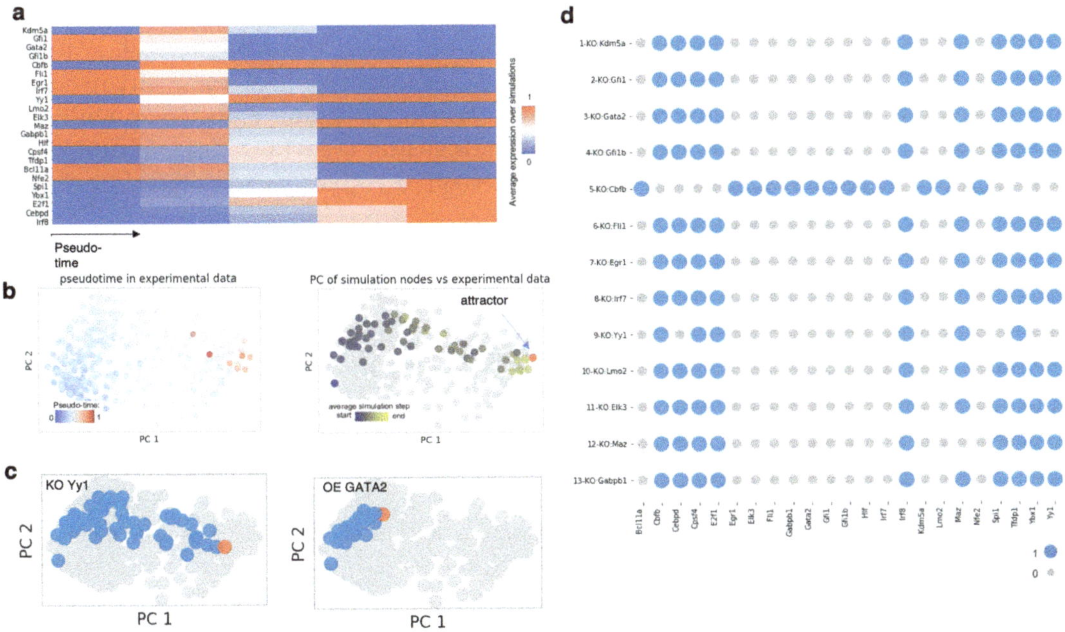

Fig. 5 GRN simulation in normal and perturbed conditions. (**a**) State transition of gene along the simulation time steps. (**b**) The PCA plot of scRNA-seq vs. simulation data. (**c**) Example of perturbation simulations. (**d**) IQCELL generates outputs for all perturbations

4 Notes

1. The sufficient sequencing depth is subjective to a particular cellular process. A general approach is to be able to recover the expression of TFs that are known to be active in particular processes in the scRNA-seq data; for a more general approach for determining the optimum sequencing depth, please see [2].
2. TFs usually have less copy numbers of mRNA in comparison to housekeeping genes. Because of this lower copy number, TFs are susceptible to dropout effects (meaning that low mRNA counts in cells can result in zero reads in scRNA-seq).
3. You can re-perform the dimensionality reduction analysis on the cells that are included in the selected clusters. This generates the dimensionality reduction representation only based on the cells of interest (Fig. 2e).
4. In some datasets, after dimensionality reduction and pseudotime inference, the user may notice an inhomogeneous distribution of number of cells along the differentiation axis (Fig. 2h). In this case we can resample (down sample) the cells along the pseudo-time axis to get a more uniform distribution. This will help to eliminate the bias of GRN inference in future steps toward the populated regions in pseudo-time.

5. The number of bins (N) has a default value of 5 for this example. But this can be changed into higher values in case that there are some genes that have high variability inside the bins. In this case the variability of gene along the pseudo-time cannot be captured sufficiently. Generally, a safe option will be to have more bins than 5 (which works in this example) (Fig. 3c).
6. Here we use the elbow method (Fig. 3e) to select the genes with distinctively highest GDS. However, it is important to note that the GRN inference will be most efficient if the number of genes is kept below 25 [6].
7. Here we assume that the genes of interest are not determined beforehand. It might, however, be the case that the user already has a set of genes that would like to use instead of the suggested genes or use it in addition to the autogenerated list of genes.

   In the first case, it is possible to skip the gene selection steps (Notebook2a and Notebook2b) and move to Notebook3. In the second case, the user should execute the gene selection pipeline fully and replace the genes of interest with the lower raked gens of the pipeline (genes with lower GDS). Note that in both cases, having more than 25 genes will slow down the pipeline analytics
8. The value of "knn_value" parameter for dropout algorithm is best to be between 2 and 10, depending on the gene expression count depth on the TFs of interest; bigger value will impute and correct more but might cause artifacts and unreal correlation between genes.
9. Occasionally, the automated sorting of genes becomes inaccurate because of noisy transitioning points. This might need to be manually checked out by the user to make sure the hierarchy is made correctly and modify it if needed.
10. The value of "sign_threshold" identifies how many interactions to be included in the initial gene–gene interaction network (Fig. 4f). The default value is 0.9 times the standard deviation of the Pearson correlation between gene pairs "pearson_std." If this produces not enough interactions (at least few for each gene), decrease the value of the cut-off.
11. This step is the most time-consuming step in IQCELL. Given the number of available CPUs, number of genes, and number of cells, this step may take from tens of minutes to hours. When the possible update rules for all genes is inferred and exported, this step is done (based on the outputs in the output box in the notebook). You can change the number of processors varying the value of "recommended_number_of_jobs."

12. The parameter "step" with the default value of 100 should be updated based on the number of cells and genes. With 1000 cells, the step value of 100 is sufficient.
13. The default value of maximum activators of the genes is set to 4; this can be decreased if needed. The default value of maximum inhibitor of the genes is set to 2 (this can be increased up to 4 if needed).
14. Here we only demonstrate single-gene perturbations, but iqcell can perform any combination of genetic perturbations.

Acknowledgments

Funding: TH and PWZ were supported by the Canadian Institutes of Health Research (CIHR), Foundation Grant FRN 154283, and the Natural Sciences and Engineering Research Council of Canada (NSERC), Discovery Grant RGPIN-2020-06496), to PWZ. PWZ is the Canada Research Chair in Stem Cell Bioengineering (https://www.chairs-chaires.gc.ca). The funders had no role in study design, data collection and analysis, decision to publish, or preparation of the manuscript. The funders had no role in study design, data collection and analysis, decision to publish, or preparation of the manuscript. The authors declare that no competing interests exist.

References

1. Aibar S, González-Blas CB, Moerman T, Huynh-Thu VA, Imrichova H, Hulselmans G, Rambow F, Marine J-C, Geurts P, Aerts J, van den Oord J, Atak ZK, Wouters J, Aerts S (2017) SCENIC: single-cell regulatory network inference and clustering. Nat Methods 14:1083–1086. https://doi.org/10.1038/nmeth.4463
2. Bass AJ, Robinson DG, Storey JD (2019) Determining sufficient sequencing depth in RNA-Seq differential expression studies. bioRxiv: 635623
3. Boveri T (1907) Zellen-Studien: Die Entwicklung dispermer Seeigel-Eier. G. Fischer
4. Britten RJ, Davidson EH (1969) Gene regulation for higher cells: a theory. Science 165: 349–357
5. Davidson EH (2002) A genomic regulatory network for development. Science 295:1669–1678. https://doi.org/10.1126/science.1069883
6. Heydari T, Langley MA, Fisher CL, Aguilar-Hidalgo D, Shukla S, Yachie-Kinoshita A, Hughes M, M. McNagny K, Zandstra PW (2022) IQCELL: a platform for predicting the effect of gene perturbations on developmental trajectories using single-cell RNA-seq data. PLOS Comput Biol 18:e1009907. https://doi.org/10.1371/journal.pcbi.1009907
7. Kathiriya IS, Rao KS, Iacono G, Devine WP, Blair AP, Hota SK, Lai MH, Garay BI, Thomas R, Gong HZ, Wasson LK, Goyal P, Sukonnik T, Hu KM, Akgun GA, Bernard LD, Akerberg BN, Gu F, Li K, Speir ML, Haeussler M, Pu WT, Stuart JM, Seidman CE, Seidman JG, Heyn H, Bruneau BG (2021) Modeling human TBX5 haploinsufficiency predicts regulatory networks for congenital heart disease. Dev Cell 56:292–309.e9. https://doi.org/10.1016/j.devcel.2020.11.020
8. Lipsitz YY, Timmins NE, Zandstra PW (2016) Quality cell therapy manufacturing by design. Nat Biotechnol 34:393–400. https://doi.org/10.1038/nbt.3525
9. Longabaugh WJR, Zeng W, Zhang JA, Hosokawa H, Jansen CS, Li L, Romero-Wolf M, Liu P, Kueh HY, Mortazavi A, Rothenberg EV (2017) Bcl11b and

combinatorial resolution of cell fate in the T-cell gene regulatory network. Proc Natl Acad Sci 114:5800–5807. https://doi.org/10.1073/pnas.1610617114
10. Morgan TH, others (1934) Embryology and genetics. Embryol Genet
11. Nestorowa S, Hamey FK, Pijuan Sala B, Diamanti E, Shepherd M, Laurenti E, Wilson NK, Kent DG, Göttgens B (2016) A single-cell resolution map of mouse hematopoietic stem and progenitor cell differentiation. Blood 128: e20–e31. https://doi.org/10.1182/blood-2016-05-716480
12. Peter IS, Davidson EH (2015) Genomic control process: development and evolution. Academic Press, San Diego
13. Peter IS, Faure E, Davidson EH (2012) Predictive computation of genomic logic processing functions in embryonic development. Proc Natl Acad Sci 109:16434–16442. https://doi.org/10.1073/pnas.1207852109
14. Small S, Kraut R, Hoey T, Warrior R, Levine M (1991) Transcriptional regulation of a pair-rule stripe in Drosophila. Genes Dev 5:827–839
15. Street K, Risso D, Fletcher RB, Das D, Ngai J, Yosef N, Purdom E, Dudoit S (2018) Slingshot: cell lineage and pseudotime inference for single-cell transcriptomics. BMC Genomics 19:477. https://doi.org/10.1186/s12864-018-4772-0
16. Takahashi K, Yamanaka S (2006) Induction of pluripotent stem cells from mouse embryonic and adult fibroblast cultures by defined factors. Cell 126:663–676. https://doi.org/10.1016/j.cell.2006.07.024
17. Tewary M, Shakiba N, Zandstra PW (2018) Stem cell bioengineering: building from stem cell biology. Nat Rev Genet 19:595–614. https://doi.org/10.1038/s41576-018-0040-z
18. Van de Sande B, Flerin C, Davie K, De Waegeneer M, Hulselmans G, Aibar S, Seurinck R, Saelens W, Cannoodt R, Rouchon Q, Verbeiren T, De Maeyer D, Reumers J, Saeys Y, Aerts S (2020) A scalable SCENIC workflow for single-cell gene regulatory network analysis. Nat Protoc 15:2247–2276. https://doi.org/10.1038/s41596-020-0336-2
19. Wolf FA, Hamey FK, Plass M, Solana J, Dahlin JS, Göttgens B, Rajewsky N, Simon L, Theis FJ (2019) PAGA: graph abstraction reconciles clustering with trajectory inference through a topology preserving map of single cells. Genome Biol 20:59. https://doi.org/10.1186/s13059-019-1663-x
20. Yachie-Kinoshita A, Onishi K, Ostblom J, Langley MA, Posfai E, Rossant J, Zandstra PW (2018) Modeling signaling-dependent pluripotency with Boolean logic to predict cell fate transitions. Mol Syst Biol 14. https://doi.org/10.15252/msb.20177952
21. Yuh C-H, Moore JG, Davidson EH (1996) Quantitative functional interrelations within the cis-regulatory system of the S. purpuratus Endo16 gene. Development 122:4045–4056

Methods in Molecular Biology (2024) 2767: 263–273
DOI 10.1007/7651_2023_489

Published online: 24 May 2023

ShapeMetrics: A 3D Cell Segmentation Pipeline for Single-Cell Spatial Morphometric Analysis

Ceren Pajanoja and Laura Kerosuo

Abstract

There is a growing need for single-cell level data analysis in correlation with the advancements of microscopy techniques. Morphology-based statistics gathered from individual cells are essential for detection and quantification of even subtle changes within the complex tissues, yet the information available from high-resolution imaging is oftentimes sub-optimally utilized due to the lack of proper computational analysis software. Here we present ShapeMetrics, a 3D cell segmentation pipeline that we have developed to identify, analyze, and quantify single cells in an image. This MATLAB-based script enables users to extract morphological parameters, such as ellipticity, longest axis, cell elongation, or the ratio between cell volume and surface area. We have specifically invested in creating a user-friendly pipeline, aimed for biologists with a limited computational background. Our pipeline is presented with detailed stepwise instructions, starting from the establishment of machine learning–based prediction files of immuno-labeled cell membranes followed by the application of 3D cell segmentation and parameter extraction script, leading to the morphometric analysis and spatial visualization of cell clusters defined by their morphometric features.

Key words 3D cell segmentation, Cell image, Cell membranes, Cell morphology, Cell shape, Cell size, Feature extraction, Ilastik, MATLAB, ShapeMetrics, Spatial localization, User-friendly code

1 Introduction

Within the past decade, the biological and medical imaging fields have advanced profoundly. High-resolution visualization of cell structures and high-content screening of tissues are being used to study functional and developmental processes, as well for improved disease diagnostics [1, 2]. Accurate and robust segmentation of cells in a three-dimensional manner within complex tissues is the first step to study cellular dynamics. Various computational methods are being developed to produce reliable results [3, 4]. Granting methods with machine learning or deep learning algorithms provide highly efficient results; however, their usage requires advanced computational skills creating a disconnect between the researchers that produce the images and the ones with the ability to extract the most meaningful information out of the data. Hence, we have established a 3D cell segmentation pipeline ShapeMetrics [5], a MATLAB-based script that utilizes a machine learning algorithm via the Ilastik software [6] combined with detailed step-by-step instructions, making it convenient for a broad range of researchers (Fig. 1). Our user-friendly pipeline requires fluorescent biomarkers to label the cell membrane to segment individual cells and to perform morphometric measurements. Users will then assign pixel annotations on cell borders and to inside/outside of cells by training a random forest classifier by using the Ilastik Software, which results in the creation of prediction maps that are used in our custom MATLAB script (modified from [7]). Once the cells in the 3D image are segmented, the script allows the user to extract morphological parameters such as longest axis of the cell, elongation, ellipticity, or ratio of cell volume to area (a readout for membrane protrusions), thus providing quantified information on the cell morphometrics within the original tissue samples. According to unbiased clustering of the cells according to differences in the cell shapes and sizes, the results are displayed as heatmaps, from which the cells within different subclusters (specific representation of certain morphometric features) are pseudo-colored and mapped back into their original locations in the tissue image (Fig. 1). We provide example results from both epithelial and mesenchymal tissues including the chick neural tube and embryonic mouse kidney. The examples show not only successfully segmented cells but also statistically significant separation of the cells based on their volumetric features [5].

2 Materials

2.1 Software (All Publicly Available with No Cost)

The "read me" file with detailed instructions is found together with the code from https://github.com/KerosuoLab/Shapemetrics.

1. Ilastik (no earlier than version 1.3.0, https://www.ilastik.org/).
2. MATLAB (no earlier than version R2020a, https://se.mathworks.com/products/matlab.html) (*see* **Note 1**).
3. Fiji / ImageJ (https://fiji.sc/).

3 Methods

3.1 Immunostaining of Cell Membranes

1. Strong and clear fluorescent signal (low signal to noise ratio) of the cell membranes is required for this pipeline to successfully segment all cells within the image, which can be obtained by immunostaining or genetic lines. This is a critical step, and before moving forward, it is necessary to test your cocktail of membrane-binding antibodies or genetic lines in your tissue of interest to make sure they equally stain all cells in the tissue.

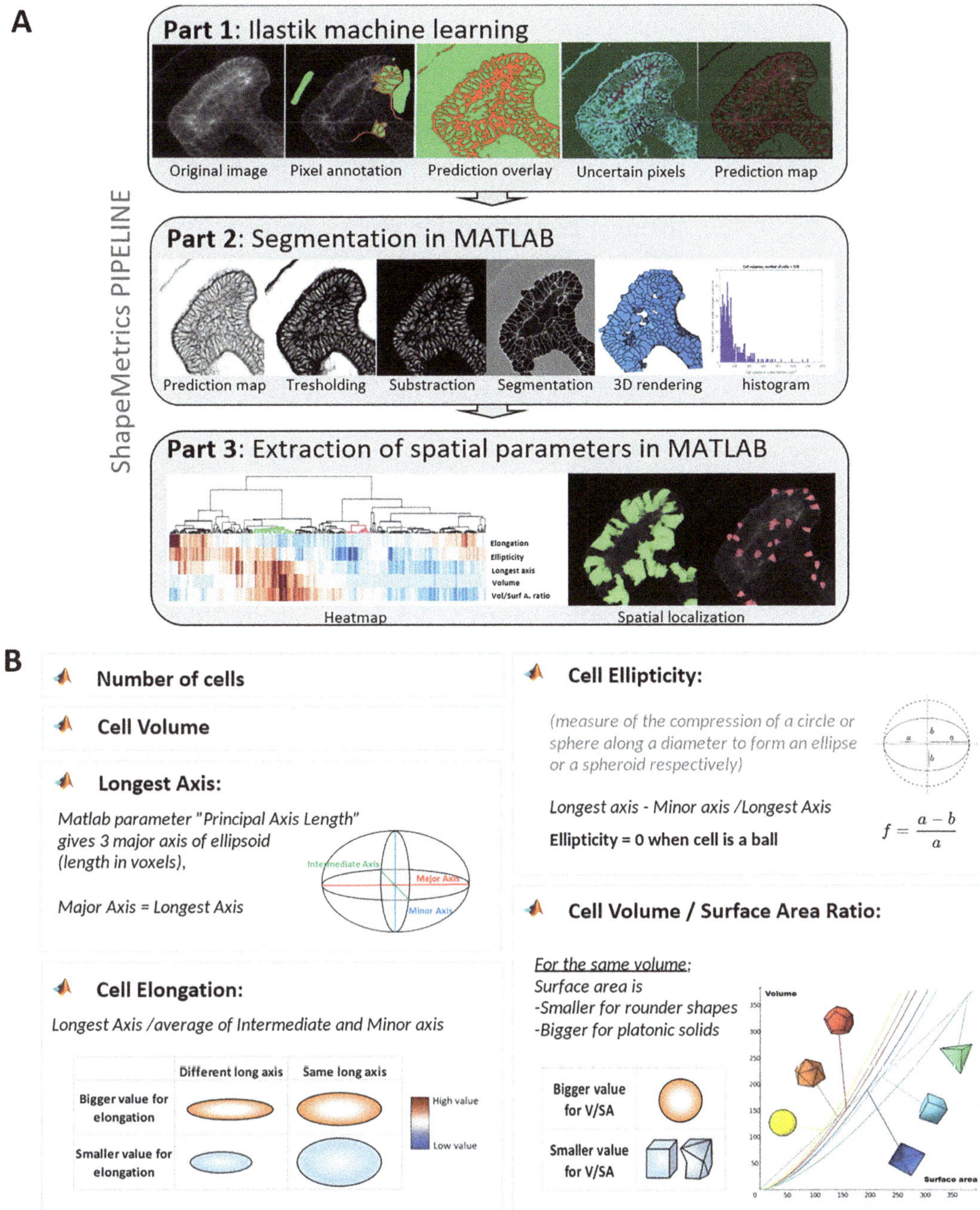

Fig. 1 (**a**) Overview of ShapeMetrics segmentation and volumetric analysis pipeline. Part 1 starts with the user drawing pixel annotation in the Ilastik software (membrane label shown with red and the inside/outside of the cell label shown with green) on the original image according to immunostaining that marks the cell membranes. Prediction overlay shows trained areas on the sample with user's labels. Software also provides an overlay of cyan color indicating uncertain labeling. Once training is complete, the prediction map is exported. Part 2 shows representative images for each step during the fine tuning of cell segmentation as described in Subheading 3.4. At the end of the segmentation, users can view the segmentation via the Matlab Volume Viewer App as 3D rendering after which the spatial parameters are extracted. Part 3 represents an example of an unbiased clustering of the cells according to their morphological parameters. The cells included in the selected branches of the heatmap are mapped back into the original image and pseudo-colored. (**b**) The full list and definitions of the volumetric parameters that are provided in the ShapeMetrics pipeline

2. In our example experiments, we used a cocktail of membrane-binding antibodies: E-cahderin (BD Transduction 610181), B-catenin (Abcam ab6391), and N-cadherin (MNCD2, Developmental Studies Hybridoma Bank, AB_528119). Primary antibodies were visualized by secondary antibodies against multiple proteins in the same channel to enhance the signal and to avoid gaps in the membrane expression. All antibodies were incubated overnight at +4 °C.

3.2 Image Acquisition

1. High-quality images (high resolution, low background) are necessary for this segmentation pipeline, which can be captured via confocal or epifluorescent microscopes.
2. To achieve successful segmentation, the images should be captured with 40×, 63×, or 100× objective and with 0.5 μm intervals in Z-axis. For example, the chick neural tube images presented in our examples were acquired by using the Andor Dragonfly Spinning disc confocal microscope with a 63× water immersion objective (NA 1.7), and the kidney samples were imaged using a Leica TCS SP8 X confocal microscope with a 63× oil-immersion objective (NA 1.4).

3.3 Creating a Prediction Map with Ilastik (Fig. 1, Part 1)

1. Open the Ilastik software and choose the Pixel Classification workflow in order to create a new project from the startup screen. Save the project file into the folder of your choice. It is recommended to save it in the same folder with the image files that will be used for this project.
2. Upload the data for segmentation (.tiff file format with *z*-stacks of the membrane immunostaining) from "Add New" > "Add Separate Image(s)" options on the right panel.
3. From "Feature Selection" applet on the left panel, click on "Select Features" button to bring up the feature selection dialog. For the ShapeMetrics pipeline, it is advised to select all pixel feature types and scales that are shown in the dialog.
4. After selecting the features, switch to the "Training" applet. Add two labels by pressing the "+Add Label" button. Each label corresponds to a pixel class to be used for separation. Use Label 1 to mark the cell membrane and use Label 2 to mark the background and the cell interiors.
5. To start the machine learning algorithm and the prediction results, press the "Live Update" button. The Random Forest classifier will be trained from the marked pixel annotations.
6. When the live update is done, predictions of each label will be displayed as an overlay on the image. Examine the results for corrections and errors. If needed, more annotations can be

added by drawing more labels to the image to improve the result (*see* **Note 2**).

7. In order to export the pixel probability map for the segmentation, switch to the "Prediction export" applet and choose "Export all." This will generate .h5 file in the same folder as the input image used for the training.

3.4 MATLAB Watershed Segmentation Algorithm (Fig. 1, Part 2)

1. Open the segmentation3D_script.m in MATLAB and set the path for the folder that holds all input files that will be used for the analysis (*see* **Note 3**).
2. Load the ilastik prediction map file by typing the name of the . h5 format file within the brackets (the prediction map consists of continuous pixel values assigned from 0 (black) to 1 (white) based on the machine learning training done by the user).
3. Run the first section of the script by pressing the "Run Section" button. This section will visualize the prediction map as a *z*-projection.
4. Similarly, in the next section, load the original image file and visualize it.
5. Add the pixel size information in order to obtain the volumes in cubed microns. Pixel size (μm/pixel) can be calculated by dividing the physical length of a pixel on the CCD camera of the microscope you are using by the total magnification (example: an image taken with 60x magnification using a CCD camera with a pixel size of 16 μm, pixel size of the image is $16 / 60 = 0.266$ μm).
6. The next section of the code creates a blurred version of the membrane z-stack image visualized as a summed z-projection. This will help to connect any gaps in the staining.
7. Pixel Size Thresholding: In order to reach high accuracy of the segmentation, restrict the size of the grouped pixels by thresholding pixel values of the prediction map. Four different threshold values and their visualizations are provided in this section for users to choose from (*see* **Note 4**).
8. Cell Size Thresholding: Users need to test and visualize the threshold values to determine the minimum and maximum limits for cell size. The aim here is to exclude the background to only extract the individual cells. Choose the optimal threshold value and set the chosen variable as equal to seg_final variable (*see* **Note 5**).
9. Next, use the blurred version of the image together with the thresholded prediction map (via MATLAB built-in function "imimposemin") to create a seed for the watershed algorithm.

10. Now, apply watershed segmentation on this seed to create a label matrix that contains information and labels of each individual cell. This way, we use the machine learning segmentation tool (Ilastik) together with the watershed algorithm to provide the most accurate and unbiased segmentation possible.
11. Lastly, apply fine-tuning to the segmentation results by applying yet another threshold value to remove connected components with fewer than the certain amount of pixels from the image, similar to **step 8** (we recommend using similar min/-max values).
12. Visualize the final label matrix for the cell segmentation via Volume Viewer in the MATLAB App section where the app provides a variety of viewing approaches for evaluating the results (*see* **Note 6**).

3.5 Extraction of Morphological Features (Fig. 1, Part 3, and Fig. 2)

1. Extract the volumetric and spatial parameter values from the label matrix using MATLAB's built-in function "regionprops3". In our script, we have made a shortcut access to 8 different parameters in total (utilizing regionprop3 function's own parameters as well as custom ones). All the extracted parameters are saved in a structure array we named as "stats" (*see* **Note 7**).

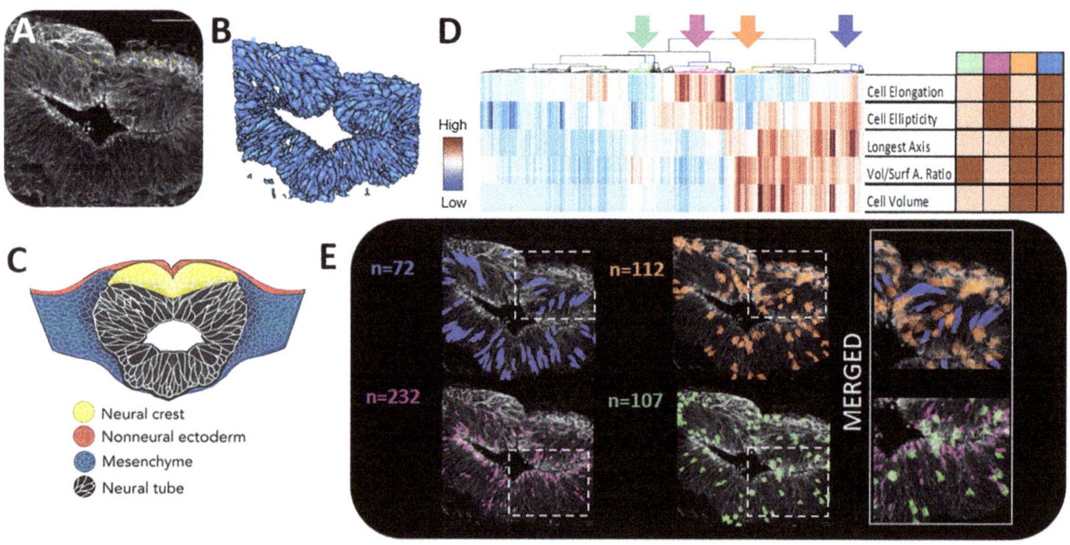

Fig. 2 (**a**) Immunostaining of cell membranes in a Hamburger-Hamilton Stage 9 (HH9) chick midbrain cross-section (scale bar 40 μm). (**b**) The 3D rendition image after segmentation of the neural tube and non-neural ectoderm. (**c**) A cartoon depicting different tissue types in the chick embryo. (**d**) A heatmap demonstrates morphological differences of the subpopulations according to the chosen parameters, the arrows mark the branches of interest. For example, cells in the pink population are small, elliptical, and elongated. (**e**) Pseudo-coloring of the cell groups that were marked in (**d**) shows that the pink cells align the lumen of the neural tube

2. By running the histogram plot, you can explore the volume distribution of the segmented cells within the image.
3. Next, we have provided several matrices that consists of different parameter combinations for clustering and heat map visualizations (*see* **Note 8**).
4. Using these matrices, create heatmaps where cells are hierarchically clustered according to all or selected parameters (using MATLAB's built-in function "clustergram").

3.6 Mapping of Cell Features Back into Spatial Context in the Original Images (Fig. 2)

1. Once you determine the branch of interest (cluster of cells) on the heatmap, move your mouse over the branch node until you see a red marker appearing. Click and hold the red marker to display the branch number. Type this number into the script where you see the variable "branches." Make sure that, within the same section of the script, you also type the correct heatmap name for the selected branch number (*see* **Note 9**).
2. Next, cut the selected group of cells (the branch) from the original label matrix by typing the branch number. Then, assign a color of your choice to this new label matrix of selected cells (*see* **Note 10**).
3. The script will create a representative visualization by placing color-coded centroids on top of the selected cells. The final sub-label matrix file is saved in .tif format.
4. In order to pseudo-color whole cells in the original membrane image, open the image in Fiji (ImageJ) together with this newly created sub-label matrix in .tif format. Overlay the images by using the following steps, Image >Stacks >Tools>Interleave, and create an average intensity *z*-projection.

3.7 Pooling of Multiple Samples into One Heatmap (Fig. 3)

1. Follow the segmentation steps carefully and save the segmented label matrix with morphological parameter information for each image (variables named "stats_matrix_all" and "stats") in separate folders.
2. Start by typing the correct path for where those folders are located on your computer so that all selected folders will be pooled into a structure array.
3. The rest of the analysis is performed by following the same steps for creating a selection of parameters and plotting the heatmap, as described above in Subheading 3.6.

3.8 Statistical Comparisons (Fig. 3)

Follow the steps for pooling multiple samples into one heatmap, then plot your heatmap of interest and choose two clusters that you would like to compare.

1. Plot each embryo separately according to chosen parameters (*see* **Note 11**). In our example, we have shown five different

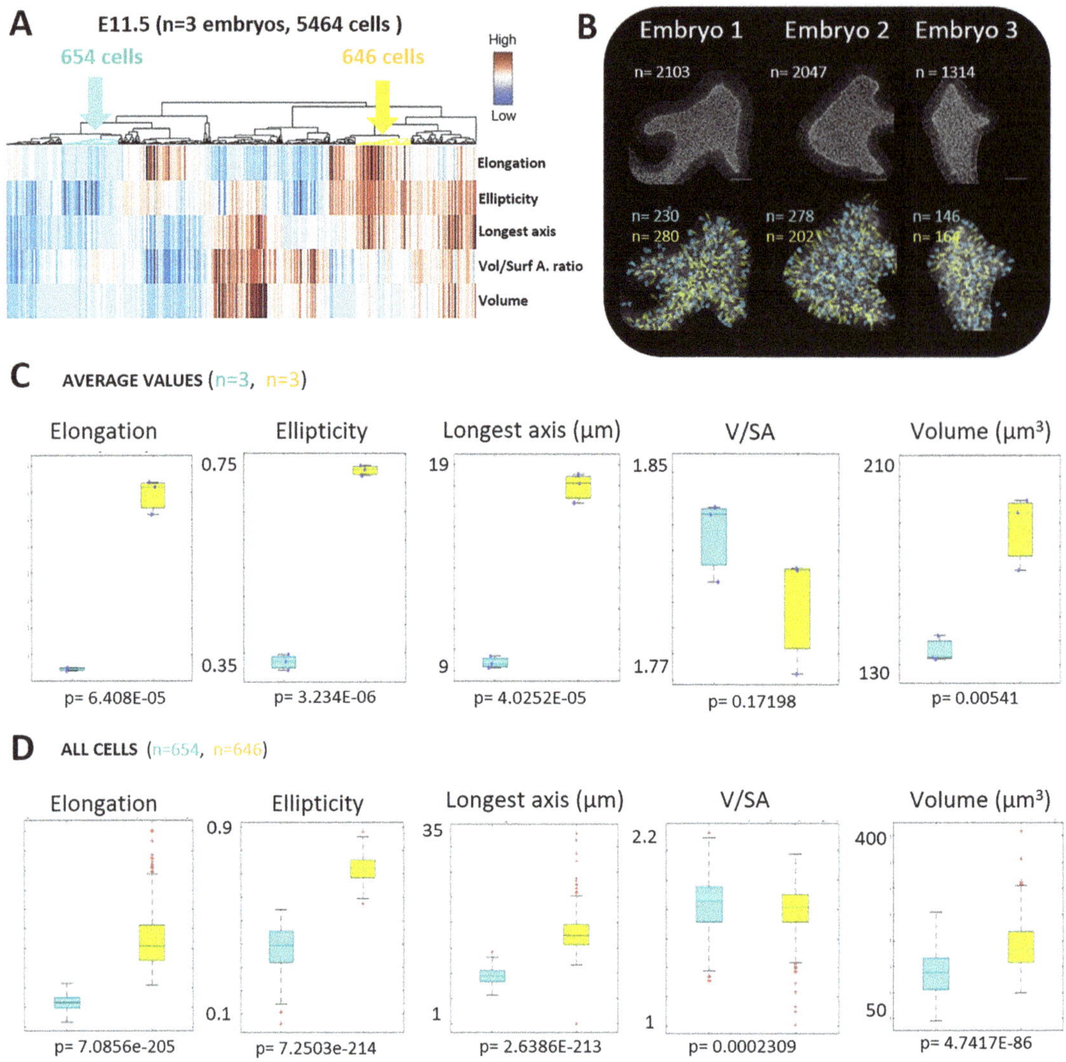

Fig. 3 An example of pooling multiple samples (Subheading 3.7) for statistical comparison (Subheading 3.8). (**a**) The heatmap shows pooled data from three developing kidney samples (E11.5) with a total of 5,464 cells. Two different morphological populations are selected (yellow represents elongated, elliptical intermediate size cells with longest axis, while the blue branch consists of cells that are small and round with short axis). (**b**) The original membrane staining of each embryo and their pseudo-colored visualization indicated by yellow and cyan groups. (**c**) Statistical (t-test) comparison of the yellow and cyan cell types shown by averaging the values per embryo (**d**) and by plotting individual cells, as described in Subheading 3.8

parameters for three embryos as box plots (shown as yellow vs blue cluster).

2. Next, using individual data points coming from biological replicates, you can apply a statistical test of your choice (t-test was used in our example) for the same parameter values between two different clusters (*see* **Note 12**).

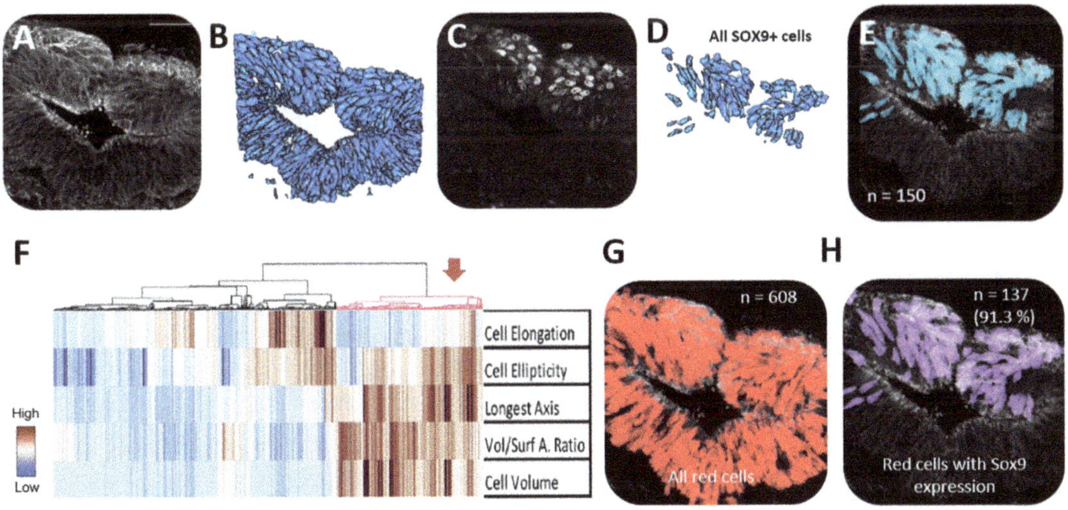

Fig. 4 An example of cell segmentation combined with biomarker analysis. (**a**) HH9 chick midbrain cross-section with membrane staining (Scale bar 40 μm). (**b**) 3D rendering of the segmented cells. (**c**) Sox9 nuclear staining on the same sample. (**d**) 3D rendering of the Sox9-labeled cell segmentation. (**e**) Pseudo-coloring of the cells that are positive for Sox9 staining. (**f**) A heatmap representing all the cells that are segmented in the image. One of the two main branches that consists of all the large cells of the sample that also are elliptical is marked with red. (**g**) Pseudo-coloring of the cells in the red branch. (**h**) Pseudo-coloring of Sox9-positive cells within the red population. 91% of the Sox9-positive cells belong to the red branch

3.9 Extracting Morphometric Features of the Cells That Are Marked with Biomarkers (Fig. 4)

1. Using the ShapeMetrics pipeline, you can combine volumetric information with a biomarker expression. For this section, cell segmentation based on membrane staining and fluorescent biomarker staining is required (*see* **Note 13**).
2. Once you have followed the steps for cell segmentation for both images (membrane and biomarker staining) individually, drag the resulting "stats" files to your MATLAB workspace. Rename them as "stats_fluor1" and "stats_fluor2," respectively.
3. In this section, we are taking advantage of the fact that each cell has been assigned an ID number (defined by the membrane staining based segmentation). First, loop through each cell and their respective stats file that contains the centroid array that belongs to each of the respective fluorescent signal labels. If the centroid of the fluorescent signal is found inside the cell's stats list, it is labeled as "true" for that cell (Boolean value of 1). All other cells are labeled as "false" (0 value).
4. At the end of the loop there are two variables: "Npositives_-fluor1," which includes the number of cells within fluorescent signal 1 (in our example, membrane), and "Npositives_fluor2," which includes the number of cells with fluorescent signal 2 (in our example, Sox9). Additionally, the IDs of the cells

containing each fluorescent signal will be saved in the comparison matrix.

5. For visualization of the cells with the fluorescent signal, follow the section of the script where another structure array named "positives" is created, and assign a color the same way as in the basic pipeline instructed in Subheading 3.6.

4 Notes

1. Make sure that all software required for this pipeline are up to date in order to avoid possible errors.
2. The Ilastik results will be more accurate when the marking (drawing) of the pixels is done precisely. Press the eye-icon next to the "Uncertainty" button to check the uncertainty values in the overlay image. This will mark the pixels with uncertain segmentation result in turquoise color. Reduce the amount of uncertainty by introducing more labels.
3. One of the common user errors that occur while running the MATLAB script is having an incorrect path or wrong file name. The path to the correct folder can be changed from the small arrow on the top line of the MATLAB window that shows the current path, or by moving the script itself to the right folder.
4. In the prediction map in MATLAB, 0 corresponds to the pixel label for cell membrane and 1 for the cell interior/background.
5. We suggest first trying the values of min 100 and max 15,000. If the image does not show any cells, reduce the max value by 1/10. Run the sections 4 and 4.3 again. In case of the opposite scenario where both the cells and the background are visible, increase the max value by 10 times and run again the sections 4 and 4.3.
6. For detailed instructions on how to use the Volume Viewer App, see MathWorks website (https://se.mathworks.com/help/images/ref/volumeviewer-app.html).
7. You can learn about additional parameters provided by regionprops3 function from the MathWorks website (https://se.mathworks.com/help/images/ref/regionprops3.html). To extract more parameters to the script, follow the example of lines provided in our script.
8. Once you display all the provided heatmap variations and get familiar with the pipeline, you can manipulate the parameter combinations according to your needs. Segmented cells and their parameter information ("stats" variable) are saved as MATLAB and .csv form at this stage of the script. Thus, in case you want to elaborate your investigation later, running the analysis from the beginning will not be needed.

9. For guidance on how to perform the visualization step, we have provided detailed instructions within the script and in GitHub. The user needs to type the variable name to define which branch in which heatmap they want to visualize. One needs to write a new file name for the output image (.tif format).
10. You can change the color codes within the "cmp" variable. Use RGB triplet color codes.
11. All information regarding each parameter within each sample are saved in the heatmap specific structure array. You can click on the variable in the MATLAB platform and copy the values or access this information by saving the structure array as an excel file.
12. In this step, you can either use the average value for a parameter for each embryo, or you can use each segmented cell as one data point.
13. For this section, if possible, we recommend using a fluorescent marker that stains nuclei (and not cytoplasm). Non-unified cytoplasmic expression can create a challenge for labeling them in the Ilastik software and may thus not result in a single segmented object per cell.

Acknowledgments

This research was supported by the Intramural Research Program of the NIH, NIDCR, NIH ZIA DE000748 (to LK), and Väre Foundation and Emil Aaltonen Foundation (to CP).

References

1. Lemke SB, Nelson CM (2021) Dynamic changes in epithelial cell packing during tissue morphogenesis. Curr Biol 31(18):R1098–R1110. https://doi.org/10.1016/j.cub.2021.07.078
2. Mattiazzi Usaj M, Styles EB, Verster AJ, Friesen H, Boone C, Andrews BJ (2016) High-content screening for quantitative cell biology. Trends Cell Biol 26(8):598–611. https://doi.org/10.1016/j.tcb.2016.03.008
3. Hodneland E, Kogel T, Frei DM, Gerdes HH, Lundervold A (2013) CellSegm - a MATLAB toolbox for high-throughput 3D cell segmentation. Source Code Biol Med 8(1):16. https://doi.org/10.1186/1751-0473-8-16
4. Wang Z, Li H (2017) Generalizing cell segmentation and quantification. BMC Bioinformatics 18(1):189. https://doi.org/10.1186/s12859-017-1604-1
5. Takko H, Pajanoja C, Kurtzeborn K, Hsin J, Kuure S, Kerosuo L (2020) ShapeMetrics: a userfriendly pipeline for 3D cell segmentation and spatial tissue analysis. Dev Biol 462(1): 7–19. https://doi.org/10.1016/j.ydbio.2020.02.003
6. Berg S, Kutra D, Kroeger T, Straehle CN, Kausler BX, Haubold C, Schiegg M, Ales J, Beier T, Rudy M, Eren K, Cervantes JI, Xu B, Beuttenmueller F, Wolny A, Zhang C, Koethe U, Hamprecht FA, Kreshuk A (2019) Ilastik: interactive machine learning for (bio)-image analysis. Nat Methods 16(12): 1226–1232. https://doi.org/10.1038/s41592-019-0582-9
7. Lignell A, Kerosuo L, Streichan SJ, Cai L, Bronner ME (2017) Identification of a neural crest stem cell niche by Spatial Genomic Analysis. Nat Commun 8(1):1830. https://doi.org/10.1038/s41467-017-01561-w

Methods in Molecular Biology (2024) 2767: 275–292
DOI 10.1007/7651_2023_488

Published online: 08 June 2023

Guinea Pig Preimplantation Embryos: Generation, Collection, and Immunofluorescence

Jesica Canizo, Savana Biondic, Katherine Vandal Lenghan, and Sophie Petropoulos

Abstract

Studying various animal models is important for comparative biology and to better understand evolutionary development. Furthermore, when aiming to translate findings to human development, it is crucial to select an appropriate animal model that closely resembles the specific aspect of development under study. The guinea pig is highlighted as a useful platform for reproductive studies due to similarities in in utero development and general physiology with the human. This chapter outlines the methods required for guinea pig mating and collection of embryos for in vitro culture and molecular characterization. Specifically, this chapter provides detailed guidance on monitoring the estrus cycle to determine the mating time, performing a vaginal flush and smear to confirm successful mating, performing euthanasia of the guinea pig, and flushing in vivo embryos. Once collected, the embryos can be utilized for numerous downstream applications. Here we will cover embryo culturing and processing embryos for immunofluorescence.

Key words Estrus detection, Euthanasia, Flushing in vivo embryos, Guinea pig, Immunofluorescence of embryos, In vitro culture, Mating

1 Introduction

The utilization of suitable animal models is essential in comprehending the mechanistic foundations behind the impacts of early-life perturbations on long-term health and well-being. While epidemiological studies alone are insufficient to establish causality or define underlying mechanisms [1], employing animal model systems with shorter reproductive timelines enables the investigation of long-term outcomes. Additionally, it bypasses ethical and legal restrictions associated with human cohorts [2].

Guinea pigs (*Cavia porcellus*) are a valuable experimental model for Developmental Origins of Health and Disease (DOHaD) studies, with several advantages such as the ease of handling and housing, relatively low cost when compared to larger animal models, and the similarity of their reproductive and general physiology to humans. Studies in this species have contributed significantly to our understanding of DOHaD [2]. The guinea pig has a longer gestation period (69–71 days) than the mouse and rat (19–21 days), but possesses a relatively shorter gestation period than sheep (145–150 days) [3], bovine (280–291 days), porcine (112–120 days) [4], and non-human primates (150–180 days) [5]. The gestational window in the guinea pig enables enhanced temporal resolution of developmental plasticity and identification of critical time periods for vulnerability to insults [2]. It is noteworthy that the timing of development of critical organ systems affected by in utero exposures or perinatal complications, leading to adverse health outcomes in humans, is similar in the guinea pig [6]. Despite their widespread use in DOHaD and in vivo studies, little is known about the preimplantation period of guinea pigs, as well as the techniques related to reproductive biotechnology, such as superovulation, in vitro maturation, in vitro fertilization, and embryo culture. We have conducted an in-depth characterization of the preimplantation embryo in guinea pigs and have devised laboratory protocols that we wish to share, to facilitate the work of others interested in studying this model.

2 Materials

2.1 Guinea Pig Estrus Cycle Monitoring and Mating

1. Guinea pigs (Hartley): females of 500 g, 2–3 months; males of 500–700 g, 2–3 months.
2. N95 mask and PPE (personal protection equipment) (*see* **Note 1**).
3. Gloves.
4. Sterile swabs.
5. Sterile water or saline.
6. Plastic transfer pipettes.
7. Microscope slides (Leica, 3800380) and cover slips (VWR, 48366067).
8. JorVet Dip Quick Stain (or any other similar to Wright-Giemsa stain).
9. Tweezers.
10. Permanent marker, different colors, to identify the females.
11. Inverted microscope.

2.2 Euthanasia and Flushing In Vivo Embryos

1. Needles of 21 G and syringes of 10 mL.
2. Compresses.
3. Towels.
4. Rat housing cage for transport.
5. Alcohol (isopropyl or 70% ethanol).
6. Anesthesia machine and isoflurane.

7. Induction box.
8. Euthanasia kit: sterile scissors and tweezers of two sizes, large and for microdissection.
9. Heating plate (SnuggleSafe ® heating pad).
10. 100 mm culture dish
11. 35 mm culture dishes (embryo tested: Sarstedt, 83.3900 and Nunc IVF dish Life Technologies, 150255)
12. Plastic transfer pipettes (Pasteur).
13. 15 mL sterile centrifuge tubes
14. Sterile tips of p1000 and p200.
15. Pipettes p1000 and p200.
16. Mouth pipette assembly (Sigma Aldrich, A5177-5EA) and pre-pulled capillary (Chemglass Life Sciences, CG-1840-02) or handling pipettes (Vitrolife, 15533).
17. M2 medium (Sigma Aldrich, M7167) with antibiotic: flushing and manipulation medium.
18. PBS with antibiotic: 9.9 mL of PBS plus 100 μL of 100× penicillin/streptomycin.
19. 100× penicillin/streptomycin solution (Life Technologies, 15140122): Aliquot into 1 mL portions and freeze at −20 °C until ready to use. Once thawed, store at 4–8 °C for up to 14 days.
20. Humidified CO_2 incubator.
21. Stereomicroscope.

2.3 Processing Embryos: Immunofluorescence of Preimplantation Embryos

1. Sterile PBS.
2. Triton X-100.
3. BSA.
4. 4% PFA: diluted from 32% stock (Electron microscopy Science, 15714). Check before using, if precipitation is seen, make a new aliquot. Keep at 4–8 °C protected from light.
5. Permeabilization: 0.25% Triton X-100 in PBS. Prepare fresh every time.
6. Blocking solution: 3% BSA in PBS. Prepare a stock of 10 mL, sterilize it using a 0.22 μm membrane, and aliquot 1 mL into sterile Eppendorf tubes. Store at 4–8 °C until use. Discard the solution once it has been opened and is no longer sterile.
7. Primary antibodies are generally diluted to 1/200 for nuclear proteins and 1/50 for membrane proteins. It is important not to vortex the antibodies but to mix them by shaking up and down the tube. We recommend optimizing the antibody concentration for each type of sample.

8. Secondary antibodies: generally diluted 1/500.
9. Hoechst or DAPI. Prepare a 10,000× dilution. Keep aliquots of 10 μL at −20 °C protected from light.
10. Mouth pipette assembly and pulled capillary to move embryos (*see* Subheading 2.2 for details).
11. 72-well mini trays (VWR, CA62409-296)
12. Micropipette p20.
13. Tips 200 μL.
14. Round cover slips: 1.5, 18 mm D, 0.17 mm thickness (Harvard Apparatus, 64-0714).
15. Rectangular cover slips: 1.5, 24 × 50 mm (Fisher Scientific, 12544E).
16. Spacers (Grace Biolabs, Secure Seal, 470352).
17. Tweezers.
18. Confocal microscope.

2.4 In Vitro Culture from Morula to Blastocyst

1. Sterile PBS.
2. Sterile N2B27 medium (*see* methods Subheading 3.4 for preparation details).
3. Mineral oil (Sigma Aldrich, ES-005-C).
4. 35 mm embryo culture dishes (embryo tested: Sarstedt, 83.3900 and Nunc IVF Dish Life Technologies, 150255)
5. Pipettes p1000, p200, p20, p2.
6. Sterile tips.
7. Membrane filter 0.22 μm.
8. Syringe.
9. Mouth pipette assembly and pre-pulled capillary or handling pipettes (*see* Subheading 2.2 for details).
10. Humidified incubator with 5% CO_2.
11. Sterile hood.

3 Methods

3.1 Guinea Pig's Estrus Cycle Monitoring and Mating

3.1.1 Monitoring the Vaginal Membrane

1. Remove enrichment items from the cage to facilitate the capture of a female, (*see* **Note 1**) Fig. 1a.
2. For new females, examine the vagina every day in the afternoon until the vaginal opening is visible (*see* **Note 2**), Fig. 1b. The vaginal opening corresponds to estrus, which is day 0 of the estrous cycle and the period during which the female is in heat.

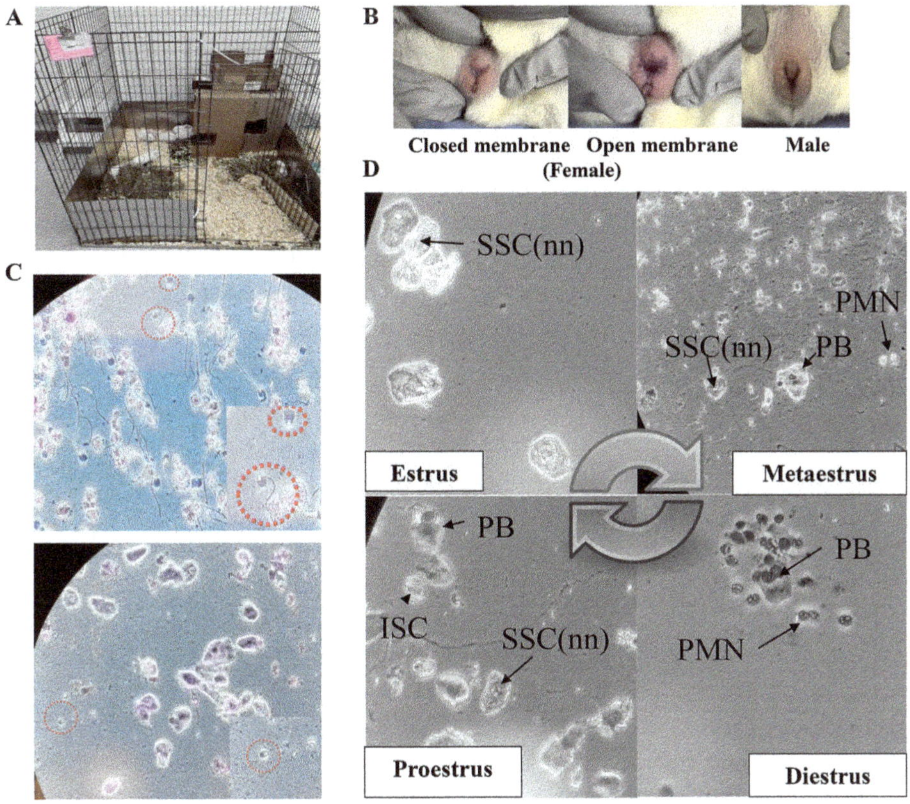

Fig. 1 Estrous cycle monitoring. (**a**) Females group-housed with enrichment consisting of cardboard boxes that serve as hiding spots. (**b**) The status of the female membrane during the estrous cycle is shown; the left-hand image shows a closed membrane, while the center image shows an open membrane. The right-hand image corresponds to a male. (**c**) Representative smear samples of sperm, showing the entire spermatozoa and head only. (**d**) The image shows the different types of cells found in each phase of the estrus cycle, such as non-nucleated superficial squamous cells (SSC (nn)), parabasal cells (PB), polymorphonuclear leukocytes (PMN), nucleated superficial squamous cells (SSC), and intermediate squamous cells (ISC)

The female will allow mating in the hours following this opening (*see* **Note 3**).

3. For females whose previous estrus date (i.e., vaginal opening) is known, examine the vagina every day starting on day 14. The opening will occur approximately every 15–19 days.

3.1.2 Mating

As soon as a vaginal opening is observed, place the female with a male in a cage. If multiple males are housed together, you must isolate the male/female pair in another cage (*see* **Note 4**). It is also possible to house the animals in a harem, with one male and several females (*see* **Note 5**).

3.1.3 Vaginal Flush and Smear

1. Early the next morning (7:00–7:30 am), perform a vaginal flush and smear on the coupled female. Take a transfer pipette containing 0.2–0.3 mL of sterile saline or water. Gently insert

the tip of the pipette 3–5 mm into the vagina. Gently press on the pipette to release a small amount of sterile saline or water and reaspirate the fluid. Ensure to use light pressure to not wash the vagina. Place the fluid on a well-labeled slide and add a cover slip.

2. For the smear, moisten a sterile swab with sterile water or saline. Insert the swab tip 1 cm into the vagina and rotate it. Roll the swab onto a well-labeled slide. It is important to roll the swab and not rub it. Allow it to air dry.
3. Return the female to the male.
4. Stain the smear with JorVet Dip Quick Stain (*see* **Note 6**). Dip the slide in the fixative solution (cyan) and incubate for 5 s, drain excess fluid onto the paper towel. Repeat the procedure with solution 2 (red) and 3 (purple). Rinse the slide under gently running tap water (touching the back of the slide) until the slide is pink and the water runs clear. Let it dry.
5. Identify the presence of sperm (Fig. 1c) and the stage of the female's estrous cycle (Fig. 1d) (*see* **Note 7**). This procedure will help determine the exact day of fecundation. E1.5 is considered next day at noon, after the presence of a sperm positive smear. Fecundation usually occurs by midnight, *E0*.
6. Once the sperm is confirmed in a smear, perform an additional vaginal swab, usually the following afternoon, to monitor female's progression to metestrus.

3.1.4 Determining the Different Stages of the Estrous Cycle

1. Check the vaginal flush sample and the smear using an inverted microscope. Generally, a 20 × objective magnification is good to identify cells and sperm.
2. Classify the sample according to Table 1 and Fig. 1d.
3. Following the vaginal opening, non-nucleated superficial squamous cells (resembling nucleus-free corn flakes) should be observed in the vaginal rinses and smears. The presence of spermatozoa should be observed the day after mating. Only the presence of spermatozoa will confirm mating (*see* **Note 8**). The appearance of neutrophils on the morning or afternoon smear will confirm the transition from estrus to metestrus.

3.2 Euthanasia and Flushing In Vivo Embryos

3.2.1 General Anesthesia of Guinea Pigs by Inhalation of Isoflurane

1. Connect the induction chamber to the anesthesia machine and adjust the oxygen flow rate to approximately 1–3 L/min and the isoflurane to 3–5%. Fill the induction chamber with the oxygen-isoflurane mixture before placing the guinea pig in the chamber.
2. Once the guinea pig is placed inside the induction chamber (Fig. 2a, b), wait a few minutes until it becomes unconscious (*see* **Note 9**).

Table 1
Guinea pigs estrus cycle characteristics

Cycle stage	Characteristics	Duration
Estrus	Presence of predominantly non-nucleated superficial squamous cells (SSC). The vaginal epithelium has rapidly proliferated and thickened to withstand the mechanical insult of copulation	Short. Usually occurs during the night, between 6 p.m. and 6 a.m.
Metestrus	Presence of superficial squamous cells (SSC) with and without nuclei, parabasal cells (PB), and especially the marked arrival of neutrophils (PMN)	2–4 days
Diestrus	Presence of nucleated cells and neutrophils (PMN). Disappearance of non-nucleated superficial squamous cells	8–10 days
Proestrus	Increase in the proportion of non-nucleated superficial squamous cells (SSC) and intermediate squamous cells (ISC); gradual decrease in leukocytes	2–4 days

3. Connect the oxygen/isoflurane tube and the purge tube from the anesthesia machine to the main circuit tube connected to a mask at the end (Fig. 2c). Transfer the animal to the mask and adjust the isoflurane vaporizer to 1.5–2% or higher depending on the animal size.

3.2.2 Exsanguination by Cardiac Puncture

1. Place the unconscious animal in dorsal recumbency.
2. Check the depth of anesthesia. There should be no withdrawal reflex. With a toothed forceps, press firmly on the interdigital region of a hind limb. If there is any limb contraction, adjust the depth of anesthesia and repeat the withdrawal reflex.
3. Use a 20 G 1″ to 23 G 1″ needle and a syringe that can hold 10 to 20 mL as needed.
4. Palpate the landmark before inserting the needle, either the tip of the sternum and the depression to its left of the animal (similar to the rat).
5. Create a slight negative pressure by pulling back on the plunger of the syringe to draw in some air.
6. Insert the needle fully into the depression and bevel up, at an angle of 30 to 45°. As soon as blood enters the syringe, hold still and withdraw the maximum amount of blood. If the blood stops flowing during the collection, adjust the location of the needle (deeper or shallower, change the angle, etc.) (*see* **Note 10**).

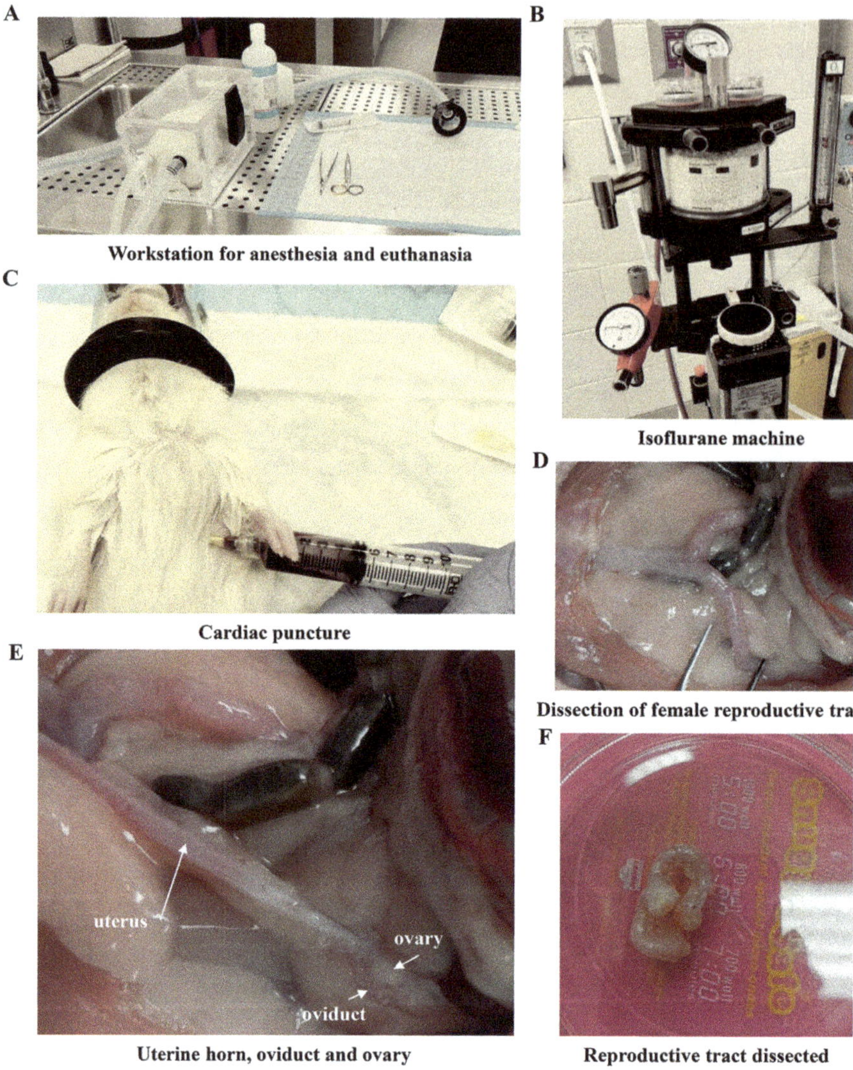

Fig. 2 Anesthesia and euthanasia. (**a**) Workstation setup includes an anesthesia induction box, face mask, tweezers, scissors, syringe, and 70% alcohol. (**b**) An anesthesia machine is used with isoflurane. (**c**) Cardiac puncture is performed laterally. (**d**) Ventral dissection is performed to isolate the reproductive tract. (**e**) The image shows the uterine horn, oviduct, and ovary. (**f**) The dissected reproductive tract is placed in a dish that is heated by a warming plate

7. Confirm the animal's death by observing vital signs before proceeding the dissection and, if necessary, apply a second euthanasia method (bilateral pneumothorax) (*see* **Note 11**).

3.2.3 Dissection

1. Following confirmation that the animal no longer exhibits any vital signs, dissection may commence. 70% ethanol or isopropanol can be sprayed onto the abdomen until it is thoroughly moistened (*see* **Note 12**).

2. With the aid of forceps and scissors, a midline incision should be made, followed by a transverse incision, taking care to incise only the skin.
3. Make a central incision and carefully open the abdominal cavity, taking care to avoid cutting the intestines.
4. Using microdissection scissors and smooth microforceps separate the fat from the uterus and ovaries (Fig. 2d, e). Isolate the uterus/ovaries.
5. Place the uterus, oviducts, and ovaries in a 100 mm Petri dish on a heating plate if available (Fig. 2f).
6. Transport to the laboratory for the flushing procedure.

3.2.4 Flushing

1. Prior to flushing the embryos, equilibrate and pre-warm the PBS and M2 culture media by placing each tube with 7–10 mL of each for at least 30 min in a humidified incubator at 37 °C and 5% CO_2 in air. Prepare the culture dish by placing drops (10–100 μL each) of M2 manipulation media in the bottom of a 35 mm plastic dish (Nunc embryo tested, *see* Subheading 2.2) and cover with embryo-tested light mineral oil. Pre-warm as well.
2. Prepare three 35 mm petri dishes (Sarstedt dishes) making straight lines with a blunt needle (Fig. 3a). Make perpendicular straight lines as well, this facilitates the search of embryos (Fig. 3b).
3. After sacrificing a pregnant female and removing the entire uterus plus ovaries, place them in p100 dish and wash twice with pre-warmed PBS with antibiotic (Fig. 3c).
4. Place the dish under a stereomicroscope and dissect the ovaries/oviduct from the uterus. Look for the corpus luteum in each ovary to have an idea of how many embryos could be gotten from each side (Fig. 3d).
5. Depending on the embryonic day/stage of the embryos, flush the oviduct or uterus (Table 2).
6. Use a 1 mL syringe with a 26-gauge needle for the uterus, or a blunt 30-gauge needle for the oviduct (insert the needle in the oviduct infundibulum) using a tweezer to keep the oviduct or uterus in position. Gently flush the uterus or oviduct in the prepared 35 mm dish that already has pre-warmed M2 (*see* Fig. 3e, f).
7. Flush each side of the uterus or oviduct two times (1 mL each) and collect the resulting 2 mL in one 35 mm dish (Fig. 3g).
8. Look for the embryos in the flushed media (Fig. 3h) and transfer them into a previously prepared 35 mm culture petri dish, using a mouth pipette attached to manipulation pipette or

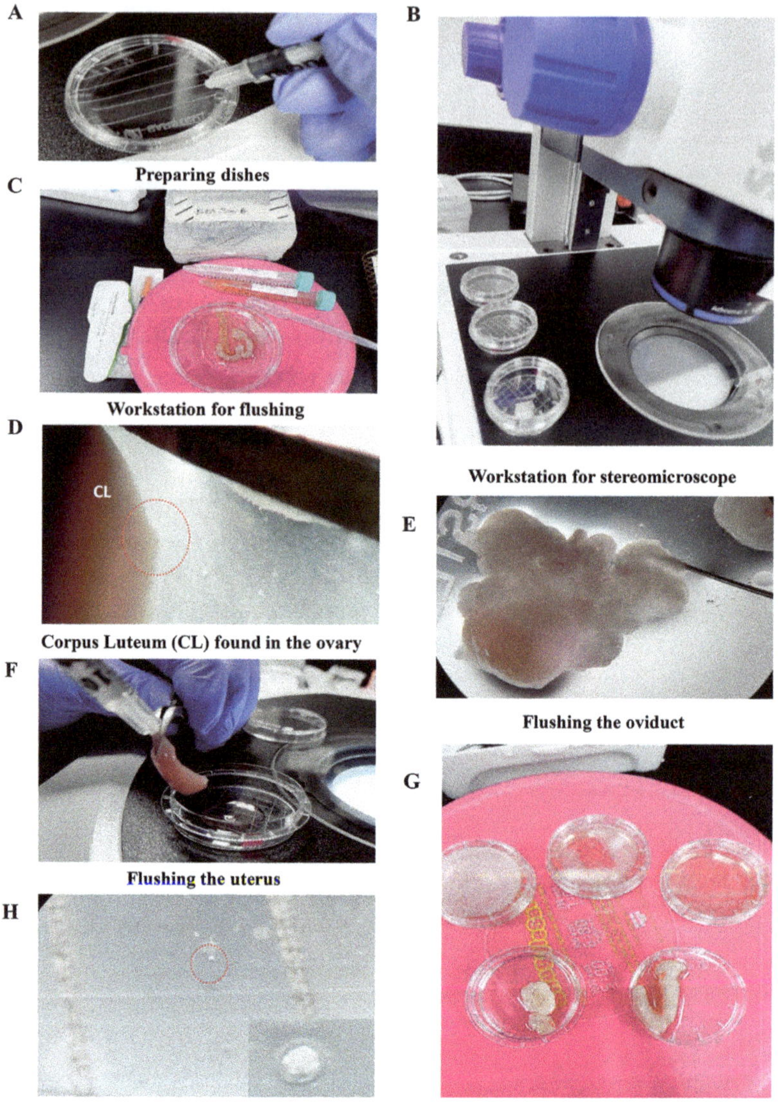

Fig. 3 Flushing guinea pigs. (**a**) Preparation of dishes consists of drawing lines on the bottom of the plate with a broken insulin needle. The pattern facilitates an organized search for the embryos. (**b**) Stereomicroscope used to locate the embryos. A final magnification of 25–40× is necessary. (**c**) The working station includes a hot plate, insulin needle, mouth pipette, sterile microdissection kit, plastic Pasteur pipette, Falcon tubes with PBS and M2 with antibiotics. (**d**) The image shows an ovary with corpus luteum surrounded in red (CL). Typically, 1–5 CLs are found between both ovaries. (**e**) Flushing of the oviduct under the stereomicroscope. (**f**) Uterine flushing. (**g**) Dissected ovaries and uterine horns after flushing. (**h**) Representative image of an embryo found after flushing, a compacted morula

Table 2
In vivo developmental dynamics of guinea pigs' embryos

Embryonic day	Number of cells	Stage	Flushing
E0.5–1	1	Zygote	Ampulla
E1.5–2	2	Cleavage stages	Oviduct
E3	4	Cleavage stages	Oviduct
E3.5	8	Cleavage stages	Oviduct
E4	16	Morula	Uterus
E4–5	16–32	Compacted morula	Uterus
E5–5.25	32	Early blastocyst	Uterus
E5.25–5.5	44	Mid blastocyst (MB)	Uterus
E5.75–6	54–62	Late blastocyst (LB)	Uterus

a pre-pulled microcapillary. Move the embryos into the micro-drops under oil and transfer through several drops to rinse off the flushing M2 media. Ideally, culture several embryos together, as a higher density of embryos enhances development [7].

9. After flushing each side (first and second dishes), perform a third flush of each and collect 1 mL per side in the same third petri dish. Look for more embryos, especially if the number of embryos obtained is inferior to the number of corpus luteum.
10. Embryos in M2 media should only be kept for up to 6 h at 37 ° C and 5% CO_2, prior to using the embryos for any downstream protocol. To culture them in vitro for a prolonged time, *see* Subheading 3.4.

3.3 Immuno-fluorescence of Preimplantation Embryos

3.3.1 Fixation and Permeabilization

1. Fix embryos in 17 μL of 4% PFA. Make 2 drops and move embryos into the first and then the second drop. Incubate for 15 min at room temperature (*see* **Note 13**). Embryos can be kept in PBS at 4–8 °C (*see* **Note 14**) or proceeds with the next steps.
2. Move embryos to 17 μL of permeabilization buffer (in the 'wash drop') and then to the second drop of 17 μL of permeabilization buffer. Incubate for 10–20 min at room temperature (the time depends on the size of the embryo). For guinea pig embryos, incubate for 10 min.
3. Move embryos to the first drop of 17 μL of blocking solution, then to the second drop and incubate for 1 h at room temperature.

Table 3
Primary antibodies used in the representative guinea pig embryo

Protein	Marker	Catalogue number	Host (type)	Brand	Dilution
SOX2	Epiblast	14981182	Rat (monoclonal)	Thermofisher Scientific	1/50 in guinea pigs' embryos
Active YAP	TE	ab205270	Rabbit (monoclonal)	Abcam	1/200
GATA3	TE	CM405A	Mouse (monoclonal)	Biocare Medical	1/250

3.3.2 Primary Antibody Incubation

1. Make a master mix of all primary antibodies in blocking solution (Table 3) (*see* **Note 15**). Combine antibodies raised in different species.
2. Set up 2 droplets of 17 μL of the master mix and move embryos into the first and then the second drop.
3. Incubate overnight in the fridge. Seal the plate with parafilm and cover with aluminum foil.
4. The next day, wash the embryos with 17 μL of sterile PBS (or 1% BSA in PBS for embryos without zona pellucida) 4 times at room temperature, 5 min each wash.

3.3.3 Secondary Antibody Incubation and Mounting

1. Make a master mix of each secondary antibody in 500 μL of blocking buffer (the concentration depends on each antibody to detect and the signal/background ratio). Make 2 drops and incubate in the second for 1–2 h at room temperature, always protected from light with aluminum foil.
2. Incubate the embryos in Hoechst or DAPI for 10 min at room temperature in the dark (make two drops and incubate in the second one) (*see* **Note 16**).
3. Wash embryos in 3 sequential drops of 17 μL of sterile PBS or 0.2% triton/PBS for 5 min each at room temperature.
4. Mount the embryos in PBS between a rectangular coverslip and round coverslip using spacers. The spacers can be layered and should be adjusted depending on the size of the embryo. Use as many spacers as necessary based on the size of the embryo. Usually for guinea pig embryos, one spacer is sufficient. For mounting, first put 17 μL of PBS in the well formed by the spacer, and then dispense the embryo gently. Put the round coverslip on top with the help of a tweezer.
5. Image the embryos as soon as possible using a confocal microscope (*see* **Note 17**), Fig. 3. Keep mounted slides in fridge for short-term storage.

Table 4
N2B27 formulation (blastoids media used in guinea pigs for ex vivo culture)

Component	Catalogue	Brand	Volume
Neurobasal	21103-045	Thermo Fisher	23.45 mL
DMEM/F12 home made	11330057	Life Technologies	23.45 mL
N2 100 ×	17502048	Life Technologies	250 μL
B27 50 ×	12587010	Life Technologies	500 μL
Glutamax 100 ×	35050061	Life Technologies	500 μL
100 mM sodium pyruvate	11360070	Life Technologies	500 μL
MEM nonessential amino acids	11140050	Life Technologies	500 μL
1 M HEPES	15630-080	Gibco	500 μL
50 mM β mercaptoethanol	31350010	Life Technologies	100 μL
Penicillin/streptomycin 100×	15140122	Life Technologies	250 μL
BSA	A3311-10G	Sigma	0.5 g

3.4 In Vitro Culture from Compacted Morula to Blastocyst

1. Prepare a homemade DMEM/F12 medium to obtain a base medium with a final concentration of 29 mM $NaHCO_3$ and 15 mM HEPES. Then, prepare the N2B27 medium by mixing the components listed in Table 4 [8]. Sterilize the mixture using a membrane filter, aliquot 1 mL into 1.5 mL Eppendorf tubes, and freeze them at −20 °C. Once an aliquot is thawed, it can be stored at 4–8 °C for up to 2 weeks.
2. Prepare the culture dish by placing drops (10–100 μL each, usually 50 μL) of N2B27 culture media in the bottom of a 35 mm plastic dish and cover with embryo-tested light mineral oil (Fig. 4a).
3. Place the dish for at least 30 min in a humidified incubator at 37 °C and 5% CO_2 in air, to allow for equilibration.
4. Transfer the embryos into the micro drops under oil. Transfer through several drops, to rinse off residual M2 manipulation media. Ideally, culture several embryos together, as a higher density of embryos enhances development.
5. Under these conditions, embryos can develop from a compacted morula E4–4,25 to the late blastocyst E6,5, Fig. 4b. If the collected embryos are at an earlier stage than expected, they cannot progress in N2B27 media.

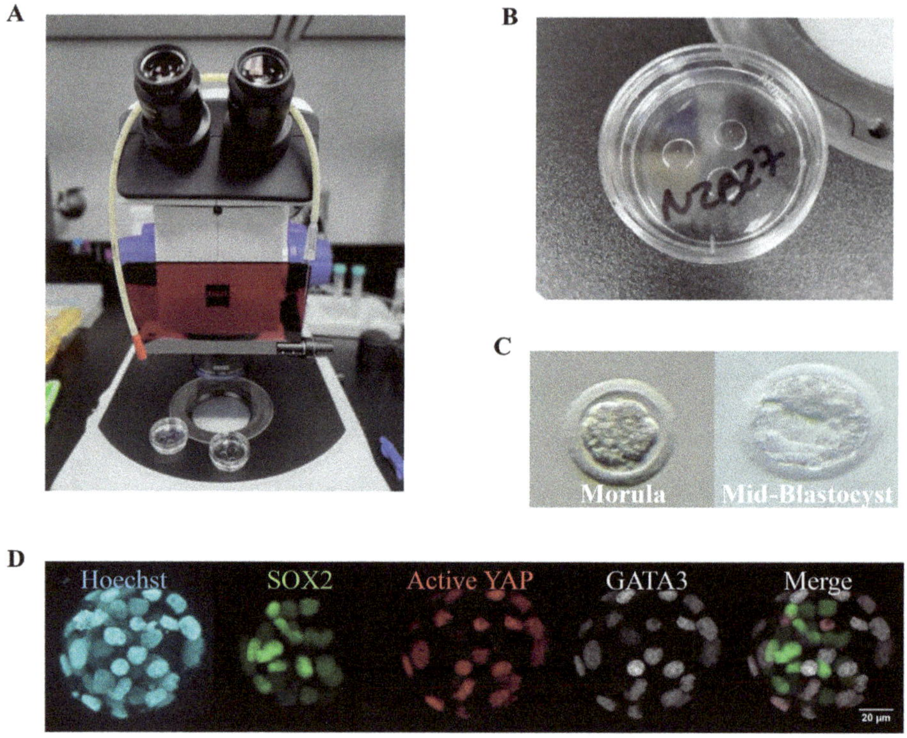

Fig. 4 In vitro culture of guinea pig embryos in N2B27 and immunofluorescence. (**a**) Workstation for culturing embryos includes a stereomicroscope, mouth pipette, and culture dishes. (**b**) Culture dish prepared with N2B27 medium drops covered with mineral oil. (**c**) Embryos cultured from morula to blastocyst in N2B27 medium. (**d**) A representative immunofluorescence image of a guinea pig embryo. Maximum projection of a z-stack (step size: 1 μm); Hoechst (nucleus); SOX2 (green), an epiblast marker; Active YAP (red); and GATA3 (gray) for the trophectoderm. Scale bar: 20 μm

4 Notes

Monitoring the vaginal membrane

1. To work with the guinea pigs, wearing scrubs, a hair cover, gloves, and an N95 mask is mandatory. Safety goggles are recommended. After handling guinea pigs, it is important for individuals to change their clothes promptly. This recommendation applies to everyone in order to prevent allergies.
2. The estrous cycle of guinea pigs typically lasts 15–19 days [9]. Ovulation is spontaneous, and the luteal phase is prolonged. During ovulation, 1 to 4 oocytes are released. Litter size ranges between 2 and 5 with an average of 2 embryos per female achieved with Hartley strain [10]. The female has a membrane that covers the opening of her vagina. This membrane disappears a few hours before estrus [11, 12]. It is essential to follow

the vaginal opening (estrus cycle monitoring) to mate the females [9] and, after the mating, confirm fecundation by smear. When a new cohort of guinea pigs arrive, we recommend following them through an entire cycle before mating them. This provides a better understanding of how quickly the membrane breaks and how long it stays open, as well as how long each female's estrus lasts.

3. Some females may have a non-synchronized cycle, which means that estrus does not occur the night after the membrane breaks. In these cases, we check the female's smear twice a day while she is housed with the male, until we get a sperm-positive sample.
4. Keep the pair together until the day of the female's euthanasia. If the males are all housed together and are only separated for mating, they can be returned to their original group with caution. The males may develop aggressive behavior, so they should be monitored. If this occurs, it is advisable to separate them and house each male individually.
5. If animals are housed in a harem, one male with up to four females is recommended. Females should have staggered estrous cycles so that the male can mate them on different days. Once the female's vaginal membrane is open, vaginal swabbing should be done daily, twice a day, until semen is present in the sample. The harem option is a good choice if there is no previous experience in mating guinea pigs. Sometimes, identifying an opening can be difficult because the female may not have a clear breakage/closure of the membrane. If this occurs, we recommend checking the female daily and record when she is first open and how long she remains open. We recommend taking smear samples to confirm estrus and start counting the days until her next cycle.
6. Use gloves or tweezers to avoid staining your fingers.
7. Most of the time, we find sperm heads instead of entire spermatozoa (*see* Fig. 1c). Sometimes, small particles similar in appearance to heads can be seen in the smear, but they can be differentiated based on their size.
8. We have found females that have had sperm in the second smear performed in the afternoon. For these we consider that mating was delayed around 6–12 h and therefore E1–1.25 is considered the following day at noon.

Euthanasia and flushing in vivo embryos

9. In general, for female guinea pigs, it is recommended to use 2 L/min of 5% isoflurane for a duration of 5 minutes. It is important to be attentive when using anesthetic gases as guinea pigs have a tendency to hold their breath.
10. An alternative method to cardiac puncture is to palpate the chest to feel the heartbeat and insert the needle directly from the side of the chest cavity, making sure it enters the heart without passing through it and protruding out. Note that the plunger must be pulled out beforehand to induce a vacuum, Fig. 2c.
11. For a bilateral pneumothorax, clean the surgical area with an antiseptic solution and make a small incision in the skin over the thorax using a surgical scalpel. Insert a sterile 25-gauge needle attached to a 1-mL syringe into the pleural space on one side of the chest. The needle should be inserted between the ribs and advanced until it punctures the pleural space. Pull back on the plunger of the syringe to remove the air from the pleural space. Repeat the procedure on the other side of the chest to create a bilateral pneumothorax. Observe the animal for any signs of distress or respiratory distress [13].
12. M2 media with HEPES allows collection and handling of preimplantation stage guinea pig embryos for prolonged periods outside CO_2 incubator.

Immunofluorescence

13. Make sure that the PFA (paraformaldehyde) does not have any precipitation. If it does, change the aliquot or prepare a fresh dilution. Prepare fresh 4% biweekly.
14. Perform immunofluorescence as soon as possible, although you can store embryos at 4–8 °C in PBS, in a parafilm-sealed plate protected from light with aluminum foil before making the permeabilization step.
15. If fluorescence quantification is required, prepare a fresh dilution each time to avoid having intensity variations in the images later. If only the number of cells needs to be counted, you can use a primary aliquot within a week.
16. We prepare a Hoechst solution at a concentration of 10,000× using molecular biology-grade water. To incubate the embryos, create a serial dilution of 1/100 twice to achieve a final concentration of 1× in PBS.
17. Images were acquired using an Olympus FV1000 upright microscope (Olympus, Japan). All images were acquired using a XLUM Plan FL N 20×/1.00 Water objective. Samples were positioned between two coverslip (#1.5, 0.17 mm) with

spacers (one for small embryos like guinea pigs and mice). For excitation, 405 nm (solid state), 488 nm (Argon laser), 543 nm (solid state) and 635 nm (solid state) lasers were used for DAPI/Hoechst, Alexa Fluor (AF) 488, AF555 or AF594 and AF647, respectively. For detection, photomultiplier tube (PMT) detectors were set as follows: a first SDM490 was positioned in front of the first PMT associated with a BA 430–470 for DAPI detection; then a SDM560 was positioned in front of the second PMT associated with a BA 535–565 for AF488 detection; a SDM640 was positioned in front of the third PMT associated with a BA 560–660 for AF594 detection; and finally, a mirror was positioned in front of the last detector with a BA 655–755 for AF647 detection. All images were acquired sequentially (frame mode) as followed: AF488 and AF647 simultaneously in first, AF594 or AF555 in second, and DAPI/Hoechst at the end of the sequence. Images were acquired in either a 512 × 512 pixel format at zoom 3 (with a pixel resolution of 414 nm) or an 800 × 800 pixel format at zoom 4, depending on the protein of interest. The acquisition speed was set at 4 μs per pixel with a line Kalmann of 3. Z-stack images were acquired with a 1 μm step size.

Acknowledgments

This work was supported by the Canadian Institutes of Health Research (PJT-178082) and NSERC PGS D scholarship to KV. SP holds the Canada Research Chair in Functional Genomics of Reproduction and Development (950-233204). We thank the CRCHUM imaging facility for allowing us to use the confocal microscope.

References

1. Dickinson H, Moss TJ, Gatford KL, Moritz KM, Akison L, Fullston T, Hryciw DH, Maloney CA, Morris MJ, Wooldridge AL, Schjenken JE, Robertson SA, Waddell BJ, Mark PJ, Wyrwoll CS, Ellery SJ, Thornburg KL, Muhlhausler BS, Morrison JL (2016) A review of fundamental principles for animal models of DOHaD research: an Australian perspective. J Dev Orig Health Dis 7(5):449–472. https://doi.org/10.1017/S2040174416000477
2. Morrison JL, Botting KJ, Darby JRT, David AL, Dyson RM, Gatford KL, Gray C, Herrera EA, Hirst JJ, Kim B, Kind KL, Krause BJ, Matthews SG, Palliser HK, Regnault TRH, Richardson BS, Sasaki A, Thompson LP, Berry MJ (2018) Guinea pig models for translation of the developmental origins of health and disease hypothesis into the clinic. J Physiol 596(23):5535–5569. https://doi.org/10.1113/JP274948
3. Sisk DB (1976) Chapter 7 - Physiology. In: Wagner JE, Manning PJ (eds) The biology of the Guinea Pig. Academic Press, San Diego, pp 63–98
4. EFSA Panel on Animal Health and Welfare (AHAW), More S, Bicout D, Botner A, Butterworth A, Calistri P, Depner K, Edwards S, Garin-Bastuji B, Good M, Gortazar Schmidt C, Michel V, Miranda MA, Saxmose Nielsen S, Velarde A, Thulke H-H, Sihvonen L, Spoolder H, Stegeman JA, Raj M, Willeberg P, Candiani D, Winckler C (2017) Animal welfare aspects in respect of the slaughter or killing of

pregnant livestock animals (cattle, pigs, sheep, goats, horses). EFSA J 15(5):e04782–e04782. https://doi.org/10.2903/j.efsa.2017.4782

5. Tarantal AF, Noctor SC, Hartigan-O'Connor DJ (2022) Nonhuman primates in translational research. Annu Rev Anim Biosci 10:441–468. https://doi.org/10.1146/annurev-animal-021419-083813
6. Briscoe TA, Rehn AE, Dieni S, Duncan JR, Wlodek ME, Owens JA, Rees SM (2004) Cardiovascular and renal disease in the adolescent Guinea pig after chronic placental insufficiency. Am J Obstet Gynecol 191(3):847–855. https://doi.org/10.1016/j.ajog.2004.01.050
7. Piliszek A, Kwon GS, Hadjantonakis A-K (2011) Ex utero culture and live imaging of mouse embryos. Methods Mol Biol 770:243–257. https://doi.org/10.1007/978-1-61779-210-6_9
8. Kagawa H, Javali A, Khoei HH, Sommer TM, Sestini G, Novatchkova M, Scholte op Reimer Y, Castel G, Bruneau A, Maenhoudt N, Lammers J, Loubersac S, Freour T, Vankelecom H, David L, Rivron N (2022) Human blastoids model blastocyst development and implantation. Nature 601(7894):600–605. https://doi.org/10.1038/s41586-021-04267-8
9. Lilley KG, Epping RJ, Hafner LM (1997) The guinea pig estrous cycle: correlation of vaginal impedance measurements with vaginal cytologic findings. Lab Anim Sci 47(6):632–637
10. Shomer NH, Holcombe H, Harkness JE (2015) Biology and diseases of Guinea pigs. Lab Anim Sci 247. https://doi.org/10.1016/B978-0-12-409527-4.00006-7
11. Stockard CR, Papanicolaou GN (1919) The vaginal closure membrane, copulation, and the vaginal plug in the Guinea-pig, with further considerations of the œstrous Rhythm. Biol Bull 37(4):222–245. https://doi.org/10.2307/1536207
12. de Matos AS, Kugelmeier T, Guimarães DA de A, da Silva KSM (2022) Early puberty in short-haired Guinea pigs kept in laboratory animal facilities. Anim Reprod 19(1):e20210068. https://doi.org/10.1590/1984-3143-AR2021-0068
13. Salerno FG, Paré P, Ludwig MS (1998) A comparative study of elastic properties of rat and Guinea pig parenchymal strips. Am J Respir Crit Care Med 157(3):846–852. https://doi.org/10.1164/ajrccm.157.3.9705078

Methods in Molecular Biology (2024) 2767: 293–305
DOI 10.1007/7651_2023_494

Published online: 08 July 2023

Manual Dissociation of Mammalian Preimplantation Embryos for Single-Cell Genomics

Katherine Vandal, Savana Biondic, Jesica Canizo, and Sophie Petropoulos

Abstract

Single-cell genomics allow the characterization and quantification of molecular heterogeneity from a wide variety of tissues. Here, we describe the manual dissociation and collection of single cells, a method adapted for the characterization of precious small tissues like preimplantation embryos. We also describe the acquisition of mouse embryos by flushing of the oviducts. The cells can then be used in multiple sequencing protocols, for example, Smart-seq2, Smart-seq3, smallseq, and scBSseq.

Key words Preimplantation embryo, Single-cell, Sequencing, Mouse, Human, Embryo collection

1 Introduction

Single-cell genomics have become increasingly popular to characterize the transcriptome and genome heterogeneity in a wide variety of tissues [1]. High-throughput methods of single cell isolation have been developed to capture a large numbers of cells [2]. However, these methods are not appropriate for very small tissues of precious samples, such as preimplantation embryos, given the small number of cells in the starting material and the low capture rates. Manual isolation of single cells is better adapted to preimplantation embryos with minimal loss of cells. Here, we describe the acquisition of mouse or human embryos, their manual dissociation into single cells, and their collection with micropipetting. We have successfully used the method described here to sequence the full-length transcriptome using Smart-seq2, smallseq, and scBS-seq, but this method could potentially be used for many other sequencing pipelines by adapting the cell collection medium [3–8].

Embryos are acquired differently depending on the species. Human embryos, usually provided by a collaborating fertility clinic, must be handled according to the approved ethics protocol. Mouse embryos are acquired from flushing the oviducts of a previously

---

Supplementary Information The online version contains supplementary material available at https://doi.org/10.1007/7651_2023_494.

mated female. To increase the number of embryos collected per mouse and decrease the number of animals used, the females are super-ovulated using a combination of pregnant mare serum gonadotropin (PMSG) and human chorionic gonadotropin (hCG) [9]. Embryos from all species are then cultured in standard conditions until their desired developmental stage is reached. Dissociation of the embryos into single cells is then performed under a microscope, and the cells are stored into a cell collection master mix adapted to the sequencing protocol that will be used.

2 Materials

2.1 Acquisition of Embryos

1. Hormones (PMSG and hCG): 25 IU/mL in filtered and autoclaved PBS 1×. Store aliquots (500–1000 μL) at −20 °C.
2. M2 mouse embryo culture media (Sigma-Aldrich, MR-015-D; *see* **Note 1**).
3. KSOM mouse embryo culture media (Sigma-Aldrich, MR-121-D).
4. Filtered mineral oil, suitable for embryo or stem cell culture (EMD Millipore, ES-005-C).
5. 10 × 35 mm tissue culture plastic petri dishes (Sarstedt, 83.3900).
6. Injection needles: 1 mL syringe (BD, 309659) and 27G needle (BD, 305109). Fill the syringe before attaching the needle to minimize the formation of air bubbles in the syringe.
7. Flushing needles: 30G Insulin syringe with permanent needle (Covidien, 8881601600). Blunt the tip of the needle using a rough surface, for example, a knife sharpening stone.
8. Sterile forceps and scissors for mouse dissection (Fig. 2a).
9. Aspirator tube assemblies for calibrated microcapillary pipettes (Millipore Sigma, A5177).
10. Pulled capillaries of size 120 μm. Pull 0.8–1.1 mm I. D × 0.25 mm wall × 100 mm length melting point tubes using a vertical micropipette puller (We pull at 700 heat level, 300 pull strength, and 4.85 μm trip point on a P-30 by Sutter Instrument), and then cut the end of the formed capillaries using a micro forge to obtain the desired size (Fig. 1).
11. Stereo microscope equipped with a transillumination base and a magnification capacity of 20× to 30×. The working distance should be at least 5 cm to allow tissue manipulation while under the microscope.

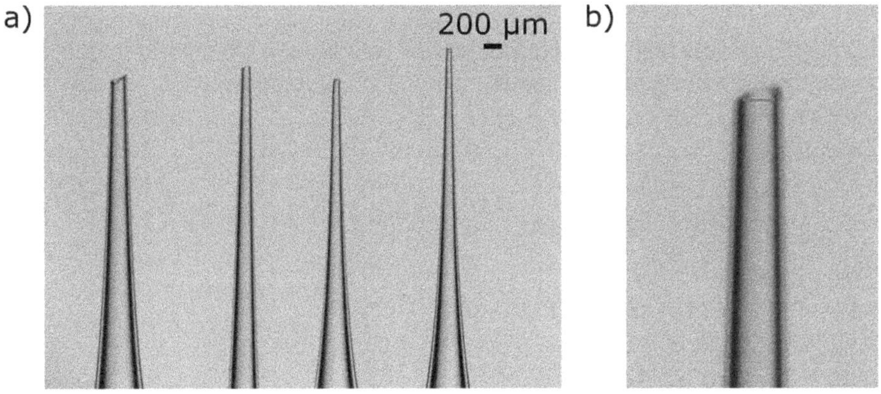

Fig. 1 Size of capillaries useful for embryo manipulation and cell dissociation and picking. (**a**) Most commonly used sizes for embryo manipulation (120 μm), cell dissociation (120, 80, and 60 μm), and picking (40 μm). From left to right: 120 μm, 80 μm, 60 μm, and 40 μm. (**b**) Red line shows how to measure capillary dimension. Measure the opening of the capillary, not including the glass walls

2.2 Dissociation of Embryos into Single Cells

1. TrypLE express enzyme without phenol red (Gibco, 2323459).
2. Filtered mineral oil, suitable for embryo or stem cell culture (Mouse: EMD Millipore, ES-005-C; Human: Vitrolife, 10029).
3. Acidic Tyrode's solution (Sigma-Aldrich, T1788).
4. 1× PBS.
5. 10 × 35 mm tissue culture plastic petri dishes (Sarstedt, 83.3900).
6. 8-tube PCR strips, free of DNA, DNase, and RNase (Eppendorf, 951010022).
7. Pulled capillaries of size 120 μm, 80 μm, 60 μm, 40 μm, and 20 μm if dissociating larger blastocysts (Fig. 1). Prepare at least one capillary of each size for each embryo that will be dissociated.
8. Stereo microscope equipped with a transillumination base and a magnification capacity of at least 40×, 60× being ideal. The working distance should be at least 5 cm to allow embryo manipulation while under the microscope.
9. Nunc IVF Petri Dish 35 mm × 10 mm (Thermo Scientific, 150255).

3 Methods

3.1 Acquisition of Embryos

The method of acquisition for human embryos is dependent on the country, the fertility clinic, and the procedure outlined in your specific ethical approval (*see* **Note 2**). The following describes the acquisition of mouse preimplantation embryos.

Table 1
Hormone injection schedule and the timing of mouse embryo development to help guide collection and embryo staging. The expected location of embryos in the reproductive organs according to their developmental stage

Day	Manipulation	Embryo stage	Location in reproductive organs
−3 (12 PM)	PMSG injection		
−1 (12 PM)	hCG injection and mating		
0 (12 AM)		Fertilization	
0.25 (6–7 AM)	Vaginal plug check		
0.375 (9 AM)		Zygote	Ampulla
1.375 (9–10 AM)		2 cell	Oviduct
1.75 (4–6 PM)		4 cell	Oviduct
2.25 (6–8 AM)		8 cell (compaction)	Oviduct
2.75 (4–6 PM)		16 cell (morula)	Oviduct
3.25 (6–10 AM)		32 cell (early blastocyst)	Uterine horn
3.75 (4–8 PM)		64 cell (late blastocyst)	Uterine horn

3.1.1 Superovulation Schedule

1. Inject the female intraperitoneally (IP) with 5.0 IU of PMSG between 11 AM and 1 PM, this will be considered day −3 (*see* Table 1 and **Note 3**).
2. 46–48 h after PMSG injection, inject the female IP with 5.0 IU of hCG between 11 AM and 1 PM (*see* **Note 4**), and then place the female with the male for mating (*see* **Note 5**).
3. *Optional*: The following morning (day 0.25), around 7 AM, check the female for a mating plug (*see* **Notes 6** and 7).

3.1.2 Flushing

See Fig. 2.

1. Prepare a culture dish by placing 30–50 μL KSOM droplets on a 35 mm dish (Thermo Scientific, 150255) and then by covering with mineral oil. Equilibrate in a 5% CO_2/37 °C incubator for at least 4 h before flushing or overnight.
2. Prepare a 1.5 mL microcentrifuge tube with approximately 500 μL of 37 °C M2 medium, one per mouse to flush.
3. Euthanize the female according to your ethics protocol at the time corresponding to the developmental stage desired (*see* Table 1).

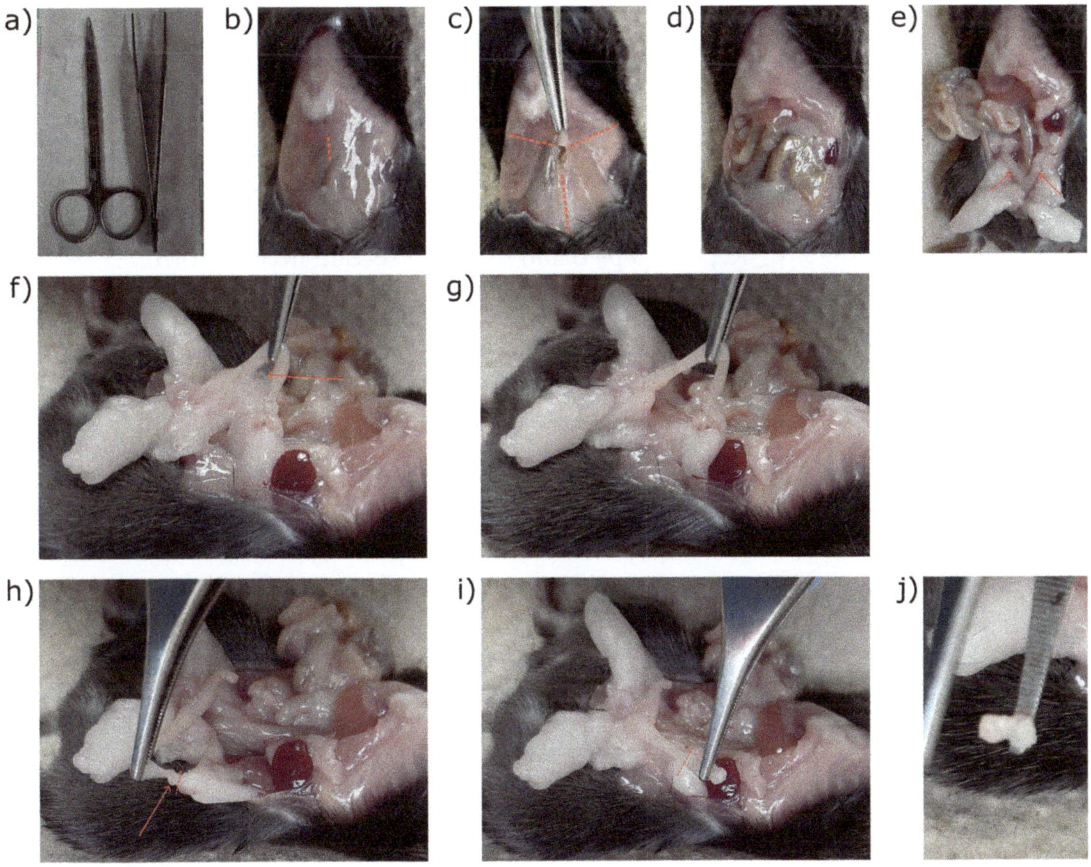

Fig. 2 Dissection of a mouse oviduct. (**a**) Tools needed for the dissection. (**b**) Opened abdomen with peritoneum intact. Red dashed line shows location of first cut of the peritoneum. (**c**) Red dashed lines show the cuts to make in the peritoneum to reveal the abdominal cavity. (**d**) Abdominal cavity. (**e**) Location of the reproductive organs. Red arrows point to both uterine horns. Panel (**f**) shows how to pick up the uterine horn with forceps to facilitate the removal of the connective tissue and fat, shown by the red arrow. (**g**) Uterine horn after removal of connective tissue and fat. Panel (**h**) shows the position of the forceps to pull the oviduct away from the ovary. The red dashed line shows where to cut, between the oviduct and the ovary. Panel (**i**) shows the location of the cut along the uterine horn to release the oviduct. (**j**) Oviduct and part of uterine horn cut out, ready to be flushed

4. Open the abdomen by making a small incision at the midline, then pulling the skin away from the incision, toward the head and tail (Fig. 2b).
5. Cut the peritoneum to reveal the abdominal cavity (*see* Fig. 2c, d). The reproductive organs should be located in the dorsal region, under the digestive organs (*see* Fig. 2e).
6. Pick up one uterine horn using forceps, then gradually remove the connective tissue and fat attached to the uterine horn and the oviduct by inserting the tip of the scissors between the uterine horn and the mesometrium and by opening the scissors while pulling them down (Fig. 2f, g).

7. Reposition the forceps to grab the uterine horn close to the oviduct and pull gently so that the oviduct is lightly pulled away from the ovary. Cut between the ovary and the oviduct, making sure to keep the whole oviduct (*see* Fig. 2h). Then cut the uterine horn 3–5 mm away from the oviduct (*see* Fig. 2i, j, **Note 8**). Immediately place the cut-out oviduct/uterine horn in the microcentrifuge tube containing warmed M2 medium.
8. Repeat **steps 6** and 7 for the second uterine horn.

   The following steps are done under a microscope.
9. Place a small drop of 21–37 °C M2 medium in the cover of a 35 mm petri dish (Sarstedt, 83.3900). Then, using forceps, place one oviduct in the drop.
10. Fill a blunt 1 mL needle with 21–37 °C M2 medium.
11. Locate the infundibulum (Fig. 3a, b and Video 1), use the forceps to immobilize it (Fig. 3c), and insert the needle into

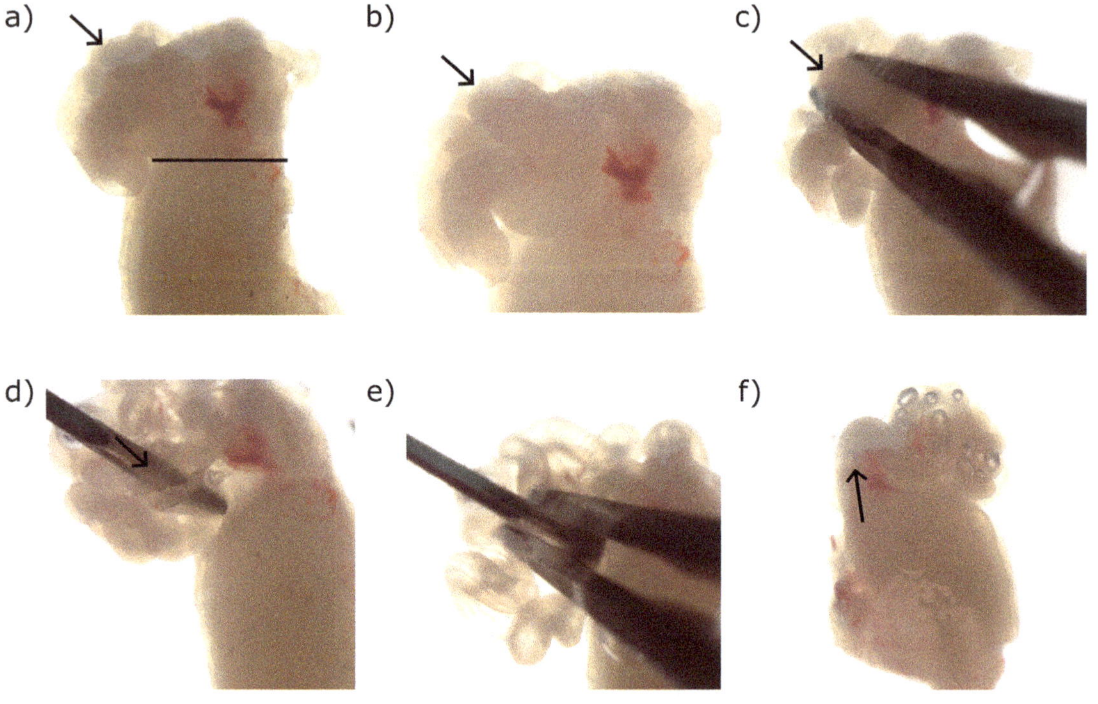

Fig. 3 Flushing of a mouse oviduct. (**a**) Oviduct and part of uterine horn cut out. The black line shows the separation between the uterine horn (below) and the oviduct (above). The black arrow points the opening of the infundibulum. (**b**) The black arrow points the opening of the infundibulum. (**c**) Infundibulum grabbed by forceps, ready for needle insertion. The black arrow points the opening of the infundibulum. (**d**) Needle inserted into the infundibulum. The black arrow points the opening of the infundibulum. (**e**). Needle inserted into the infundibulum and secured by the forceps. (**f**) Overview of the oviduct and part of uterine horn cut out. The black arrow points the location and angle of insertion of the needle when too many holes have been made in the oviduct

the infundibulum, making sure not to pierce a hole in the oviduct (*see* **Note 9**, Fig. 3d).

12. Grab the oviduct around the needle to stabilize it (Fig. 3e), and then gently flush the oviduct and uterine horn by injecting M2 medium until the medium that comes out at the cut end of the uterine horn contains no more debris (*see* **Note 10**).
13. Let the embryos settle at the bottom of the dish for at least 2 min (*see* **Note 11**).
14. Place a 30–50 μL drop of M2 medium in a new 35 mm dish (Sarstedt, 83.3900).
15. Using a 120 μm pulled capillary attached to an aspirator tube, use mouth pipetting (*see* **Note 12**) to collect all the embryos from the flushing drop and move them to the clean drop of M2 medium made in **step 15** (*see* **Note 13**).
16. To wash the embryos, place 5 30–50 μL drops of warm equilibrated KSOM medium on the same 35 mm dish, and then sequentially move the embryos to those drops.
17. Move the embryos to the culture dish prepared in **step 1**.
18. Culture in a 5% CO_2 incubator at 37 °C until ready to dissociate into single cells.

3.2 Dissociation of Embryos into Single Cells

This procedure can be applied to all mammalian embryos.

1. Prepare the cell lysis buffer according to the sequencing protocol desired, then aliquot into 8-well strips and store on wet ice.
2. Prepare two Styrofoam boxes, one containing wet ice and one containing dry ice.
3. Prepare a 35 mm dish (Sarstedt, 83.3900) by pipetting 5–7 30–50 μL drops of TrypLE and covering with mineral oil.
4. On the underside of the lid of a 35 mm dish (Sarstedt, 83.3900), pipette a single 30–50 μL drop of Tyrode's acid solution. On the underside of the lid of a second 35 mm dish, pipette a single 30–50 μL drop of PBS.
5. If present, remove the zona from the embryo by placing it in the Tyrode's acid drop, using a 120 μm capillary. Keep the embryo moving and watch it closely (*see* **Note 14** and Video 2).
6. As soon as the zona is removed, move the embryo to the PBS droplet to wash the Tyrode's, and then move it to a clean TrypLE droplet (*see* Fig. 4a, b, **Note 15**).
7. Switch to an 80 μm capillary. Pipette up and down the embryo to help loosen the cells (*see* Fig. 4e, **Note 16** and Videos 3 and 4).

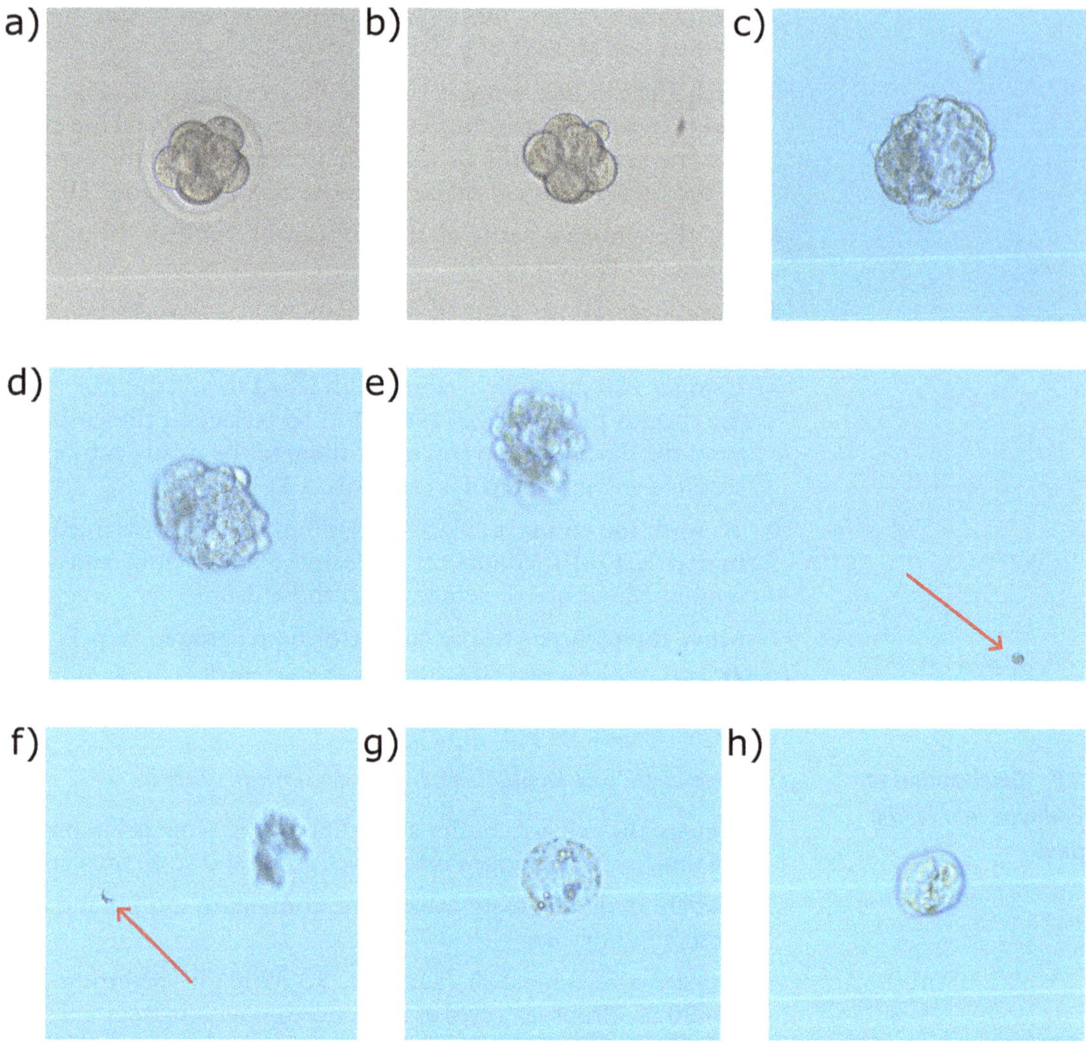

Fig. 4 Dissociation of mouse embryos. Panels (**a**) and (**b**) show a morula before and after zona removal in Tyrode's acid solution, respectively. Panel (**c**) shows a blastocyst after zona removal, and panel (**d**) shows the same blastocyst after cavity collapse. (**e**) Blastocyst during the process of cell dissociation. The red arrow shows a single cell, ready to be picked. (**f**) Blastocyst during the process of cell dissociation. The red arrow shows a debris. Notice how it is elongated, and not round. Panels (**g**) and (**h**) show a single cell at a higher magnification than panel (**e**). Notice how they have a round shape, but are not perfectly round, and have some texture to them

8. When enough single cells have detached to fill one full strip, switch to a 40 μm pulled capillary and pick one cell up. Make sure it is in the capillary by blowing it out and sucking it up again (*see* **Note 17** and Video 4).
9. Carefully lower the capillary into the well of a strip about half way into the lysis buffer and completely empty the capillary.

Make sure not to break the capillary against the well and to not make bubbles (*see* **Note 18**).

10. Repeat **steps 8** and **9** to fill one 8-well strip, and then immediately cap the strip and transfer it to dry ice to flash freeze (*see* **Note 19** and Video 5).
11. Repeat steps 7–10 until no more cells are available or it has been around 30 min (*see* **Notes 20** and **21**). Store filled 8-well strips at −80 °C.

4 Notes

1. M2 medium should not be equilibrated, as its pH is adjusted for use in ambient conditions. Use the medium either at room temperature or at 37 °C.
2. Human embryos are generally acquired either fresh or cryopreserved. If obtained cryopreserved, inquire about the method of freezing used by the clinic to use the corresponding thaw kit. This will increase survival rates. If the embryos need to be cultured further after acquisition, use IVF quality plastic ware, as well as culture medium and oil specifically made for human embryo culture. Place in an incubator with conditions of 5% CO_2/5% O_2 at 37 °C. Human embryos are very sensitive; therefore, opening of the incubator door should be reduced to a minimum, as well as taking out the embryos for checking development under the microscope.
3. PMSG provides follicle stimulating hormone (FSH), which stimulates follicle maturation. Optimal female age is between 6 and 12 weeks. Our experience has shown that after 12–16 weeks, mating probability and fertilization rate is decreased.
4. hCG mimics the surge of luteinizing hormone (LH), resulting in ovulation approximately 12 h after injection, around midnight, denoted in this protocol as embryonic day (E) 0.0 or day 0.0. Mice also tend to mate around midnight [10].
5. Place the female in the male cage and not vice versa to avoid aggression from the female toward the male. Fertilization rates are increased if the male has mated between 1 and 3 weeks prior. If the male has not mated for more than 3 weeks or has never mated before, it can be useful to mate him with a practice female, that is, a female from which no embryos are needed. This can be a female in excess from a transgenic colony maintenance, for example. Each male's breeding history should be monitored. Some males are not good breeders, shown either by low fertilization rates or by low embryo quality. In addition,

age can impact fertilization rate and embryo quality, and males need to be routinely replaced.

6. For some strains, the vaginal plug is hard to find or disintegrates quickly after mating and is therefore not a good indicator of successful mating. To check if the presence of a vaginal plug is a good indicator of successful fertilization for your strain, you can check for the presence/absence of a vaginal plug following mating and record whether embryos can be collected or only unfertilized oocytes. This will allow you to calculate false positive and false negative rates.
7. The female can be separated from the male as early as the next morning following copulation. It is also possible to keep them together until the time of flushing, depending on what is more convenient.
8. If the embryos are collected at the zygote stage, instead of flushing, locate the swollen ampulla and tear it using a needle to release the cumulus mass. Wash the cumulus mass in warmed hyaluronidase solution to remove the cumulus cells. Proceed with **step 16**.
9. If embryo collection is performed at the blastocyst stage (day 3.25 and later), the whole uterine horn should be kept and flushed.
10. Inject the M2 medium slowly, using a low pressure. High pressure can pierce the oviduct and create holes. If a hole is accidently created, there are alternatives to flushing through the infundibulum. First, the needle can be inserted lower down the oviduct, by piercing a hole, then injecting M2 medium. Second, the needle can be inserted at the end of the oviduct, close to the uterine horn, and the M2 medium can be injected so that it passes through the oviduct and exits from the infundibulum (*see* Fig. 3f). As a last resort, the whole oviduct can be shredded by using a sharp needle or scissors. This last method is not ideal as it leads to a decreased recovery of embryos, as they can stay stuck in unshredded sections of the oviduct or lost in the high number of debris resulting from this method.
11. The second oviduct can be flushed during that time.
12. Mouth pipetting is a technique where small amounts of liquid are transferred from one place to another using pressure (suction/blowing) produced by the mouth. Capillaries are attached to an aspirator tube at one end, which is held, while the other end possesses a mouth piece used to generate the pressure differences. While this is our preferred method of moving embryos between drops, this can be replaced with whatever technique the user may find preferable during flushing. However, in the subsequent step of embryo dissociation and single-cell collection (Subheading 3.2) mouth pipetting is

required as it allows for smaller volumes to be taken up and blown out at more precise pressures, and also utilizes capillaries which are small enough to dissociate clumps of cells.

13. Adjusting the contrast of the microscope can be useful to better locate the embryos by the presence of the zona pellucida. Embryos are often found under the reproductive organs.
14. The zona pellucida usually takes 10–60 s to disintegrate. We find that keeping the embryo moving by dispensing it at the top of the Tyrode's drop and then letting it fall down to the bottom before picking it up again minimizes embryo manipulation while preventing the embryo from sticking to the bottom of the dish. You should watch the embryo at all times during zona removal and should see the zona start to thin out. The zona is almost entirely removed once you begin to observe the definition of the outer embryo cells. The embryo should appear "bumpy," clearly seeing the cell edges, versus smooth as with the zona.
15. When passing through the oil layer with a capillary, lightly blow to avoid oil from entering the capillary.
16. Use a capillary with an opening approximately 3/4 the size of the mass of cells to be dissociated. A bigger capillary will not dissociate the cells efficiently, while a smaller capillary can lead to the cells disintegrating. Depending on the stage of the embryo, it can take more or less time before the first cells start to detach from the embryo. Non-compacted embryos (8 cells or less) will require minimal manipulation to dissociate. In contrast, blastocysts can take up to 10 min before cells start to dissociate from the embryo. When an embryo takes a long time to start dissociating, take short breaks (30–60 s) from pipetting the embryo up and down to allow the TrypLE to loosen the cells. When working with a blastocyst, it can be useful to pipette up and down with an 80 μm capillary until the cavity collapses (*see* Fig. 4c, d); then switch to a small capillary (60–70 μm) to help loosen the cells. The first cell that dissociates, usually well before the others, is usually the polar body. Often, groups of cells will detach from the embryo. When that happens, switch to a smaller capillary and start to pipette the group of cells up and down to dissociate the cells further. Cells sometimes dissociate as doublets, which can be hard to separate. The use of a smaller capillary can help. Try to work as quickly as possible to avoid mRNA degradation, but not at the cost of losing the embryo.
17. It is helpful to scratch the bottom of the plate lightly to orient yourself and know where to focus the microscope. Generally, you want to pick cells that have settled to the bottom of the plate to ensure that only cells are collected and not debris or an

air bubble. Debris are not as round as cells, and usually have a more elongated shape (*see* Fig. 4e, f). Cells will settle to the bottom of the dish, are not perfectly round, and have some texture to them (*see* Fig. 4g, h), while air bubbles tend to float to the oil layer, are perfectly round, and don't have any texture inside of them. The cell should be picked up with a very small volume of TrypLE (almost negligible, 0.1–0.2 μL) as a high volume can affect the cell lysis reaction. To minimize the volume of TrypLE picked up with the cell, pipette down until a very low volume of liquid is in the capillary, and then hold the pressure when moving close to the cell. Release the pressure, but do not create suction, and take the capillary out of the dish quickly. It is also possible to hold the pressure while taking the capillary out of the dish.

18. To validate that the cell has been dispensed in the well and is not still in the capillary, move the capillary to the TrypLE droplet and pipette some liquid up, then gently empty the capillary completely, but without making bubbles. The formation of bubbles can lead to loss of cells or the whole embryo.
19. Fill the strip as quickly as possible so that the lysis reaction is not completed before the flash freeze. We usually try to fill a strip in under 5 min. This is why it is recommended to have at least 8 single cells ready to be picked up before starting to fill a strip. However, do not dissociate too many single cells at once, as this will risk some cells lysing before they are ready to be collected.
20. Use fresh capillaries and fresh Tyrode's acid and TrypLE drops for each embryo to avoid cross-contamination.
21. When dissociating human embryos, the first cells that dissociate are usually trophectoderm cells and the last cells are from the inner cell mass. Therefore, to make sure both types of cells are collected, do not collect only the first cells that dissociate. Alternatively, your clinical collaborators can laser-dissect the trophectoderm cells and you can pick from both populations (inner cell mass + polar trophectoderm and mural trophectoderm), or immunosurgery can be performed to enrich for the inner cell mass.

Acknowledgments

This work was supported by the Canadian Institutes of Health Research (PJT-178082) and NSERC PGS D scholarship to K.V. S.P. holds the Canada Research Chair in Functional Genomics of Reproduction and Development (950-233204).

References

1. Macaulay IC, Voet T (2014) Single cell genomics: advances and future perspectives. PLoS Genet 10:e1004126
2. Kulkarni A, Anderson AG, Merullo DP et al (2019) Beyond bulk: a review of single cell transcriptomics methodologies and applications. Curr Opin Biotechnol 58:129–136
3. Picelli S, Faridani OR, Björklund ÅK et al (2014) Full-length RNA-seq from single cells using Smart-seq2. 9:171–181
4. Zhao C, Biondic S, Vandal K et al (2022) Single-cell multi-omics of human preimplantation embryos shows susceptibility to glucocorticoids. Genome Res 32:1627–1641
5. Chen G, Schell JP, Benitez JA et al (2016) Single-cell analyses of X Chromosome inactivation dynamics and pluripotency during differentiation. 26:1342–1354
6. Hagemann-Jensen M, Abdullayev I, Sandberg R et al (2018) Small-seq for single-cell small-RNA sequencing. Nat Protoc 13:2407–2424
7. Clark SJ, Smallwood SA, Lee HJ et al (2017) Genome-wide base-resolution mapping of DNA methylation in single cells using single-cell bisulfite sequencing (scBS-seq). Nat Protoc 12:534–547
8. Hildebrand S, Hultin S, Subramani A et al (2017) The E-cadherin/AmotL2 complex organizes actin filaments required for epithelial hexagonal packing and blastocyst hatching. 7: 9540
9. Behringer R, Gertsenstein M, Nagy KV et al (2018) Administration of gonadotropins for superovulation in mice. Cold Spring Harb Protoc 2018:pdb.prot092403
10. Behringer R, Gertsenstein M, Nagy K et al (2014) Manipulating the mouse embryo: a laboratory manual. Cold Spring Harbor Laboratory Press

Methods in Molecular Biology (2024) 2767: 307–320
DOI 10.1007/7651_2023_490

Published online: 02 June 2023

Whole-Mount RNA, Single-Molecule RNA (smRNA), and DNA Fluorescence In Situ Hybridization (FISH) in Mammalian Embryos

Jesica Canizo, Katherine Vandal, Savana Biondic, and Sophie Petropoulos

Abstract

Fluorescence in situ hybridization (FISH) provides a valuable tool for studying the spatial localization of and expression level of genes and cell function in diverse biological contexts. In this chapter, we describe a protocol for the simultaneous detection of RNA (including single-molecule (sm)RNA) and DNA in mammalian embryos using FISH. RNA FISH is a technique that enables the detection and visualization of specific RNA molecules within cells. With advancements in technology, the sensitivity and specificity of RNA FISH has been improved to allow the detection of individual mRNA molecules. Both RNA and smRNA are detected using a set of fluorescent-labeled probes, which are complementary to a specific nucleotide sequence corresponding to the gene of interest. These probes hybridize to the target RNA molecules, enabling the simultaneous detection of multiple RNAs within the same cell or tissue. DNA FISH is performed using probes directed at the DNA sequence to detect the genome region of interest. In this chapter, we provide a protocol to process mammalian embryos for FISH with probe examples specifically for studying X-Chromosome activity. By utilizing other probe designs, this protocol can be adapted for the visualization and quantification of other genes and chromosomal regions of interest.

Key words DNA FISH, Guinea pig, Human, Mammalian embryos, Mouse, RNA FISH, smRNA FISH

1 Introduction

RNA/smRNA and DNA FISH are powerful molecular cytogenetic techniques used to visualize RNA and DNA molecules within cells or tissues [1, 2]. RNA FISH is based on the hybridization of fluorescently labeled probes with complementary RNA sequences, followed by fluorescence microscopy imaging. smRNA FISH is a subtype of RNA FISH that uses a set of fluorescently labeled short probes (18–22 nucleotides long) [3], which hybridize to target RNA molecules, enabling the detection of mRNA at single-molecule resolution [4]. By detecting the localization and abundance of specific mRNAs, smRNA FISH provides a valuable tool for studying gene expression patterns, mRNA localization, and cell function in diverse biological contexts. Further, smRNA FISH can detect other sub-types of RNA such as lncRNA and miRNAs [5] and is compatible with diverse sample types, experimental conditions, and has the potential for multiplexing [6]. Using the same principle of hybridization to complementary sequences, DNA FISH allows for the precise mapping of genetic material within cells and is commonly used to track the inheritance patterns of specific chromosomes, identify chromosomal abnormalities, and monitor the progression of meiosis during embryonic development [2]. DNA and RNA/smRNA FISH are highly specific, making them useful tools for both basic research and clinical applications. They offer unique advantages to be used in a wide range of biological contexts. Examples of their application obtained with the protocol described here can be viewed in our previous publications [7, 8].

2 Materials

2.1 RNA/smRNA FISH of Embryos

1. Probes can be designed in-house or purchased predesigned by Stellaris. Further, they can be commercially synthetized or prepared in-house.
2. 96 % Ethanol.
3. Nitric acid.
4. Silanization solution: add 2 mL of 3-amino-propyltriethoxysilane (Sigma, 440140-100ML) to 98 mL of acetone. Use fresh solution each time. Discard in a proper waste container (*see* **Note 1**).
5. Molecular Biology water, RNAse, DNAse free (VWR, 02-0201-1000).
6. Fixing solution: fresh 4 % PFA in PBS diluted from a 32 % commercial solution (Electron microscopy Science, 15714).
7. 1× PBS, Ca^{2+} and Mg^{2+} free.
8. Methanol.
9. 20× SSC (Ambion, AM9763).
10. TE, pH 8 (Ambion, AM9849).
11. Hoechst or DAPI.
12. Prolong Gold (Molecular probes, P36961).
13. Formamide (Ambion, AM9342). Aliquot and store in −20 °C protected from light. Use a fresh aliquot each time.
14. Dextran sulfate (Sigma, D6001-10G).
15. *E. coli* tRNA (Roche, 10 109 541 001).
16. Ribonucleoside vanadyl complex (RVC) 200 mM (Sigma, R3380 or NEB, S1402S).

17. RNAse away or any other cleaning solution that avoid RNAse contamination.
18. Rectangular coverglass, 24 × 50 mm, #1.5 (Fisher Scientific, 12544E) (these need to be silanized prior to mounting embryos).
19. Super Frost rectangular glass slides.
20. Tool to move embryos with minimal volume (10 μL pipette, denudation pipette or pulled glass capillary).
21. Tissue wipers (VWR Light duty, 82003-820 US or 115-0202 Europe)
22. Micropipettes p1000, p200.
23. RNAse free filter tips.
24. Petri dish, 100 mm and 60 mm diameter.
25. Wash and dry® Coverslips rack (Sigma, Z743684).
26. Beakers of multiple sizes.
27. Glass graduated cylinders.
28. LoBind microcentrifuge tubes (Eppendorf, 022431048).
29. Parafilm.
30. Scissors.
31. Rubber cement or other sealer.
32. Plastic tweezers.
33. Surgical tweezers.
34. Paint brush.
35. Alcohol-waterproof marker.
36. 75-well mini trays (VWR, CA62409-296).
37. Water bath at 70 °C.
38. Stereomicroscope.
39. Incubator that goes to 38.5 °C or hybridization chamber (VWR, 24000 V).
40. Chemical hood.
41. Wide-field microscope or spinning disk (*see* **Note 2**).

2.2 DNA FISH

1. PBS without Ca^{2+} and Mg^{2+}, warmed to 37 °C.
2. Fixing solution: 987.86 μL of 4 % PFA in PBS, 7.14 μL of tergitol 70 % and 5 μL of triton x-100.
3. Denature solution (formamide 70 %): 7 mL formamide, 1 mL 20 × SSC, 2 mL RNAse DNAse free water. Prepare in a chemical hood fresh every week and store protected from light. Warm to 73 °C in water bath before use.

4. Washing solution (formamide 50 %): 5 mL formamide, 1 mL 20 × SSC, 4 mL RNAse DNAse free water (*see* **Note 3**). Prepare in a chemical hood fresh every week and store at 4 C protected from light. Warm to room temperature 1 h before use.
5. Triton HCl: 40 mL of 0.2 N HCl and 200 μL of triton X-100. Prepare fresh in a chemical hood.
6. RNAse solution: 30 μL of RNAse A and 970 μL of PBS Ca^{2+} and Mg^{2+} free. Prepare fresh.
7. 70 %, 80 %, and 100 % Ethanol.
8. Set of Probes (*see* **Note 4**).
9. Prolong Gold mounting medium.
10. Petri dish, 100 mm and 60 mm diameter.
11. Plastic and metal tweezers.
12. Laboratory desiccator.
13. Water bath.
14. Heat block.
15. Hybridization chamber or incubator.
16. Chemical hood.
17. Micropipettes p1000 and p200.
18. Tips without filter.
19. Circular coverslips, 15 mm diameter, #1 1/2 thickness (Electron Microscopy Sciences).
20. LoBind microcentrifuge tubes (Eppendorf, 022431048).
21. Wide-field microscope or spinning disk (same characteristics than for smRNA FISH).

3 Methods

3.1 RNA/mRNA FISH of Embryos

3.1.1 Preparing Hybridization Buffer Stock

1. Prepare a 50 mg/mL stock of BSA in 2× SSC. Aliquot and freeze at −20 °C (*see* **Note 5**).
2. Prepare a 20 mg/mL stock of *E. coli* tRNA in water, aliquot and freeze at −20 °C.
3. Thaw the commercial stock of RVC, aliquot into 1.5 mL microcentrifuge tubes and freeze at −20 °C (*see* **Note 6**).
4. Mix dextran with water according to Table 1 and incubate for 30 min at room temperature to allow hydration and dilution.
5. Add other components (Table 1) one by one. Mix each one by pipetting up and down (do not vortex).
6. Aliquot 330 or 660 μL of hybridization buffer into low binding microcentrifuge tubes and freeze at −20 °C until use (*see* **Note 7**). Prepare fresh every 4 months.

Table 1
mRNA hybridization buffer composition

Component	Final concentration	5 mL final volume
Deionized formamide	10 %	500 μL
Dextran sulfate	10 %	0.5 g
SSC	2×	500 μL of a 20 × stock
E. coli tRNA	1 mg/mL	250 μL
RVC	2 mM	50 μL of 200 mM stock
BSA	0.2 mg/mL	20 μL of 50 mg/mL stock
RNAse free water		3.65 mL

3.1.2 Coverslip Cleaning

1. Empty a box of coverslips (#1.5) into a beaker. Rinse coverslips in distilled water two times for 20 min each.
2. Incubate the coverslips in concentrated nitric acid (contained in a parafilm-sealed beaker) under the chemical hood for 24 h at room temperature.
3. Rinse coverslips in water 4 times for 1 h each.
4. Fill a beaker with distilled water and sterilize the coverslips by autoclaving.
5. Wash coverslips once with 96 % ethanol and store them indefinitely in 96 % ethanol.

3.1.3 Silanization of Coverslips

1. Place the coverslips in a rack under a chemical hood, and air dry until all the ethanol has evaporated.
2. Immerse the rack with coverslips in the silanizing solution using plastic tweezers to manipulate the rack and incubate for 60 s (*see* Fig. 1).
3. Shake off excess liquid and transfer the rack of slides to a new beaker with water. Agitate and incubate for 60 s.
4. Remove the washed slides and put them in a desiccator until dry (generally overnight).
5. Store the coverslips in the glass box inside a dry chamber until use (~1 month).

3.1.4 Preparing Wash Solution

Wash solution (20 % formamide): mix 2 mL formamide, 1 mL 20 × SSC and 7 mL of RNAse free H_2O. It can be kept at 4 °C for a week protected from light. Take it out at room temperature 1 h before use.

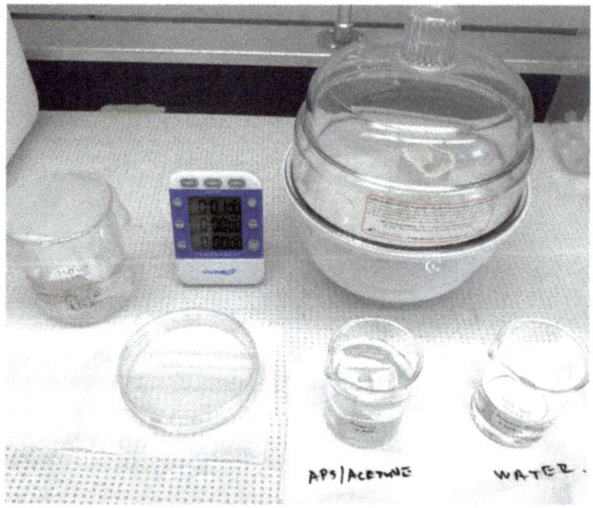

Fig. 1 Silanization of coverslips. The workstation consists of precleaned coverslips stored in 96 % ethanol, a laboratory desiccator for drying and storing silanized coverslips, silanization solution, water, coverslips rack, a timer, and a petri dish to hold the plastic tweezers (not shown in this picture)

3.1.5 RNA/sm RNA FISH

1. Preheat a water bath to 70 °C, preheat an incubator or hybridization chamber to 38.5 °C and cool an aliquot of methanol at −20 °C.
2. Fill two wells of a 75-well mini tray with 17 μL of 4 % PFA (in PBS). Fix embryos by washing them in the first well of PFA briefly, then incubating them in the second well for 15 min at room temperature.
3. Fill three subsequent wells of the 75-well mini tray with PBS. Wash the embryos by moving them into these PBS-containing wells, incubating them for 5 min in the last well at room temperature (*see* **Note 8**).
4. Place embryos onto a silanized coverslip using a mouth pipette attached to a prepulled capillary under stereomicroscope (*see* Fig. 2a, b) (you can arrange embryos in groups on one coverslip) (*see* **Note 9**), transferring the least amount of liquid possible. During this time, make marks around the embryos using an alcohol-water proof marker to visualize their locations (*see* Fig. 2c). After marking the location of the embryo, we may carefully remove excess liquid by aspiration using a mouth pipette attached to a prepulled capillary, taking care to avoid aspirating the embryo back onto the pipette. This process helps the embryo to attach to the silanized coverslip. Allow the embryo to dry onto the coverslip for approximately 2–5 min, just before seen signs of crystallization.

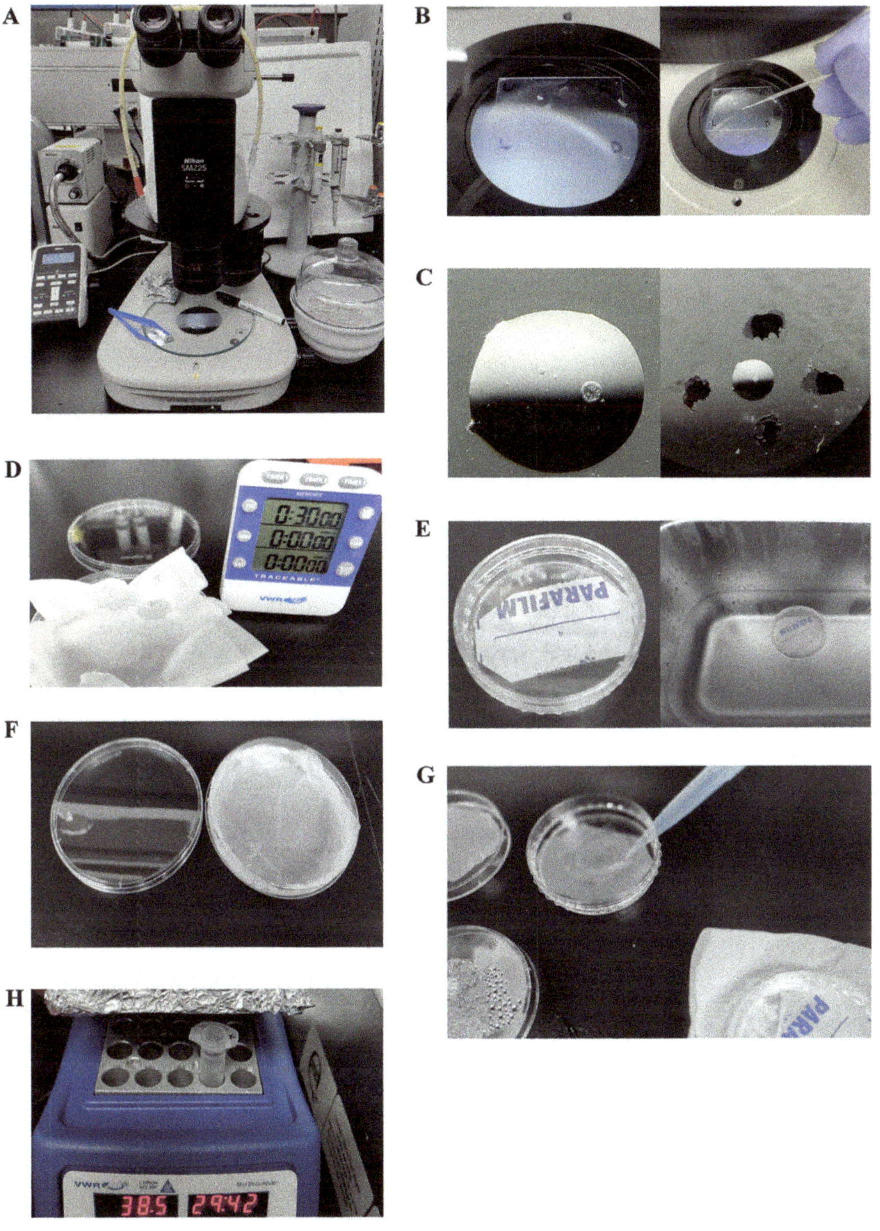

Fig. 2 RNA/smRNA FISH steps. (**a**) The workstation for mounting embryos which is comprised of a stereomicroscope, plastic tweezers, mouth pipette, and silanized coverslips (**b**). Image depicting the transfer of embryos to the silanized coverslips. (**c**) Image depicting the dotted marks used to localize the sample on the coverslips. (**d**) Image of the coverslips drying. (**e**) Image depicting the setup for heat shock in TE. (**f**) Image of the wet chamber for incubations. (**g**) Image of the wash step performed in a petri dish with parafilm. (**h**) Image of the prewarmed hybridization buffer plus probes at 38.5 °C.

5. Using plastic tweezers, move the coverslip onto a 60 mm dish with a square of non-stretched parafilm lining the bottom (*see* **Note 10**). Add about 3 mL of prechilled methanol carefully, avoiding pouring it directly onto the embryos, and incubate them for 10 min at −20 °C (*see* **Note 11**).
6. Remove the coverslip from methanol and air dry for 30 min by placing it on a Kimwipe (*see* Fig. 2d). At this point, attached embryos can be checked under stereomicroscope using a piece of parafilm to prevent the coverslips from sticking to the glass.
7. Meanwhile, prepare a 60 mm petri dish with parafilm (*see* **Note 12**), add 2–3 mL of TE, and heat at 70 °C in a pre-warmed water bath (*see* Fig. 2e).
8. During this time, prepare the wet chamber (*see* Fig. 2f). In a 100 mm petri dish, add wet paper towels and parafilm over the paper towels.
9. Using plastic tweezers, gently slide the coverslip (sample side up) onto the warm TE prepared in **step 7**. Cover the dish with the lid and heat shock at 70 °C for 10 min.
10. In the meantime, prepare the probes. Add each desired probe set to 300 μL of hybridization buffer in a low binding microcentrifuge tube (*see* **Note 13**).
11. Move the coverslip to a fresh 60 mm dish with parafilm using plastic tweezers and wash by pipetting 1 mL of 2× SSC gently onto one corner of the coverslip, avoiding bubbles (*see* Fig. 2g). Incubate for 1 min.
12. Add 300 μL of prewarmed hybridization + probe mix (*see* Fig. 2h) onto the parafilm in the 100 mm dish and place the coverslip on top of it (sample side down). Incubate for 5 min at room temperature protected from light, and then seal the dish with parafilm. Take caution to avoid any air bubbles.
13. Incubate for 2–6 h at 38.5 °C in the hybridization chamber or an incubator protected from light (*see* **Note 14**).
14. Move the coverslip to a fresh 60 mm dish with parafilm and wash with 1 mL of prewarmed 20 % formamide/2× SSC for 4 × 15 min each at 38.5 °C. The last wash should be with 1: 10000 Hoechst diluted in the wash solution (*see* **Note 15**).
15. Move the coverslip using plastic tweezers to a new 60 mm dish with parafilm and wash with 1 mL of 2× SSC at room temperature for 1 min.
16. Add a drop of Prolong Gold onto a Superfrost Plus slide and lay the coverslip on top (sample side down), avoiding bubbles. Wipe any excess Prolong Gold that leaks when coverslip is placed.
17. Dry at room temperature for 24 h, protected from light.

18. Seal with rubber cement.
19. Image (*see* **Note 16**).
20. Keep slides in the fridge until performing DNA FISH the following day or as soon as possible.

3.2 DNA FISH

Day 1 DNA FISH (using the same embryos that were imaged above for RNA/smRNA FISH)

1. Preheat the slide chamber to 42 °C. Put 0.2 N HCl/Triton X-100 and 70 % ethanol on ice. Prewarm 70 % formamide/2× SSC to 73 °C ± 1 in a water bath. Preheat PBS to 37 °C.
2. Remove rubber cement from the slide using a razor blade or pulling it off with tweezers.
3. Loosen the Prolong Gold by immersing the slide into 37 °C PBS for 15–30 min (*see* Fig. 3a). Using surgical metal tweezers, slide gently the coverslip off the slide. The embryos should still be attached to the coverslip. Presence of embryos can be confirmed with a stereomicroscope.
4. Wash the coverslip with 1 mL PBS for 5 min at room temperature (*see* **Note 17**), (*see* Fig. 3b).
5. Re-fix samples on coverslips by pipetting 1 mL of DNA fixing solution at room temperature and incubating for 10 min in a petri dish with parafilm (*see* Fig. 3b).
6. Wash with 1 mL PBS at room temperature for 5 min.
7. Prepare a wet chamber as indicated in **step 8** for the RNA FISH protocol.
8. Blot any excess liquid off the coverslip and place it face down into a 300 μL drop of RNase solution placed over a piece of parafilm in a humidity chamber. Incubate at 37 °C for 30 min.
9. Wash with 1 mL of PBS at room temperature for 5 min.
10. Place a 60 mm petri dish with parafilm on wet ice. Move the coverslip with plastic tweezers to that dish and add 1 mL of 0.2 N HCl/Triton X-100. Incubate on ice for 10 min (*see* Fig. 3c).
11. During this time, start preparing the probes by warming the probes and buffer to room temperature and making the dilution in hybridization buffer (*see* **Note 18**).
12. Wash coverslips with 1 mL PBS at room temperature for 5 min.
13. Denature samples in a petri dish with parafilm containing 1 mL 70 % formamide/2 × SSC. Dip the coverslip gently and incubate for 7 min at 73 °C ± 1, sample side up.
14. Dehydrate the sample immediately by moving the coverslip to a new 60 mm petri dish with parafilm and adding 1 mL of prechilled 70 % ethanol. Incubate on ice for 2 min (*see* Fig. 3c).

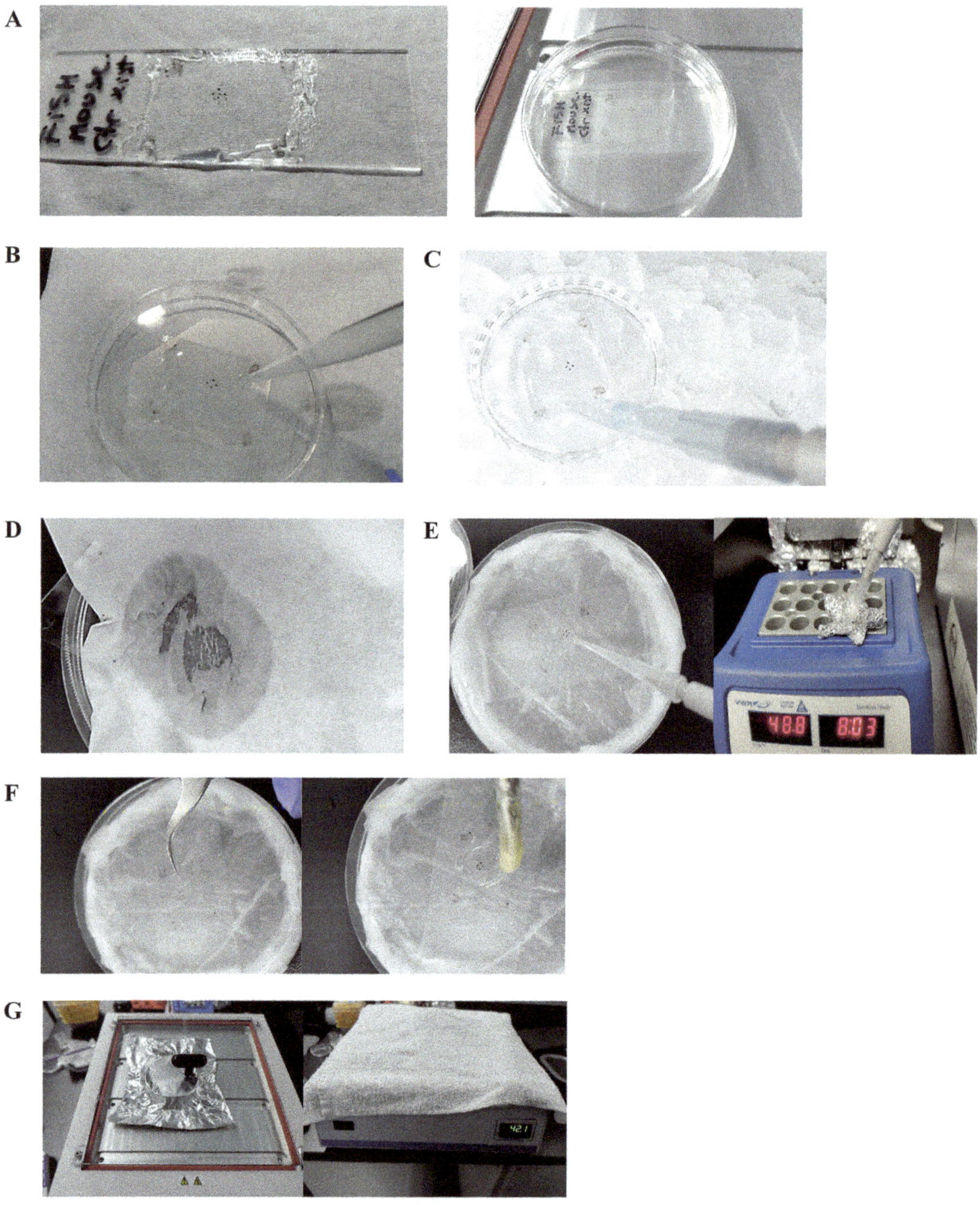

Fig. 3 DNA FISH steps. (**a**) Image depicting the steps associated with sample recovery. (Left) A slide containing the sample mounted for RNA/smRNA FISH and (right) the slide immersed in PBS at 37 °C. (**b**) Setup for washing and re-fixing the sample, which includes a petri dish with parafilm and the sample (face up). (**c**) Setup for HCl solution and 70 % ethanol incubations, which includes a petri dish with the sample on ice. (**d**) Image depicting the air-drying of coverslips. (**e**) Image depicting the pipetting of probes. (**f**) Imaging depicting the mounting of samples for hybridization. (Left) The application of round coverslips following the addition of hybridization solution and (right) sealing the coverslip with rubber cement. (**g**) Image depicting the overnight the incubation of samples with DNA probes

15. Initiate probe denaturation by incubating the solution prepared in **step 11** at 73 °C ± 1 for 5 min while ensuring that the tube is covered with aluminum foil to protect it from light.
16. Remove the probe solution from the water bath and place it on a 45–50 °C heat block until use.
17. Dehydrate the sample immediately by moving the coverslip to a new 60 mm petri dish with parafilm and adding 1 mL of 80 % ethanol for 2 min at room temperature followed by 1 mL of 100 % ethanol at room temperature for 2 min.
18. Keep the coverslips in 100 % ethanol until all the coverslips and the probe solution are ready.
19. Blot excess liquid from the coverslip and air dry on a paper towel. Do not exceed 2 min of drying (*see* Fig. 3d).
20. Apply 20 μL probe solution onto the coverslip. Keep the probe solution on the heat block while pipetting (*see* Fig. 3e). Immediately place a 15 mm circular coverslip on top, avoiding air bubbles (*see* Fig. 3f).
21. Apply rubber cement around the coverslip (*see* Fig. 3f).
22. Incubate at 42 °C overnight (up to 16 h) in a wet chamber (*see* Fig. 3g).

Day 2 DNA FISH

1. One hour before finishing the incubation step with the probes, preheat the 50 % formamide washing solution and the 2× SSC solution to 46 °C ± 1.
2. Remove the rubber cement from the coverslip and the circular coverslip with a razor blade.
3. Immediately transfer the coverslip to a petri dish with parafilm and add 1 mL of prewarmed 50 % formamide washing solution, sample side up, (*see* **Note 19**). The incubation is done in a hybridization chamber at 46 °C ± 1 for 5 min.
4. Repeat **step 3**.
5. Then move the coverslip to a new petri dish with parafilm and wash with 1 mL of 2× SSC at 46 °C ± 1 for 5 min.
6. Repeat **step 5**.
7. Move the coverslip into a new petri dish with parafilm and stain the sample by adding 1 mL of 1:10000 Hoechst solution in 2× SSC for 15 min at room temperature.
8. Wash with 1 mL 2× SSC for 5 min at room temperature.
9. Blot excess liquid from the coverslip and add a drop of Prolong Gold. Immediately apply another coverslip on top. Dry for approximately 1 h at room temperature, protected from light.

10. Image (similar setup to RNA/smRNA FISH) (*see* **Note 20**).
11. For long-term storage, use a microscope slide storage box with silica beads and keep at −20 °C.

4 Notes

RNA/smRNA FISH

1. Excess silanization solution can be prepared. We usually prepare a batch of 5–8 coverslips at a time.
2. Images for RNA FISH are acquired using a Zeiss AxioObserver Z1 Yokogawa CSU-X1 spinning disk confocal inverted microscope (Zeiss, Germany) equipped with a motorized stage, Piezo objectives, and an Evolve EMCDD (512 × 512, 16 bit, 16 μm pixel size) monochrome camera (Photometrics). We use an alphaPlan Apo 100×/1.46 Oil DICIII (UV) M27 objective (Zeiss, Germany) to acquire the images, resulting in a final pixel size of 133 nm. For DAPI excitation, we use a 405 nm (solid-state) laser associated with a double bandpass filter in emission (480/22 + LP530). For AF568 excitation, we use a 561 nm (solid-state) laser associated with a double bandpass filter in emission (460/30 + 590/30). We use the wide-field microscope configuration on the same microscope for smRNA FISH.
3. We prepare 10 mL of washing solution to wash two coverslips at the same time.
4. For detection of the X- and Y-Chromosome, we use the following probes: Vysis CEP X (DXZ1) SpectrumGreen Probe (Abbott Molecular, 05J10-023) and Vysis CEP Y (DYZ1) SpectrumOrange Probe (Abbott Molecular, 05J08-024) for human; MCENXY-10-GROR (Empire Genome Corp) for mouse.
5. Clean all work surfaces, pipettes, and plastic tweezers with RNAse away solution. Use gloves and laboratory coat throughout the procedure.
6. Each time a RVC aliquot is used, it needs to be prewarmed at 38.5 °C for 30 min. Mix by inversion before use.
7. We use 300 μL of hybridization buffer for one coverslip; therefore, we aliquot 330 μL. We also prepare 660 μL aliquots for when we manipulate two coverslips at the same time.
8. Wait until all embryos are ready before transferring to the silanized coverslip.
9. Label the coverslip in each corner with the following space references: U, R, D, L for upper, right, down, and left. This will help to identify the side of the coverslip where embryos are (sample side up).

10. Parafilm prevents the coverslip from sticking to the bottom of the petri dish and makes it easier to handle.
11. We first move the petri dish with the coverslip inside the freezer and pipette the prechilled methanol directly there. Put the lid before incubating for 10 min.
12. At this step, we do not remove the paper covering the parafilm as it prevents the coverslip from sticking to the warmed parafilm, which becomes adhesive at 70 °C.
13. To study X-Chromosome inactivation, we have successfully used human XIST Q570 (Stellaris, SMF-2038-1), human ATRX Q670 (Stellaris, VSMF-2019-5), and mouse Xist Q570 (Stellaris, SMF-3011-1). We have also used human IL6 CAL Fluor Red 610 (VSMF-20534-5; BioSearch Technologies) in another study. We reconstitute the probe set according to the manufacturer's recommendations. Briefly, we prepare a concentrated stock of 100 μM by adding 50 μL of TE to 5 nmol of dry probes. Then, we prepare working aliquots of 12.5 μM (16 × 10 μL) by adding 20 μL of the concentrated stock to 140 μL of TE. These aliquots are stored at −20 °C and protected from light. Finally, at the time of hybridization, we prepare a solution of probes ranging between 125 and 250 nM in the hybridization buffer [6–8]. Custom probes can also be designed for genes of interest, especially if different species are being studied. Generally, it is recommended to design 25–48 probes per gene target.
14. We have tested 2 h for XIST, and it worked, for other mRNAs up to 6 h. When 1 h is left, warm 20 % formamide solution to 37 °C and Prolong Gold to room temperature.
15. All wash steps should be performed protected from light.
16. High-magnification, high-numerical-aperture objective lenses are required and long exposure times (up to 1–2 s) should be used, depending on the probe and channel. High laser power (50–100 %) is also used to increase the signal. Resolution should also be high. For XIST, we use a resolution of 1088 × 1024, which generally appears as a dense cloud. However, for smRNAs, where the signal appears as tiny dots, a higher resolution of 2304 × 2304 is required. Due to the short working distance of these high-magnification/NA objectives, it is important to mount the samples on coverslips instead of regular microscope slides. To image several nuclei, Z-stacks were performed with a 220 nm step size.

DNA FISH

17. The DNA FISH protocol uses 60 mm petri dishes with parafilm for all washing steps, as indicated for the RNA/smRNA FISH in the previous section.

18. We prepare a 10 μL solution for 3–4 embryos by mixing 7 μL of CEP hybridization buffer, 1 μL of Y-Chromosome probe, 1 μL of X-Chromosome probe, and 1 μL of RNAse free H_2O in a low binding PCR tube. Then centrifuge 1–3 s, vortex, and centrifuge again. Keep the solution in the dark at 38.5 °C until ready to use.
19. Following addition of probe, all downstream steps are carried out protected from light.
20. Imaging the same region of interest (ROI): For XIST FISH followed by DNA FISH of X- and Y-Chromosomes, it is crucial to image the same area every time; otherwise, the measurements of the same site will not overlap with the signals. It is important to select ROI in every round of the experiment using recognizable landmarks in the tissue to ensure consistency.

Acknowledgments

This work was supported by the Canadian Institutes of Health Research (PJT-178082) and NSERC PGS D scholarship to K.V. SP holds the Canada Research Chair in Functional Genomics of Reproduction and Development (950-233204). We thank the CRCHUM imaging facility for use of the Zeiss spinning disk/ wide-field microscope.

References

1. Lea R, Bonev B, Dubaissi E, Vize PD, Papalopulu N (2012) Multicolor fluorescent in situ mRNA hybridization (FISH) on whole mounts and sections. In: Hoppler S, Vize PD (eds) Xenopus protocols: post-genomic approaches. Humana Press, Totowa, pp 431–444
2. Cui C, Shu W, Li P (2016) Fluorescence in situ hybridization: cell-based genetic diagnostic and research applications. Front Cell Dev Biol 4:89
3. Orjalo A, Johansson HE, Ruth JL (2011) Stellaris™ fluorescence in situ hybridization (FISH) probes: a powerful tool for mRNA detection. Nat Methods 8(10):i–ii. https://doi.org/10.1038/nmeth.f.349
4. Young AP, Jackson DJ, Wyeth RC (2020) A technical review and guide to RNA fluorescence in situ hybridization. PeerJ 8:e8806. https://doi.org/10.7717/peerj.8806
5. Raj A (2013) Single-Molecule RNA FISH. In: Roberts GCK (ed) Encyclopedia of biophysics. Springer, Berlin Heidelberg, Berlin, Heidelberg, pp 2340–2343
6. Codeluppi S, Borm LE, Zeisel A, La Manno G, van Lunteren JA, Svensson CI, Linnarsson S (2018) Spatial organization of the somatosensory cortex revealed by osmFISH. Nat Methods 15(11):932–935. https://doi.org/10.1038/s41592-018-0175-z
7. Petropoulos S, Edsgärd D, Reinius B, Deng Q, Panula SP, Codeluppi S, Plaza Reyes A, Linnarsson S, Sandberg R, Lanner F (2016) Single-cell RNA-Seq reveals lineage and X chromosome dynamics in human preimplantation embryos. Cell 165(4):1012–1026. https://doi.org/10.1016/j.cell.2016.03.023
8. Zhao C, Biondic S, Vandal K, Björklund ÅK, Hagemann-Jensen M, Sommer TM, Canizo J, Clark S, Raymond P, Zenklusen DR, Rivron N, Reik W, Petropoulos S (2022) Single-cell multiomics of human preimplantation embryos shows susceptibility to glucocorticoids. Genome Res 32(9):1627–1641. https://doi.org/10.1101/gr.276665.122

INDEX

Magdalena Zernicka-Goetz and Kursad Turksen (eds.), *Embryo Models In Vitro: Methods and Protocols*, Methods in Molecular Biology, vol. 2767, https://doi.org/10.1007/978-1-0716-3686-2,

T

U

GPSR Compliance
The European Union's (EU) General Product Safety Regulation (GPSR) is a set of rules that requires consumer products to be safe and our obligations to ensure this.

If you have any concerns about our products, you can contact us on

ProductSafety@springernature.com

In case Publisher is established outside the EU, the EU authorized representative is:

Springer Nature Customer Service Center GmbH
Europaplatz 3
69115 Heidelberg, Germany

www.ingramcontent.com/pod-product-compliance
Ingram Content Group UK Ltd.
Pitfield, Milton Keynes, MK11 3LW, UK
UKHW052003051026
15624UKWH00003B/538
* 9 7 8 1 0 7 1 6 3 6 8 5 5 *